Python 程序设计

第4版·微课版·在线学习软件版 董付国 ◎ 编著

清华大学出版社
北京

内 容 简 介

全书共 18 章，主要内容如下：第 1 章介绍 Python 基础知识与概念；第 2 章讲解列表、元组、字典、集合等常用可迭代对象；第 3 章讲解 Python 选择结构与循环结构；第 4 章讲解字符串基本操作方法与正则表达式模块 re 的用法；第 5 章讲解函数设计与使用；第 6 章讲解面向对象编程有关的知识；第 7 章讲解文本文件与二进制文件的读写，以及文件级操作与目录操作，Office 文件与 PDF 文件操作；第 8 章讲解异常处理结构及 Python 程序的调试与测试方法；第 9 章讲解标准库 tkinter 的 GUI 应用；第 10 章讲解网络编程、网页内容读取；第 11 章讲解安卓平台的 Python 编程；第 12 章讲解注册表编程以及系统运维；第 13 章讲解多线程与多进程编程；第 14 章介绍 SQLite、Access、MS SQL Server、MySQL 访问方法；第 15 章讲解图形图像编程、音乐编程、语音识别及视频处理；第 16 章介绍逆向工程与软件分析原理、IDAPython 编程、Immunity Debugger 编程以及 Windows 平台软件调试原理；第 17 章讲解 NumPy、SciPy、Matplotlib、Pandas 与 statistics 在科学计算和可视化、数据处理、统计与分析中的应用；第 18 章讲解安全哈希算法、对称密钥密码算法 DES 和 AES，以及非对称密钥密码算法 RSA 与数字签名算法 DSA。

本书既可以作为计算机及相关专业的教材，也可以作为 Python 爱好者的参考书。

本书封面贴有清华大学出版社防伪标签，无标签者不得销售。
版权所有，侵权必究。举报：010-62782989，beiqinquan@tup.tsinghua.edu.cn。

图书在版编目（CIP）数据

Python 程序设计：微课版：在线学习软件版/董付国编著. —4 版. —北京：清华大学出版社，2024.5（2024.7 重印）
ISBN 978-7-302-66379-9

Ⅰ.①P⋯ Ⅱ.①董⋯ Ⅲ.①软件工具－程序设计 Ⅳ.①TP311.561

中国国家版本馆 CIP 数据核字（2024）第 107899 号

责任编辑：白立军　杨　帆
封面设计：刘　键
责任校对：李建庄
责任印制：沈　露

出版发行：清华大学出版社
　　　　网　　址：https://www.tup.com.cn，https://www.wqxuetang.com
　　　　地　　址：北京清华大学学研大厦 A 座　　　　邮　　编：100084
　　　　社　总　机：010-83470000　　　　　　　　　　邮　　购：010-62786544
　　　　投稿与读者服务：010-62776969，c-service@tup.tsinghua.edu.cn
　　　　质量反馈：010-62772015，zhiliang@tup.tsinghua.edu.cn
　　　　课件下载：https://www.tup.com.cn，010-83470236
印 装 者：三河市龙大印装有限公司
经　　销：全国新华书店
开　　本：185mm×260mm　　　印　张：24.25　　　字　数：582 千字
版　　次：2015 年 8 月第 1 版　　2024 年 6 月第 4 版　　印　次：2024 年 7 月第 2 次印刷
定　　价：69.80 元

产品编号：105039-01

前　言

　　Python 由 Guido van Rossum 于 1989 年年底开始设计与开发，第一个公开发行版本发行于 1991 年。Python 推出不久就迅速得到了各行业人士的青睐，经过 30 多年的发展，已经渗透到计算机科学与技术、统计分析、移动终端开发、科学计算可视化、逆向工程与软件分析、图形编程与图像处理、人工智能、游戏设计与策划、网站开发、数据采集、数据分析与处理、密码学、系统运维、音乐编程、视频处理、计算机辅助教育、医药辅助设计、天文信息处理、化学、生物等几乎所有专业和领域。著名搜索引擎 Google 的核心代码使用 Python 实现，迪士尼公司的动画制作与生成采用 Python 实现，大部分 UNIX 和 Linux 都内建了 Python 环境支持，豆瓣网使用 Python 作为主体开发语言进行网站架构和有关应用的设计与开发，网易公司大量网络游戏的服务器端代码超过 70% 采用 Python 进行设计与开发，易度公司的 PaaA 企业应用云端开发平台和百度云计算平台 BAE 也都大量采用了 Python 语言，美国宇航局使用 Python 实现了 CAD/CAE/PDM 库及模型管理系统，雅虎公司使用 Python 建立全球范围的站点群，微软公司的集成开发环境 Visual Studio 2015 开始默认支持 Python 语言，开源 ERP 系统 Odoo 完全采用 Python 语言开发，引力波数据是用 Python 进行处理和分析的，大量人工智能开发框架提供 Python 接口，类似的案例数不胜数。

　　早在多年前 Python 就已经成为卡内基-梅隆大学、麻省理工学院、加州大学伯克利分校、哈佛大学等国外很多大学计算机专业或非计算机专业的程序设计入门教学语言，目前国内也有不少学校的多个专业陆续开设了 Python 程序设计课程。Python 分别于 2007 年、2010 年、2018 年、2020 年、2021 年先后 5 次被 TIOBE 网站评为年度语言；2014 年 12 月在 *IEEE Spectrum* 推出的编程语言排行榜中，Python 取得了第 5 位的好名次；2017 年至 2023 年连续 7 年 *IEEE Spectrum* 把 Python 排在第一位。

　　Python 是一门免费、开源的跨平台高级动态编程语言，支持命令式编程、函数式编程，完全支持面向对象程序设计，拥有大量功能强大的内置对象、标准库和扩展库以及众多狂热的支持者，使得各领域的研发人员、策划人员以及管理人员都能够快速实现和验证自己的思路与创意。在有些编程语言中需要编写大量代码才能实现的功能，在 Python 中直接调用内置函数或标准库对象即可实现。Python 用户只需要把主要精力放在业务逻辑的设计与实现上，在开发效率和运行效率之间达到了完美的平衡，其精妙之处令人赞叹。

　　Python 是一门快乐、优雅的语言。与 C 语言系列和 Java 等语言相比，Python 大幅降低了学习与使用的难度。Python 易学易用，语法简洁清晰，代码可读性强，编程模式非常符合人类思维方式和习惯。经常浏览 Python 社区的优秀代码、Python 标准库和扩展库文档甚至源代码，适当了解其内部工作原理，可以帮助读者编写更加优雅的 Python 程序。

　　如果读者有其他程序设计语言的基础，那么在学习和使用 Python 的过程中，一定不要把用其他语言编程的习惯和风格带到 Python 中，那样不仅会使代码变得非常冗长、烦琐，还可能严重影响代码的效率。应该尽量尝试从最自然、最简洁的角度出发去思考和解决问题，这样才能写出更加优雅、更加 Pythonic 的代码。

本书内容组织

对于 Python 程序员来说,熟练运用优秀、成熟的扩展库可以快速实现业务逻辑和创意,而 Python 语言基础知识和基本数据结构的熟练掌握则是理解和运用其他扩展库的必备条件,并且在实际开发中建议优先使用 Python 内置对象和标准库对象实现预定功能。本书前 8 章使用大量篇幅介绍 Python 编程基础知识,通过大量案例演示 Python 语言的精妙与强大。从第 9 章开始介绍大量标准库和扩展库在 GUI 编程、网络编程、移动终端编程、Windows 系统编程、多线程与多进程编程、数据库编程、多媒体编程、逆向工程与软件分析、科学计算与可视化、密码学编程等多个领域的应用。全书共 18 章,主要内容组织如下。

第 1 章 基础知识。介绍如何选择 Python 版本,Python 对象模型,数字、字符串等基本数据类型,运算符与表达式,常用内置函数,基本输入输出,Python 代码编写规范,Python 文件名等。

第 2 章 Python 可迭代对象。讲解可迭代对象常用的方法和基本操作,列表基本操作与常用方法,列表推导式,切片操作,元组与生成器推导式,序列解包,字典、集合基本操作与常用方法,字典推导式与集合推导式,以及如何使用列表实现队列、栈、二叉树、有向图等复杂数据结构。

第 3 章 选择与循环。讲解 Python 选择结构,while 循环与 for 循环,带 else 子句的循环结构,break 和 continue 语句,选择结构与循环结构的综合运用。

第 4 章 字符串与正则表达式。讲解字符串编码格式,字符串格式化、替换、分隔、连接、查找、排版等基本操作,正则表达式语法、正则表达式对象、子模式与 Match 对象,以及 Python 正则表达式模块 re 的应用。

第 5 章 函数设计与使用。讲解函数定义与调用,关键参数、默认值参数、可变长度参数等不同参数类型,全局变量与局部变量,参数传递时的序列解包,return 语句,lambda 表达式,以及函数式编程、生成器函数与可调用对象等若干高级话题。

第 6 章 面向对象程序设计。讲解类的定义与使用,self 与 cls 参数,类成员与实例成员,私有成员与公有成员,特殊方法与运算符重载,继承与多态等内容。

第 7 章 文件操作。讲解文件操作基本知识,Python 文件对象,文本文件读写操作,二进制文件读写与对象序列化,文件复制、移动、重命名,文件类型检测、完整性检查,文件压缩与解压缩,文件夹大小统计、增量备份,删除指定类型的文件以及 Office 文件操作等内容。

第 8 章 异常处理结构与程序调试、测试。讲解 Python 内置异常类与自定义异常类,不同形式的异常处理结构,使用 IDLE 和 pdb 模块调试 Python 程序,Python 单元测试相关知识。

第 9 章 tkinter 应用开发。讲解如何使用标准库 tkinter 进行 GUI 编程,通过大量实际案例演示基本组件的用法。

第 10 章 网络程序设计。讲解计算机网络基础知识,UDP、TCP 编程,网页内容读取与网页爬虫。

第 11 章 安卓平台的 Python 编程。介绍 Pydroid3 和 QPython3 开发环境的应用,讲解安卓平台的 Python 程序设计。

第 12 章 Windows 系统编程。讲解注册表编程,将 Python 程序打包为 exe 可执行文件、GUI编程、系统版本判断、在Python中调用外部程序，以及Python在系统运维中的应用。

第13章 多线程与多进程编程。讲解Python标准库threading和multiprocessing在多线程编程与多进程编程中的应用，以及多线程与多进程的数据共享与同步控制。

第14章 数据库编程。介绍SQLite数据库及其相关概念，Connection对象、Cursor对象、Row对象，以及使用Python扩展库操作Access、MS SQL Server、MySQL等数据库。

第15章 多媒体编程。讲解扩展库PyOpenGL在计算机图形编程中的应用，扩展库Pillow在图像编程中的应用，pygame、SciPy在音乐编程中的应用，speech在语音识别中的应用，以及opencv-python和moviepy在视频处理中的应用。

第16章 逆向工程与软件分析。介绍逆向工程与软件分析的原理和相关插件，IDAPython与Immunity Debugger在软件分析中的应用，以及Windows平台软件调试原理。

第17章 数据分析、科学计算与可视化。讲解扩展库NumPy、SciPy、Matplotlib在科学计算与可视化领域的应用，以及扩展库Pandas与标准库statistics在数据处理、统计与分析中的应用。

第18章 密码学编程。以pycryptodome、rsa、hashlib等模块为主讲解安全哈希算法、对称密钥密码算法DES和AES，以及非对称密钥密码算法RSA与数字签名算法DSA的应用。

本书信息量大，知识点紧凑，案例丰富，实用性强。全书100多个涉及不同行业领域的实用案例，没有多余的文字、程序输出结果或软件安装截图，充分利用宝贵的篇幅来讲解尽可能多的知识。本书作者具有20年程序设计教学经验，讲授过汇编语言、C/C++/C#、Java、PHP、Python等多门程序设计语言，编写过大量的应用程序。本书内容结合作者多年教学与开发过程中积累的许多经验和案例，并将其巧妙地糅进了相应的章节。

本书对Python内部工作原理进行了一定深度的剖析，95%以上的案例均使用Python 3.9和3.10演示，代码同样适用于Python 3.6/3.7/3.8/3.11/3.12/3.13以及更高版本实现，另外还适当介绍了高版本的部分新特性。书中适当介绍了Python代码优化和安全编程的有关知识，可以满足不同层次读者的需要。

本书适用读者

本书可以作为（但不限于）：

（1）计算机专业本科生Python程序设计教材或研究生必读书目。本科生建议72学时以上，讲授本书全部章节。

（2）数字媒体技术、软件工程、网络工程、信息安全、通信工程、电子、自动化及其他工科专业本科生或研究生Python程序设计教材。建议64学时，讲授前9章，再根据专业特点与需要在其他章节中选讲3～5章。

（3）会计、经济、金融、管理、心理学、统计及其他非工科专业研究生或本科生Python程序设计教材。建议64学时，讲授前8章中不带星号的内容，第9章的案例选讲三四个，再根据专业特点与需要在其他章节中选讲两三章，其余章节由学生根据兴趣自学。

（4）非计算机相关专业本科生公共基础课Python程序设计教材。建议48学时并边讲边练,讲授前8章中不带星号的章节,再根据需要在其他章节中选讲两三章,其余章节由学生根据兴趣自学。

(5) 专科院校或职业技术学院Python程序设计教材。建议96学时,讲授前9章中不带星号的内容以及第10、13、14、17章。

(6) Python培训用书。建议时间为一周,讲授前8章,再根据需要选讲3～5章。

(7) 具有一定Python基础的读者进阶首选学习资料。

(8) 涉及Python开发的程序员、策划人员、科研人员和管理人员阅读书目。

(9) 打算利用业余时间学习一门快乐的程序设计语言并编写几个小程序来娱乐或解决工作中小任务的读者首选学习资料。

(10) 少数对编程具有浓厚兴趣和天赋的中学生课外阅读资料。

教学资源

本书包含思政元素,提供全套教学课件、源代码、电子教案、习题答案、考试题库、在线练习与考试软件(目前有4000个客观题和780个编程题)以及教学大纲,配套资源可以登录清华大学出版社官方网站(www.tup.com.cn)下载或与作者联系索取,本书编辑的电子邮箱地址为bailj@tup.tsinghua.edu.cn,作者的微信公众号为"Python小屋"。

Python程序设计课程
思政案例分享

由于时间仓促,作者水平有限,书中难免存在纰漏,不足之处还请同行指正并通过作者联系方式进行反馈与交流。作者将不定期在公众号和微信发布和更新勘误表,并通过QQ和微信答复读者的疑问。

感谢

首先感谢父母的养育之恩,在当年那么艰苦的条件下还坚决支持我读书,没有让我像其他同龄的孩子一样辍学。感谢姐姐、姐夫多年来对我的爱护以及在老家对父母的照顾,感谢善良的弟弟、弟媳在老家对父母的照顾,正是有了你们,我才能在远离家乡的城市安心工作。感谢我的妻子在生活中对我的大力支持,也感谢懂事的女儿在我工作的时候能够在旁边安静地读书而尽量不打扰我,在定稿前和妈妈一起帮我阅读全书并检查出了几个错别字。

感谢每一位读者,感谢您在茫茫书海中选择了本书,衷心祝愿您能够从本书中受益,学到您需要的知识! 同时也期待每一位读者的热心反馈,随时欢迎您指出书中的不足!

本书在编写出版过程中得到清华大学出版社的大力支持和帮助,在此表示衷心的感谢。

董付国定稿于山东烟台

2024年1月

目 录

第 1 章 基础知识 ··· 1
- 1.1 如何选择 Python 版本 ·· 1
- 1.2 Python 安装与简单使用 ·· 3
- 1.3 使用 pip 管理 Python 扩展库 ··· 4
- 1.4 Python 基础知识 ··· 5
  - 1.4.1 Python 对象模型 ··· 5
  - 1.4.2 Python 变量 ··· 6
  - 1.4.3 数字 ··· 9
  - 1.4.4 字符串 ··· 11
  - 1.4.5 运算符与表达式 ··· 11
  - 1.4.6 常用内置函数 ·· 17
  - 1.4.7 基本输入输出 ·· 25
  - 1.4.8 模块导入与使用 ··· 25
- 1.5 Python 代码编写规范 ··· 27
- 1.6 Python 文件名 ·· 28
- 1.7 Python 程序的__name__属性 ··· 29
- *1.8 编写和使用自己的包 ·· 29
- *1.9 Python 程序伪编译与打包 ··· 30
- 1.10 案例精选 ·· 31
- *1.11 The Zen of Python ··· 33
- 本章小结 ·· 33
- 习题 ··· 34

第 2 章 Python 可迭代对象 ··· 35
- 2.1 列表 ··· 35
  - 2.1.1 列表的创建与删除 ·· 36
  - 2.1.2 列表元素的增加 ··· 36
  - 2.1.3 列表元素的删除 ··· 38
  - 2.1.4 列表元素访问与计数 ··· 41
  - 2.1.5 元素存在性测试 ··· 41
  - 2.1.6 切片操作 ·· 42
  - 2.1.7 列表排序与逆序 ··· 44
  - 2.1.8 用于序列操作的常用内置函数 ·· 45
  - 2.1.9 列表推导式 ··· 46
  - *2.1.10 使用列表实现向量运算 ··· 48

2.2 元组 49
　　2.2.1 元组的创建与删除 49
　　2.2.2 元组与列表的区别 49
　　2.2.3 序列解包 50
　　2.2.4 生成器表达式 51
2.3 字典 52
　　2.3.1 字典的创建与删除 52
　　2.3.2 字典元素的访问 53
　　2.3.3 字典元素的添加与修改 54
　　2.3.4 字典应用案例 54
2.4 集合 55
　　2.4.1 集合的创建与常用操作 55
　　2.4.2 集合运算 56
　　2.4.3 集合运用案例 56
2.5 再谈内置函数 sorted() 58
*2.6 复杂数据结构 59
　　2.6.1 堆 59
　　2.6.2 队列 60
　　2.6.3 栈 63
　　2.6.4 链表 64
　　2.6.5 二叉树 65
　　2.6.6 有向图 66
本章小结 67
习题 68

第 3 章 选择与循环 70

3.1 条件表达式 70
3.2 选择结构 71
　　3.2.1 单分支选择结构 71
　　3.2.2 双分支选择结构 71
　　3.2.3 嵌套的选择结构 72
　　3.2.4 多分支选择结构 73
　　3.2.5 选择结构应用案例 75
3.3 循环结构 76
　　3.3.1 while 循环与 for 循环 76
　　3.3.2 循环结构的优化 77
3.4 break 和 continue 语句 78
3.5 案例精选 78
本章小结 84
习题 85

第 4 章 字符串与正则表达式 …… 86
4.1 字符串 …… 86
4.1.1 字符串格式化 …… 86
4.1.2 字符串常用方法 …… 89
4.1.3 字符串常量 …… 94
*4.1.4 可变字符串 …… 96
4.1.5 中文分词与拼音处理 …… 96
4.1.6 字符串应用案例精选 …… 97
4.2 正则表达式 …… 100
4.2.1 正则表达式语法 …… 100
4.2.2 re 模块主要函数 …… 102
4.2.3 直接使用 re 模块函数 …… 105
4.2.4 使用正则表达式对象 …… 107
4.2.5 子模式与 Match 对象 …… 109
4.2.6 正则表达式应用案例精选 …… 112
本章小结 …… 116
习题 …… 117

第 5 章 函数设计与使用 …… 118
5.1 函数定义与调用 …… 119
5.2 形参与实参 …… 120
5.3 参数类型 …… 121
5.3.1 默认值参数 …… 121
5.3.2 关键参数 …… 123
5.3.3 可变长度参数 …… 123
5.3.4 参数传递时的序列解包 …… 124
5.4 return 语句 …… 125
5.5 变量作用域 …… 125
5.6 lambda 表达式 …… 127
5.7 案例精选 …… 128
5.8 高级话题 …… 134
本章小结 …… 139
习题 …… 140

第 6 章 面向对象程序设计 …… 141
6.1 类的定义与使用 …… 141
6.1.1 类定义语法 …… 141
6.1.2 self 参数 …… 142
6.1.3 类成员与实例成员 …… 142
6.1.4 私有成员与公有成员 …… 143
6.2 方法 …… 144

	6.3	属性	146
*	6.4	特殊方法与运算符重载	148
		6.4.1 常用特殊方法	148
		6.4.2 案例精选	150
	6.5	继承	155
	6.6	多态	159
	本章小结		160
	习题		160

第7章 文件操作 161

- 7.1 文件对象 161
- 7.2 文本文件内容操作案例精选 162
- 7.3 二进制文件操作案例精选 165
  - 7.3.1 使用 pickle 模块 165
  - 7.3.2 使用 struct 模块 166
- 7.4 文件级操作 166
  - 7.4.1 os 与 os.path 模块 166
  - 7.4.2 shutil 模块 167
- 7.5 目录操作 168
- 7.6 案例精选 170
- 本章小结 181
- 习题 182

第8章 异常处理结构与程序调试、测试 183

- 8.1 基本概念 183
- 8.2 Python 内置异常类与自定义异常 184
- 8.3 异常处理结构语法应用 186
  - 8.3.1 try…except 186
  - 8.3.2 try…except…else 187
  - 8.3.3 try…except…except…except 188
  - 8.3.4 try…except…else…finally 189
- 8.4 断言与上下文管理 190
  - 8.4.1 断言 190
  - 8.4.2 上下文管理 191
- *8.5 使用 IDLE 调试代码 191
- *8.6 使用 pdb 模块调试程序 192
  - 8.6.1 pdb 模块常用命令 192
  - 8.6.2 使用 pdb 模块调试 Python 程序 193
- *8.7 Python 单元测试 195
- 8.8 文档测试 198
- 8.9 性能测试 199

本章小结 ………………………………………………………………………………… 200
习题 ……………………………………………………………………………………… 201

第9章　tkinter 应用开发 …………………………………………………………… 202

9.1　tkinter 基础 ………………………………………………………………………… 202
9.1.1　tkinter 常用组件 …………………………………………………………… 202
9.1.2　tkinter 应用程序开发基本流程 ………………………………………… 203

9.2　tkinter 应用案例精选 ……………………………………………………………… 205
9.2.1　用户登录界面 ……………………………………………………………… 205
9.2.2　选择类组件应用 …………………………………………………………… 206
9.2.3　简单文本编辑器 …………………………………………………………… 208
9.2.4　简单画图程序 ……………………………………………………………… 212
9.2.5　电子时钟 …………………………………………………………………… 215
9.2.6　简易计算器 ………………………………………………………………… 217
9.2.7　桌面放大镜 ………………………………………………………………… 219
9.2.8　抽奖程序 …………………………………………………………………… 219
9.2.9　猜数游戏 …………………………………………………………………… 221
9.2.10　图片查看器程序 ………………………………………………………… 224
9.2.11　在 tkinter 应用程序中使用日历选择组件 …………………………… 225

本章小结 ………………………………………………………………………………… 227
习题 ……………………………………………………………………………………… 227

第10章　网络程序设计 ……………………………………………………………… 228

10.1　计算机网络基础知识 …………………………………………………………… 228

10.2　UDP 和 TCP 编程基础 ………………………………………………………… 230
10.2.1　UDP 编程 ………………………………………………………………… 230
10.2.2　TCP 编程 ………………………………………………………………… 232

10.3　网络编程案例精选 ……………………………………………………………… 235
10.3.1　网络嗅探器 ……………………………………………………………… 235
10.3.2　多进程端口扫描器 ……………………………………………………… 236
10.3.3　查看本机所有联网程序信息 …………………………………………… 237
10.3.4　查看局域网内 IP 地址与 MAC 地址的对应关系 …………………… 238
10.3.5　查看本机网络流量 ……………………………………………………… 238
10.3.6　局域网内服务器自动发现 ……………………………………………… 238
10.3.7　多线程＋Socket 实现素数远程查询 ………………………………… 239
10.3.8　建立和使用 TCP 长连接 ……………………………………………… 240

10.4　网页内容读取与网页爬虫 ……………………………………………………… 242
10.4.1　网页内容读取与域名处理基础知识 …………………………………… 242
10.4.2　网页爬虫实战 …………………………………………………………… 243

本章小结 ………………………………………………………………………………… 245
习题 ……………………………………………………………………………………… 246

第 11 章　安卓平台的 Python 编程 ·· 247
　11.1　QPython 简介 ·· 247
　11.2　安卓应用开发案例 ·· 248
　本章小结 ·· 251
　习题 ·· 252

第 12 章　Windows 系统编程 ·· 253
　12.1　注册表编程 ·· 253
　12.2　创建可执行文件 ·· 256
　12.3　调用外部程序 ·· 257
　12.4　创建窗口 ·· 261
　12.5　判断 Windows 操作系统的版本 ··· 263
　12.6　系统运维 ·· 264
　　12.6.1　Python 扩展库 psutil ··· 264
　　12.6.2　使用 Pywin32 实现事件查看器 ··· 266
　　12.6.3　切换用户登录身份 ·· 268
　本章小结 ·· 269
　习题 ·· 270

第 13 章　多线程与多进程编程 ·· 271
　13.1　threading 模块 ·· 271
　13.2　Thread 对象 ·· 272
　　13.2.1　Thread 对象中的方法 ··· 272
　　13.2.2　Thread 对象中的 daemon 属性 ··· 273
　13.3　线程同步技术 ·· 274
　　13.3.1　Lock/RLock 对象 ·· 275
　　13.3.2　Condition 对象 ·· 276
　　13.3.3　queue 模块 ··· 278
　　13.3.4　Event 对象 ·· 279
　　13.3.5　Semaphore 与 BoundedSemaphore ··· 280
　　13.3.6　Barrier 对象 ··· 281
　13.4　多进程编程 ·· 281
　　13.4.1　创建与启动进程 ·· 282
　　13.4.2　进程间数据交换 ·· 283
　　13.4.3　进程同步 ·· 286
　　13.4.4　标准库 subprocess ·· 286
　本章小结 ·· 290
　习题 ·· 291

第 14 章　数据库编程 ·· 292
　14.1　SQLite 应用 ·· 292
　　14.1.1　Connection 对象 ·· 293

14.1.2　Cursor 对象 ·············· 293
　　　14.1.3　Row 对象 ················· 295
　14.2　访问其他类型数据库 ················ 297
　　　14.2.1　操作 Access 数据库 ········ 297
　　　14.2.2　操作 MS SQL Server 数据库 ··· 298
　　　14.2.3　操作 MySQL 数据库 ········ 299
　本章小结 ····································· 300
　习题 ··· 301

第 15 章　多媒体编程 ······················· 302

　15.1　图形编程 ························· 302
　　　15.1.1　创建图形编程框架 ········ 302
　　　15.1.2　绘制文字 ················· 303
　　　15.1.3　绘制图形 ················· 303
　　　15.1.4　纹理映射 ················· 304
　　　15.1.5　处理键盘/鼠标事件 ······· 306
　15.2　图像编程 ························· 307
　　　15.2.1　图像处理模块 Pillow 功能简介 ·· 307
　　　15.2.2　使用 Pillow 计算椭圆中心 ·· 309
　　　15.2.3　使用 Pillow 动态生成比例分配图 ·· 310
　　　15.2.4　使用 Pillow 生成验证码图片 ·· 310
　15.3　音乐编程 ························· 312
　　　15.3.1　音乐播放 ················· 312
　　　15.3.2　WAV 波形音乐文件处理 ··· 314
　15.4　语音识别 ························· 315
　15.5　视频处理和摄像头接口调用 ······· 318
　　　15.5.1　OpenCV 应用 ············· 318
　　　15.5.2　moviepy 应用 ············· 319
　本章小结 ····································· 321
　习题 ··· 322

第 16 章　逆向工程与软件分析 ·············· 323

　16.1　主流项目与插件简介 ·············· 323
　　　16.1.1　主流项目 ················· 324
　　　16.1.2　常用插件 ················· 324
　16.2　IDAPython 与 Immunity Debugger 编程 ·· 325
　　　16.2.1　IDAPython 编程 ·········· 325
　　　16.2.2　Immunity Debugger 编程 ·· 329
　16.3　Windows 平台软件调试原理 ······ 334
　　　16.3.1　Windows 调试接口 ········ 334
　　　16.3.2　调试事件 ················· 335

	16.3.3 进程调试	336
	16.3.4 线程环境	337
	16.3.5 断点	337
16.4	案例精选	339
本章小结		341
习题		341

第 17 章 数据分析、科学计算与可视化 …… 342

17.1	NumPy 数组运算与矩阵运算	342
17.2	SciPy 简单应用	348
	17.2.1 常数与特殊函数	349
	17.2.2 SciPy 中值滤波	350
	17.2.3 使用 SciPy 进行多项式计算	351
	17.2.4 数理统计与随机变量	352
17.3	Matplotlib 可视化案例精选	352
	17.3.1 绘制折线图	352
	17.3.2 绘制散点图	353
	17.3.3 绘制饼状图	354
	17.3.4 在图例中显示公式	355
	17.3.5 创建和使用子图	356
	17.3.6 绘制有描边和填充效果的柱状图	356
	17.3.7 使用雷达图展示学生成绩	358
	17.3.8 绘制三维曲面	359
	17.3.9 绘制三维曲线	359
	17.3.10 设置图例样式	361
17.4	数据分析扩展库 Pandas 用法精要	361
17.5	统计分析模块 statistics 常用函数	365
本章小结		366
习题		367

第 18 章 密码学编程 …… 368

18.1	安全哈希算法	368
18.2	对称密钥密码算法 DES 和 AES	368
18.3	非对称密钥密码算法 RSA 与数字签名算法 DSA	370
	18.3.1 RSA	370
	18.3.2 DSA	371
本章小结		372
习题		372
参考文献		373

第 1 章 基 础 知 识

Python 是一门跨平台、开源、免费的解释型高级动态编程语言,支持伪编译把 Python 源程序转换为字节码来优化程序和提高加载速度,并且支持使用 py2exe、py2app、cx_Freeze、Nuitka 或 pyinstaller 工具将 Python 程序打包为不同平台上的可执行程序,可以在没有安装 Python 解释器和相关依赖包的系统中运行;Python 支持命令式编程、函数式编程,完全支持面向对象程序设计,语法简洁清晰,并且拥有大量的几乎支持所有领域应用开发的成熟扩展库;Python 就像胶水一样,可以把使用多种不同语言编写的程序融合到一起实现无缝拼接,更好地发挥不同语言和工具的优势,满足不同应用领域的需求。

配套资源

1.1 如何选择 Python 版本

Python 官方网站多年来同时发行 Python 2.x(已停止更新,最后一个版本为 2020 年 4 月 20 日发布的 2.7.18)和 Python 3.x 两个不同系列的版本,互相之间不兼容,很多内置函数的实现和使用方式也有较大的区别,Python 3.x 对 Python 2.x 的标准库也进行了一定程度的重新拆分和整合。本书改版完成时几个主流版本的最新版本分别为 Python 3.6.15/3.7.17/3.8.17/3.9.17/3.10.12/3.11.5/3.12.0。当较新的 Python 版本推出之后,不要急于更新和替换已安装版本,而是应该在确定自己必须使用的扩展库也推出了相应版本之后再一起进行更新。

安装好 Python 以后,在"开始"菜单中选择 IDLE(Python GUI)命令,即可启动 Python 解释器并可以看到当前安装的 Python 版本号,如图 1-1 和图 1-2 所示。在 IDLE(Python GUI)和 Python(command line)两种界面中,都以 3 个大于号 >>> 作为提示符,可以在提示符后面输入要执行的语句。在本书所有章节给出的示例代码中,>>> 符号都不需要输入,仅表示该代码是在交互模式下运行,不带该提示符的代码则表示是以程序的方式运行的。本书主要使用 IDLE 来介绍 Python 程序的开发与应用,读者也可以选择 wingIDE、PyCharm、Pythonwin、VS code、Spyder 或其他开发环境。

```
Python 3.6.8 Shell
File Edit Shell Debug Options Window Help
Python 3.6.8 (tags/v3.6.8:3c6b436a57, Dec 24 2018, 00:16:47) [MSC v.1916 64 bit
(AMD64)] on win32
Type "help", "copyright", "credits" or "license()" for more information.
>>>
```

图 1-1 Python 3.6.8 主界面(3.9 及之前的版本都与此类似)

除了在启动主界面上查看已安装的 Python 版本之外,还可以使用下面的命令随时查看。

```
>>> import sys
>>> sys.version
'3.10.11 (tags/v3.10.11:7d4cc5a, Apr 5 2023, 00:38:17) [MSC v.1929 64 bit (AMD64)]'
```

```
IDLE Shell 3.11.4
File Edit Shell Debug Options Window Help
Python 3.11.4 (tags/v3.11.4:d2340ef, Jun  7 2023, 05:45:37)
 [MSC v.1934 64 bit (AMD64)] on win32
Type "help", "copyright", "credits" or "license()" for more
 information.
>>>
```

图 1-2　Python 3.11.4 主界面（Python 3.10 及更高版本都与此类似）

```
>>> sys.version_info
sys.version_info(major=3, minor=10, micro=11, releaselevel='final', serial=0)
>>> sys.winver
'3.10'
```

另外，也可以在命令提示符环境执行命令 python -V 查看版本，请读者自行验证。

有时候可能需要同时安装多个不同的版本，例如，同时安装 Python 3.10 和 Python 3.8 以及 Python 3.12，并根据不同的开发需求在多个版本之间进行切换。多版本并存一般不影响在 IDLE 中直接运行程序，只需要启动相应版本的 IDLE 即可。在命令提示符环境中运行 Python 程序时，如果无法正确运行，可以尝试在调用 Python 主程序时指定其完整路径，或者通过修改系统 Path 变量来实现不同版本之间的切换。在 Windows 10 系统下修改系统 Path 变量的步骤：进入资源管理器，右击"此电脑"并选择"属性"命令，在弹出的对话框中单击"高级系统设置"选项，切换至"高级"选项卡，单击"环境变量"按钮，然后修改系统变量 Path，如图 1-3 所示。

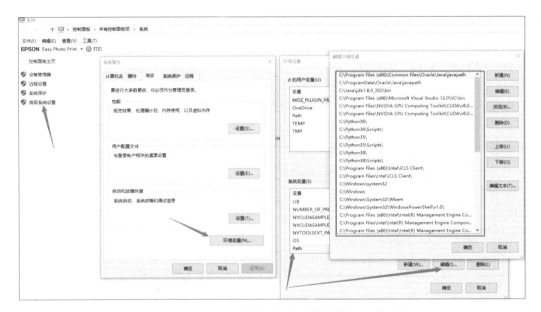

图 1-3　Windows 10 系统下修改系统变量 Path 的方法

1.2 Python 安装与简单使用

Python 的安装很简单,打开 Python 官方主页 https://www.python.org/后,进入下载页面,选择适合自己的版本下载并安装即可,注意安装路径不要太深且不要有中文和空格,最好同时选择安装 tkinter 和 pip 并选择 add to path 选项。如果使用的是 Linux 系统,例如 Ubuntu,那么很可能已经预装了某个版本的 Python,请根据需要进行升级。若未经特别说明,本书主要使用 Windows 10＋Python 3.9 进行演示,代码同样适用于 Python 3.6/3.7/3.8/3.10/3.11/3.12/3.13。

1.2

也可以根据个人爱好选择安装和使用 Anaconda3、Pycharm 或其他开发环境。本书均以官方自带的 IDLE 为例,如果使用交互式编程模式,那么直接在 IDLE 提示符>>>后面输入相应的代码并按 Enter 键执行即可,如果执行顺利,马上就可以看到执行结果,否则会抛出异常。

```
>>> 3 + 5                           #在交互模式中立即显示表达式的值
8
>>> import math                     #导入标准库 math
>>> math.sqrt(9)                    #求 9 的平方根,等价于 9**0.5
3.0
>>> 3 * (2 + 6)
24
>>> 2 / 0                           #除数不能为 0,代码抛出异常,略去了详细错误信息
ZeroDivisionError: integer division or modulo by zero
```

解决复杂问题时,往往需要编写包含多行代码的程序来实现特定的业务逻辑,同时也方便代码的保存、共享、不断完善和重复利用。在 IDLE 界面中使用菜单 File→New File 命令,创建一个程序文件,输入代码并保存(控制台应用程序扩展名为 py,GUI 程序可以保存为 pyw 文件)。可以使用菜单 Run→Check Module 命令来检查程序中是否存在语法错误;使用菜单 Run→Run Module 命令运行程序,程序运行结果将直接显示在 IDLE 交互界面上。除此之外,也可以在资源管理器中双击扩展名为 py、pyw 或 pyc 的文件直接运行;在有些情况下,可能还需要在命令提示符环境中运行 Python 程序文件。例如,假设有程序 HelloWorld.py 的内容如下:

```
def main():
    print('Hello world')
main()
```

在 IDLE 中运行该程序结果如图 1-4 所示。在命令提示符环境(或类似的 Powershell)中运行该程序的方法与结果如图 1-5 所示,该图中演示了 3 种执行 Python 程序的方法,虽然最后一种方法看上去更简单,但是应尽量使用第一种方法(前两行)来运行 Python 程序,否则可能会影响某些程序的正确运行。另外,为保证命令提示符环境中可以正常运行,应在安装时把 Python 路径添加至 Path 变量,或者也可以在安装之后配置系统的环境变量 Path。

图 1-4　在 IDLE 中运行程序

图 1-5　在命令提示符环境中运行程序

在实际开发中，如果用户能够熟练使用开发环境 IDLE 提供的一些快捷键，将会大幅提高编写速度和开发效率。在 IDLE 中，除了撤销（Ctrl＋Z）、全选（Ctrl＋A）、复制（Ctrl＋C）、粘贴（Ctrl＋V）、剪切（Ctrl＋X）等常规快捷键之外，其他比较常用的快捷键如表 1-1 所示。

表 1-1　常用的 IDLE 快捷键

快　捷　键	功　能　说　明
Alt＋P	浏览历史命令（上一条）
Alt＋N	浏览历史命令（下一条）
Ctrl＋F6	重启 Shell，之前定义的对象和导入的模块全部失效
F1	打开 Python 帮助文档
Alt＋/	自动补全前面曾经出现过的单词，如果之前有多个单词具有相同前缀，则在多个单词中循环以供选择
Ctrl＋]	缩进代码块
Ctrl＋[	取消代码块缩进
Tab	补全代码或批量缩进
Alt＋3	注释代码块
Alt＋4	取消代码块注释

1.3

1.3　使用 pip 管理 Python 扩展库

在默认情况下，安装 Python 时不会安装任何扩展库，应根据需要安装相应的扩展库。pip 是官方安装包自带的 Python 扩展库管理工具，常用 pip 子命令使用方法如表 1-2 所示。

表 1-2　常用 pip 子命令使用方法

pip 命令示例	说　　　明
pip freeze [>packages.txt]	列出已安装模块及其版本号，可以使用重定向符>把扩展库信息保存到文件 packages.txt 中
pip install SomePackage[==version]	在线安装 SomePackage 模块，可以使用方括号内的形式指定扩展库版本
pip install SomePackage.whl	通过 whl 文件离线安装扩展库
pip install -r packages.txt	读取文件 packages.txt 中的扩展库信息，并安装这些扩展库

续表

pip 命令示例	说 明
pip install --upgrade SomePackage	升级 SomePackage 模块
pip uninstall SomePackage[==version]	卸载 SomePackage 模块

对于大部分扩展库，使用 pip 工具直接在线安装都会成功，但有时候会因为缺少 VC 编译器或依赖文件而失败。在 Windows 平台上，如果在线安装扩展库失败，可以下载编译好的 whl 文件（一定不要修改下载的文件名），然后在命令提示符环境中使用 pip 命令进行离线安装。例如：

```
pip install Django-5.0.3-py3-none-any.whl
```

注意，如果计算机上安装了多个版本的 Python 开发环境，在一个版本下安装的扩展库无法在另一个版本中使用。可以在资源管理器中切换至相应版本 Python 安装目录的 scripts 文件夹中，然后在按 Shift+鼠标右键弹出的菜单中选择"在此处打开命令提示符窗口"(Windows 7)或"在此处打开 PowerShell 窗口"(Windows 10)，进入命令提示符环境执行 pip 命令。

如果由于网速问题导致在线安装速度过慢，pip 命令还支持指定国内的站点来提高速度，下面的命令用于从阿里云下载安装扩展库 jieba，其他服务器地址可以自行查阅。

```
pip install jieba -i http://mirrors.aliyun.com/pypi/simple --trusted-host mirrors.aliyun.com
```

如果遇到类似于"拒绝访问"的出错提示，可以在执行 pip 命令时增加选项--user。

如果使用 Anaconda3，安装时已经自带了几百个常用的扩展库。如果需要再安装其他扩展库，除了 pip 命令之外，也可以使用 conda 命令安装 Python 扩展库，用法与 pip 命令类似。

1.4　Python 基础知识

1.4.1　Python 对象模型

对象是 Python 语言中最基本的概念之一，Python 中的一切都是对象。Python 中有许多内置对象可直接使用，如数字、字符串、列表、字典、元组、集合、del 命令，以及 len()、id()、type()等大量内置函数。表 1-3 中列出了部分常见的 Python 内置对象，内置函数将在 1.4.6 节详细讲解。

表 1-3　Python 内置对象

对象类型	类型名称	示　　例	简 要 说 明
数字	int float complex	1234 3.14,1.3e5 3+4j	整数大小没有限制，内置支持复数及其运算
字符串	str	'swfu' "I'm a student" '"Python"' r'abc', R'bcd'	使用单引号、双引号、三引号作为定界符且可以嵌套，以字母 r 或 R 引导的表示原始字符串，见 4.1 节

续表

对象类型	类型名称	示例	简要说明
字节串	bytes	b'hello world'	以字母 b 引导，可以使用单引号、双引号、三引号作为定界符且可以嵌套
列表	list	[1, 2, 3] ['a', 'b', ['c', 2]]	所有元素放在一对方括号中，元素之间使用逗号分隔，其中的元素可以是任意类型，见 2.1 节
字典	dict	{1: 'food', 2: 'taste', 3:'import'}	所有元素放在一对花括号中，元素之间使用逗号分隔，元素形式为"键：值"，见 2.3 节
元组	tuple	(2, -5, 6) (3,)	不可变，所有元素放在一对圆括号中，元素之间使用逗号分隔，如果元组中只有一个元素，后面的逗号不能省略，见 2.2 节
集合	set frozenset	{'a', 'b', 'c'} frozenset({'d','b','c'})	所有元素放在一对花括号中，元素之间使用逗号分隔，元素不允许重复；另外，set 是可变的，frozenset 是不可变的，见 2.4 节
布尔型	bool	True, False	逻辑值，关系运算符、元素测试运算符、同一性测试运算符组成的表达式的值一般为 True 或 False
空类型	NoneType	None	空值
异常	Exception、ValueError、TypeError		Python 内置大量异常类，分别对应不同类型的异常，见第 8 章
文件		f = open('data.dat', 'rb')	open 是 Python 内置函数，使用指定的模式打开文件，返回文件对象，见第 7 章
迭代器对象		生成器对象、zip 对象、enumerate 对象、map 对象、filter 对象等	具有惰性求值的特点，其中的元素只能使用一次
编程单元		函数（使用 def 定义） 类（使用 class 定义） 模块（类型为 module）	函数（见第 5 章）和类（见第 6 章）都属于可调用对象，模块用来集中存放函数、类、常量或其他对象

1.4.2 Python 变量

在 Python 中，不需要事先声明变量名及其类型，直接赋值即可创建各种类型的变量。Python 是一种动态类型语言，也就是说，变量的类型是可以随时变化的，下面的代码演示了 Python 变量类型的变化（实际是创建了新变量）。

```
>>> x = 3                      #创建整型变量
>>> print(type(x))             #查看对象类型
<class 'int'>
>>> x = 'Hello world.'         #创建字符串变量
>>> print(type(x))
<class 'str'>
>>> x = [1, 2, 3]              #创建列表变量
>>> print(type(x))
<class 'list'>
>>> i: int = '3'               #声明变量 i 为整型，但并不真正约束和检查值的类型
```

代码中首先创建了整型变量 x，然后又分别创建了字符串和列表类型的变量 x。当创建了字符串类型的变量 x 之后，之前创建的整型变量 x 将自动失效，创建列表对象 x 之后，之前创建的字符串变量 x 将自动失效。可以将该模型理解为"状态机"，在显式修改其类型或删除之前，变量将一直保持上次的类型。

虽然不需要在使用之前显式地声明变量及其类型，但是 Python 仍属于强类型编程语言，Python 解释器会根据赋值或运算来自动推断变量类型，任何时刻任何变量都属于确定的类型。

在程序中，如果变量出现在赋值分隔符或复合赋值分隔符（例如+=、*=）的左边表示创建变量或修改变量的值，否则表示引用该变量的值，这一点同样适用于使用下标来访问列表、字典等可变对象以及其他自定义对象中元素的情况。例如：

```
>>> x = 3                      #创建整型变量
>>> print(x**2)                #引用变量的值
9
>>> x += 6                     #修改变量值，实际是创建新变量，相当于 x = x + 6
>>> print(x)                   #读取变量值并输出显示
9
>>> x = [1, 2, 3]              #创建列表对象
>>> print(x)
[1, 2, 3]
>>> x[1] = 5                   #修改列表元素值，实际是修改元素的引用
>>> print(x)                   #输出显示整个列表
[1, 5, 3]
>>> print(x[2])                #输出显示列表的指定元素
3
```

字符串和元组属于不可变序列，不能通过下标的方式来修改其中的元素值，例如下面的代码试图修改元组中元素的值时抛出异常。

```
>>> x = (1, 2, 3)
>>> print(x)
(1, 2, 3)
>>> x[1] = 5
TypeError: 'tuple' object does not support item assignment
```

在 Python 中，允许多个变量引用同一个值，例如：

```
>>> x = 3
>>> id(x)                      #查看变量 x 引用的内存地址，和自己的运行结果不一样是正常的
2603133239600
>>> y = x                      #y 和 x 引用同一个值
>>> id(y)
2603133239600                  #地址与 x 相同
```

继续上面的示例代码，当修改其中一个变量的值以后，实际修改的是变量的引用，其内存地址将会变化，但这并不影响另一个变量。例如，接着上面的代码继续执行下面的代码：

```
>>> x += 6
>>> id(x)                                              #地址发生改变
2603133239792
```

```
>>> y
3
>>> id(y)                                          #地址没有改变
2603133239600
```

在这段代码中,内置函数 id() 用于返回一个对象的内存地址。可以看出,在 Python 中修改变量值的操作,并不是修改了变量的值,而是修改了变量指向的内存地址。这是因为 Python 解释器首先读取变量 x 原来的值,然后将其加 6,并将结果存放于新的内存中,最后将变量 x 指向该结果的内存空间,如图 1-6 所示。

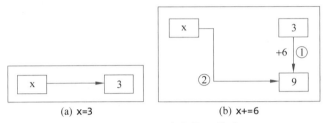

(a) x=3 　　　(b) x+=6

图 1-6　Python 内存管理模式

Python 采用基于值的内存管理方式,如果为不同变量赋值为相同值,这个值在内存中只保存一份,多个变量指向同一个值的内存空间首地址,这样可以减少内存空间的占用,提高内存利用率。Python 启动时,会对[-5, 256]的整数进行缓存。也就是说,如果多个变量的值相等且介于[-5, 256]内,那么这些变量共用同一个值的内存空间。对于区间[-5, 256]之外的整数,同一个程序或交互模式下同一个语句的同值不同名变量会共用同一个内存空间,不同程序或交互模式下不同语句不遵守这个约定。例如(双栏排的代码先看左半部分,再看右半部分,后同):

```
>>> x = -6
>>> y = -6
>>> id(x) == id(y)
False
>>> x = -5
>>> y = -5
>>> id(x) == id(y)
True
>>> x = 255
>>> y = 255
>>> id(x) == id(y)
True
>>> x = 256
>>> y = 256
>>> id(x) == id(y)
True
>>> x = 257
>>> y = 257
```

```
>>> id(x) == id(y)
False
>>> x = 3.0
>>> y = 3.0
>>> id(x) == id(y)
False
>>> x, y = 300000, 300000
>>> id(x) == id(y)
True
>>> x = [666666, 666666]
>>> y = (666666, 666666)
>>> id(x[0]) == id(x[1])
True
>>> id(y[0]) == id(y[1])
True
>>> id(x[0]) == id(y[0])
False
```

创建程序文件 memoryTest.py,编写下面的代码并运行,会发现两次输出结果都为 True。

```
x = 30000000000
y = 30000000000
print(id(x) == id(y))
x = 3.4
y = 3.4
print(id(x) == id(y))
```

综上,Python不会对实数进行缓存,交互模式下不同语句的同值不同名变量不共用同一个内存空间,交互模式下同一个语句或同一个程序的同值不同名变量会共用同一个内存空间。字符串的内存驻留机制比较复杂,与字符串长度和包含的字符种类及创建方式都有关系,一般不关注内存使用情况。

Python具有自动管理内存的功能,会跟踪所有的值,并自动删除不再使用或者引用次数为0的值。如果确定某个变量不再使用,可以使用del命令显式删除该变量,值的引用次数减1,当某个值的引用次数变为0时,将会被Python的垃圾回收机制自动删除并释放内存空间。可以使用标准库函数sys.getrefcount(x)查看对象x的值的引用次数。

最后,在定义变量名(包括函数名和类名)时,需要注意以下问题。

(1) 变量名必须以英文字母、汉字或下画线开头,但以下画线开头的名字在Python中有特殊含义,本书第6章会详细讲解。

(2) 变量名中不能有空格及中英文标点符号(括号、引号、逗号、斜线、反斜线、冒号、句号、问号等)。

(3) 不能使用关键字作为变量名,可以导入keyword模块后使用print(keyword.kwlist)查看所有关键字。

(4) 不建议使用系统内置的模块名、类型名或函数名,以及已导入的模块名及其成员名作为变量名,这会改变其类型和含义。

(5) 变量名区分英文字母的大小写,如student和Student是不同的变量。

1.4.3 数字

Python数值类型主要有整数、浮点数和复数。整数类型可以分为下面几种。

(1) 十进制整数,如0、-1、9、123。

(2) 十六进制整数,使用0、1、2、3、4、5、6、7、8、9、a、b、c、d、e、f来表示整数,必须以0x开头,如0x10、0xfa、0xabcdef。

(3) 八进制整数,使用8个数字0、1、2、3、4、5、6、7来表示整数,必须以0o开头,如0o35、0o11。

(4) 二进制整数,使用2个数字0、1来表示整数,必须以0b开头,如0b101、0b100。

在Python中,整数可以任意大。

```
>>> a = 9999999999999999999999999999
>>> a * a
99999999999999999999999999980000000000000000000000000001
>>> a ** 3
999999999999999999999999999700000000000000000000000000029999999999999999999999
9999999999
```

浮点数也称实数或小数,.3、15.0、0.37、-11.2、1.2e2、314.15e-2都是合法的浮点数。使用时

需要注意的是,浮点数的计算、存储和表示有误差,应避免直接比较两个浮点数是否相等。例如:

```
>>>0.4 - 0.3                           #有误差
0.10000000000000003
>>>0.4 - 0.3 == 0.1                    #直接比较是否相等可能会得到错误结果
False
>>> from math import isclose           #从模块中导入对象,见1.4.8节
>>> isclose(0.4 - 0.3, 0.1)            #测试两个浮点数是否足够接近
True
```

另外,在标准库 decimal 中提供了高精度实数,在标准库 fractions 中提供了分数,下面代码演示了相关的用法。

```
>>> from decimal import Decimal, getcontext
>>> Decimal('1') / Decimal('3')
Decimal('0.3333333333333333333333333333')
>>> getcontext().prec = 40             #修改有效数字位数
>>> Decimal('1') / Decimal('3')
Decimal('0.3333333333333333333333333333333333333333')
>>> from fractions import Fraction
>>> Fraction(3,5) + Fraction(3,7)      #分数相加
Fraction(36, 35)
>>> Fraction(3,5) * Fraction(3,7)      #分数相乘
Fraction(9, 35)
>>> Fraction(3,5) / Fraction(3,7)      #分数相除
Fraction(7, 5)
```

Python 中的复数与数学中复数的形式完全一致,由实部和虚部构成,使用 j 或 J 来表示虚部。

```
>>> a = 3 + 4j
>>> b = 5 + 6j
>>> c = a + b
>>> c
(8+10j)
>>> c.real                             #查看复数实部
8.0
>>> c.imag                             #查看复数虚部
10.0
>>> a.conjugate()                      #返回共轭复数
(3-4j)
>>> a * b                              #复数乘法
(-9+38j)
>>> a / b                              #复数除法
(0.6393442622950819+0.03278688524590165j)
```

Python 3.6 开始支持在数字中间位置使用单个下画线作为分隔来提高数字的可读性,类似于数学上使用逗号作为千位分隔符。

```
>>> 1_000_000
1000000
>>> 1_2_3_4
1234
```

```
>>>1_2 + 3_4j
(12+34j)
>>>1_2.3_45
12.345
```

1.4.4 字符串

在 Python 中,字符串属于不可变序列,可以使用单引号、双引号或三引号进行界定,并且单引号、双引号、三单引号、三双引号还可以互相嵌套,省去了转义字符的麻烦。例如:

`'abc'、'123'、'中国'、"Python"、"'Tom said,"Let's go"'"`

都是合法字符串,空字符串表示为''或""或"""",即一对不包含任何内容的任意字符串界定符。一对三单引号或三双引号表示的字符串支持换行,支持复杂排版格式的字符串,也可以在程序中表示较长的注释。更多关于字符串的内容在第 4 章和第 5 章中将分别进行介绍。

Python 支持转义字符,反斜线可以与紧邻后面的字符共同表示其他含义,常用的转义字符如表 1-4 所示。

表 1-4 常用的转义字符

转义字符	含 义	转义字符	含 义
\n	换行符	\"	双引号
\t	制表符	\\	一个\
\r	回车	\ooo	1~3 位八进制数对应的字符
\'	单引号	\xhh	2 位十六进制数对应的字符
\uhhhh	4 位十六进制数对应的字符	\Uxxxxxxxx	8 位十六进制数对应的字符,要求不大于\U0010FFFF

字符串界定符前面加字母 r 或 R 表示原始字符串,其中的特殊字符不进行转义,每个字符都表示字面含义,但字符串的最后不能是单个\符号。原始字符串主要用于正则表达式,也可以用于简化文件路径或 URL 的输入,可参考第 4 章的内容。

1.4.5 运算符与表达式

运算符用来表示特定类型对象支持的行为和对象之间的操作。运算符的功能与对象类型密切相关,不同类型的对象支持的运算符不同,同一个运算符作用于不同类型的对象时功能也会有所区别。常用的 Python 运算符如表 1-5 所示,大致按照优先级从低到高的顺序排列,且表格中第一列中同一个单元格内的运算符具有相同的优先级。在计算表达式时,会先计算高优先级的运算符再计算低优先级的运算符,相同优先级的运算符从左向右依次进行计算(幂运算符" ** "和条件表达式"…if…else…"除外)。不建议花太多时间记忆运算符的优先级和结合性,更建议在编写程序时使用圆括号明确表达式的计算顺序。

表 1-5 常用的 Python 运算符

运算符示例	功 能 说 明
:=	赋值运算，Python 3.8 新增，俗称海象运算符
lambda[parameter]: expression	用于定义 lambda 表达式，功能相当于函数，parameter 相当于函数形参，可以没有；expression 表达式的值相当于函数返回值
value1 if condition else value2	用于表示一个二选一的表达式，其中，value1、condition、value2 都为表达式，如果 condition 的值等价于 True 则整个表达式的值为 value1 的值，否则整个表达式的值为 value2 的值，类似于一个双分支选择结构，见 3.2.2 节
or	"逻辑或"运算，以表达式 exp1 or exp2 为例，如果 exp1 的值等价于 True 则返回 exp1 的值，否则返回 exp2 的值
and	"逻辑与"运算，以表达式 exp1 and exp2 为例，如果 exp1 的值等价于 False 则返回 exp1 的值，否则返回 exp2 的值
not	"逻辑非"运算，对于表达式 not x，如果 x 的值等价于 True 则返回 False，否则返回 True
in、not in is、is not <、<=、>、>=、==、!=	元素测试，表达式 x in y 的值当且仅当 y 中包含元素 x 时才会为 True；测试两个变量是否引用同一个对象，如果两个对象引用的是同一个对象，那么它们的内存地址相同；关系运算，用于比较大小，作用于集合时表示测试集合的包含关系
|	"按位或"运算，集合并集
^	"按位异或"运算，集合对称差集
&	"按位与"运算，集合交集
<<、>>	左移位、右移位
+ -	算术加法，列表、元组、字符串连接； 算术减法，集合差集
* @ / // %	算术乘法，序列重复； 矩阵乘法； 真除； 整除； 求余数，字符串格式化
+ - ~	正号； 负号，相反数； 按位求反
**	幂运算，指数可以为小数，例如，3**0.5 表示计算 3 的平方根
[ ] . ( )	下标，切片； 属性访问，成员访问； 函数定义或调用，修改表达式计算顺序，声明多行代码为一个语句
[ ]、( )、{ }	定义列表、元组、字典和集合，以及列表推导式、生成器表达式、字典推导式、集合推导式

除了表 1-5 中列出的运算符，Python 还有+=、-=、*=、/=、//=、**=、|=、&=、^=等复合赋值分隔符。作用于可哈希对象时其功能为运算符和等号的组合，例如，如果 x 为数值那么 x +=3 等价于 x = x +3，x |=5 等价于 x = x | 5。但作用于列表、字典、集合等不可哈希对象

时要复杂一些,例如,当 x 为列表时 x += [5] 等价于 x.extend([5]),即列表 x 的引用不变。

(1) + 运算符除了用于算术加法以外,还可以用于列表、元组、字符串的连接,但不支持不同内置类型的对象之间相加或连接。

```
>>> [1, 2, 3] + [4, 5, 6]                #连接两个列表
[1, 2, 3, 4, 5, 6]
>>> (1, 2, 3) + (4,)                     #连接两个元组
(1, 2, 3, 4)
>>> 'abcd' + '1234'                      #连接两个字符串
'abcd1234'
>>> 'A' + 1                              #不支持字符串与数字相加,抛出异常
TypeError: Can only concatenate str (not 'int') to str
>>> True + 3                             #Python 内部把 True 当作 1 处理
4
>>> False + 3                            #把 False 当作 0 处理
3
```

(2) * 运算符不仅可以用于算术乘法,还可以用于列表、字符串或元组等类型变量与整数进行 * 运算,表示重复内容并返回重复后的新对象。

```
>>> 2.0 * 3                              #浮点数与整数相乘
6.0
>>> (3 +4j) * 2                          #复数与整数相乘
(6 +8j)
>>> (3 +4j) * (3-4j)                     #复数与复数相乘
(25 +0j)
>>> 'a' * 10                             #字符串重复
'aaaaaaaaaa'
>>> [1,2,3] * 3                          #列表重复
[1, 2, 3, 1, 2, 3, 1, 2, 3]
>>> (1,2,3) * 3                          #元组重复
(1, 2, 3, 1, 2, 3, 1, 2, 3)
```

(3) Python 中的除法有两种,/ 和 // 分别表示真除法和整除运算。

```
>>> 3 / 5                                #结果为实数
0.6
>>> 3 // 5                               #结果为整数
0
>>> 3.0 / 5                              
0.6
>>> 3.0 // 5                             #实数形式的整数值
0.0
>>> 13 // 10                             #向下取整
1
>>> -13 // 10                            #结果为小于-1.3 的最大整数
-2
```

(4) % 运算符除可用于字符串格式化之外,还可以对整数和浮点数计算余数。但是由于浮点数的精度影响,计算结果可能会有误差。

```
>>> 3.1 % 2
1.1
>>> 6.3 % 2.1
```

```
2.0999999999999996
>>>5.7 % 4.8
0.9000000000000004
>>>6 % 2
0
>>> -17 % 4                        #余数与%右侧的运算数符号一致
3
>>>17 % -4                         #(17-(-3))能被(-4)整除,17-(17//-4*-4)=-3
-3
```

(5) 关系运算符可以连用,作用于集合时表示测试集合之间的包含关系。

```
>>>1 < 3 < 5                       #等价于 1 < 3 and 3 < 5
True
>>>'Hello' > 'world'               #逐个比较对应位置上的字符,得出结论立即停止
False
>>>[1, 2, 3] < [1, 2, 4]           #比较列表大小,规则与字符串比较类似
True
>>> 'Hello' > 3                    #字符串和数字不能比较
TypeError: '>' not supported between instance of 'str' and 'int'
>>>{1, 2, 3} < {1, 2, 3, 4}        #测试是否真子集
True
```

(6) 元素测试运算符 in 用于测试一个对象是不是另一个对象的元素。

```
>>>3 in [1, 2, 3]                  #测试 3 是否存在列表[1, 2, 3]中
True
>>>5 in range(1, 10, 1)            #range()是用来生成指定范围数字的内置函数
True
>>>'abc' in 'abcdefg'              #子字符串测试
True
>>>for i in (3, 5, 7):             #循环,成员遍历,见第 3 章
    print(i, end='\t')             #注意,这里按两下 Enter 键才会执行

3    5    7
```

(7) 同一性测试运算符 is 用于测试两个对象是否引用同一个地址,是则返回 True,否则返回 False。引用同一个对象的两个变量具有相同的内存地址。

```
>>>3 is 3
True
>>>x = [300, 300, 300]
>>>x[0] is x[1]                    #基于值的内存管理,同一个值在内存中只有一份
True
>>>x = [1, 2, 3]
>>>y = [1, 2, 3]
>>>x is y                          #上面形式创建的 x 和 y 不是同一个列表对象
False
```

(8) 位运算符只能用于整数,其内部执行过程:首先将整数转换为二进制数,然后右对齐,必要的时候左侧补 0,按位进行运算,最后再把计算结果转换为十进制数返回。

```
>>>3 << 2                          #把 3 左移 2 位,每左移 1 位相当于乘以 2
12
```

```
>>> 3 & 7                              #"按位与"运算
3
>>> 3 | 8                              #"按位或"运算
11
>>> 3 ^ 5                              #"按位异或"运算
6
```

(9) 集合的并集、交集、对称差集等运算借助于位运算符来实现,差集使用减号运算符实现。

```
>>> {1, 2, 3} | {3, 4, 5}              #并集,自动删除重复元素,集合中每个元素都是唯一的
{1, 2, 3, 4, 5}
>>> {1, 2, 3} & {3, 4, 5}              #交集
{3}
>>> {1, 2, 3} ^ {3, 4, 5}              #对称差集
{1, 2, 4, 5}
>>> {1, 2, 3} - {3, 4, 5}              #差集
{1, 2}
```

(10) and 和 or 具有惰性求值特点,只计算必须计算的表达式。

```
>>> 3 > 5 and a > 3                    #注意,此时并没有定义变量 a,但不会出错
False
>>> 3 > 5 or a > 3                     #3 > 5 的值为 False,所以需要计算后面表达式,出错
NameError: name 'a' is not defined
>>> 3 < 5 or a > 3                     #3 < 5 的值为 True,不需要计算后面表达式
True
>>> 3 and 5                            #最后一个计算的表达式的值作为整个表达式的值
5
>>> 3 and 5 > 2
True
>>> 3 not in [1, 2, 3]                 #逻辑非运算 not
False
>>> 3 is not 5                         #not 的计算结果肯定是 True 或 False 二者之一
True
```

(11) 逗号并不是运算符,只是一个普通分隔符。

```
>>> 'a' in 'b', 'a'
(False, 'a')
>>> 'a' in ('b', 'a')
True
>>> x = 3, 5                           #创建元组
>>> x
(3, 5)
>>> x = 3 + 5, 7
>>> x
(8, 7)
```

(12) Python 不支持++和--运算符,虽然在形式上有时候似乎可以这样用,但实际上是另外的含义。

```
>>> i = 3
>>> ++i                                #正正得正
```

```
3
>>>+(+i)                                    #与++i 等价
3
>>>i++                                      #Python 不支持++运算符,语法错误
SyntaxError: invalid syntax
>>>--i                                      #负负得正
3
>>>-(-i)                                    #与--i 等价
3
```

(13) 方括号"[]"可以用于定义列表或列表推导式;还可以用于指定整数下标或切片,访问列表、元组、字符串中的一个或一部分元素;也可以指定字典的"键"作下标,访问对应的"值"。圆点运算符"."用于访问模块、类或对象的成员。

```
>>>import random
>>>data = random.choices(range(10), k =10)  #调用random模块中的函数
>>>print(data)
[8, 6, 1, 8, 9, 6, 8, 6, 2, 1]
>>>data.sort()                              #调用列表对象的sort()方法
>>>print(data)
[1, 1, 2, 6, 6, 6, 8, 8, 8, 9]
>>>print(data[3])                           #访问列表中下标为3的元素
6
>>>print(data[1:5])                         #访问列表中下标介于[1,5)区间的元素
[1, 2, 6, 6]
>>>data.remove(8)                           #调用列表方法remove()删除第一个8
>>>print(data)
[1, 1, 2, 6, 6, 6, 8, 8, 9]
>>>data = {'red':(1,0,0), 'green':(0,1,0), 'blue':(0,0,1)}
>>>print(data['red'])                       #使用"键"作下标,访问对应的"值"
(1, 0, 0)
```

(14) 虽然很多人一直习惯把等号"="称作赋值运算符,但严格来说,Python 中的等号是不能算作运算符的,它只是变量名或参数名与表达式之间的分隔符,用于表示等号左侧的变量引用右侧表达式的计算结果。

Python 3.8 及之后的版本新增了真正的赋值运算符":=",也称海象运算符,可以在表达式中创建变量并为变量赋值,如果运用得当可以让代码更加简洁。这个运算符不能在普通语句中直接使用,如果必须使用需要在外面加一对圆括号。下面的代码演示了赋值运算符":="的用法,第一段代码中用到了随机选择,每次运行结果可能不一样,请自行运行程序并查看结果,其中用到的选择结构与循环结构见第3章,文件操作见第7章。

```
from random import choices

text =''.join(choices('01', k=100))
if (c:=text.count('0'))>50:
    print(f'0出现的次数多,有{c}次。')
else:
    print(f'1出现的次数多,有{100 - c}次。')

with open('news.txt', encoding= 'utf8') as fp:
```

```
    while (length:=len(line:=fp.readline())) >0:
        if length >30:
            print(f'第一个长度大于30的行为{line},长度为{length}')
            break
    else:
        print('没有长度大于30的行。')

for num in (data:=[1,2,3]):
    print(num)
#for循环结构中创建的循环变量在循环结束之后还可以访问
print(data, num)
```

在Python中，单个任意类型的对象或常数属于合法表达式，使用运算符连接的变量和常量，以及函数调用的任意组合也属于合法的表达式。

1.4.6 常用内置函数

内置函数不需要导入任何模块即可直接使用。例如，本书很多例子中都会用到的map()函数就属于Python内置函数，该函数在本书后面会有讲解，也可以直接跳至第5章阅读，或者使用help(map)查看该函数帮助文档进行学习。

执行下面的命令可以列出所有内置函数和内置对象：

```
>>> dir(__builtins__)                    #前后各两个下画线，共4个
```

Python常用的内置函数及其功能简要说明如表1-6所示，其中方括号表示可选参数，可有可无。

表1-6 Python常用的内置函数及其功能简要说明

函 数	功能简要说明
abs(x, /)	返回数字x的绝对值或复数x的模，斜线表示该位置之前的所有参数必须为位置参数（见5.3节）。例如，只能使用abs(-5)的形式调用，不能使用abs(x=-5)的形式进行调用
aiter(async_iterable, /)	返回异步可迭代对象的异步迭代器对象，Python 3.10新增
all(iterable, /)	如果可迭代对象（可以使用for循环逐个遍历其中元素的对象，从面向对象程序设计的角度来讲，可迭代对象是指实现了特殊方法__iter__()的类的对象）iterable为空或其中所有元素都等价于True则返回True，否则返回False
anext(…)	返回异步迭代器对象中的下一个元素，Python 3.10新增
any(iterable, /)	如果可迭代对象iterable中存在等价于True的元素就返回True；否则返回False；如果可迭代对象iterable为空，则返回False
ascii(obj, /)	返回对象的ASCII码表示形式，进行必要的转义。例如，ascii('1234')返回"'1234'"，ascii('微信公众号：Python小屋')返回"'\\u5fae\\u4fe1\\u516c\\u4f17\\u53f7\\uff1aPython\\u5c0f\\u5c4b'"，其中'\\u5fae'为汉字'微'的转义字符
bin(number, /)	返回整数number的二进制形式的字符串，参数number可以是二进制、八进制、十进制或十六进制数。例如，bin(5)的值为'0b101'，bin(-5)的值为'-0b101'，bin(511)的值为'0b111111111'

续表

函　　数	功能简要说明
bool(x)	如果参数 x 的值等价于 True 就返回 True，否则返回 False
breakpoint(*args, **kws)	启动并进入 pdb 调试器环境
bytearray(iterable_of_ints) bytearray(string, encoding 　　　　[, errors]) bytearray(bytes_or_buffer) bytearray(int) bytearray()	返回可变的字节数组，可以使用函数 dir() 和 help() 查看字节数组对象的详细用法。例如，bytearray((65, 97, 103)) 返回 bytearray(b'Aag')，bytearray('社会主义核心价值观', 'gbk') 返回 bytearray(b'\xc9\xe7\xbb\xe1\xd6\xf7\xd2\xe5\xba\xcb\xd0\xc4\xbc\xdb\xd6\xb5\xb9\xdb')，bytearray(b'abcd') 返回 bytearray(b'abcd')，bytearray(5) 返回 bytearray(b'\x00\x00\x00\x00\x00')
bytes(iterable_of_ints) bytes(string, encoding 　　　[, errors]) bytes(bytes_or_buffer) bytes(int) bytes()	创建字节串或把其他类型数据转换为字节串，不带参数时创建空字节串。例如，bytes(5) 表示创建包含 5 个 0 的字节串 b'\x00\x00\x00\x00\x00'，bytes((97, 98, 99)) 表示把若干介于 [0, 255] 区间的整数转换为字节串 b'abc'，bytes((97,)) 可用于把一个介于 [0, 255] 区间的整数 97 转换为字节串 b'a'，bytes('董付国', 'utf-8') 使用 UTF-8 编码格式把字符串 '董付国' 转换为字节串 b'\xe8\x91\xa3\xe4\xbb\x98\xe5\x9b\xbd'
callable(obj, /)	如果参数 obj 为可调用对象就返回 True，否则返回 False。Python 中的可调用对象包括函数、lambda 表达式、类、类方法、静态方法、对象方法、实现了特殊方法 __call__() 的类的对象
classmethod(function)	修饰器函数，用于把一个普通成员方法转换为类方法
chr(i, /)	返回 Unicode 编码为 i 的字符，其中 0 <= i <= 0x10ffff
compile(source, filename, 　　　　mode, flags=0, 　　　　dont_inherit=False, 　　　　optimize=-1, *, 　　　　_feature_version=-1)	把 Python 程序源码伪编译为字节码，可被 exec() 或 eval() 函数执行
complex(real=0, imag=0)	返回复数，其中 real 是实部，imag 是虚部，默认值均为 0，直接调用函数 complex() 不加参数时返回虚数 0j
delattr(obj, name, /)	删除对象 obj 的 name 属性，等价于 del obj.name
dict() dict(mapping) dict(iterable) dict(**kwargs)	创建空字典或把可迭代对象转换为字典。参数名前面加两个星号表示可以接收多个关键参数，也就是调用函数时以 name=value 形式传递的参数（见 5.3.2 节）
dir([object])	返回指定对象或模块 object 的成员列表，如果不带参数则返回包含当前作用域内所有可用对象名的列表
divmod(x, y, /)	计算整商和余数，返回元组 (x//y, x%y)，满足恒等式：div * y + mod ≡ x
enumerate(iterable, start=0)	枚举可迭代对象 iterable 中的元素，返回可生成形式为 (start, iterable[0])、(start+1, iterable[1])、(start+2, iterable[2])、…的值的 enumerate 对象（属于迭代器对象，具有惰性求值的特点，从面向对象程序设计的角度来讲，迭代器对象是指实现了特殊方法 __iter__() 和 __next__() 的类的对象），start 表示计数的起始值，默认为 0

续表

函　　数	功能简要说明
eval(source, globals=None, locals=None, /)	计算并返回字符串或字节串对象 source 中表达式的值，参数 globals 和 locals 用来指定字符串 source 中变量的值，如果二者有冲突，以 locals 为准。如果参数 globals 和 locals 都没有指定，就在当前作用域及更外层作用域中搜索字符串 source 中的变量并替换。该函数可以对任意字符串进行求值，有安全隐患，建议使用标准库 ast 中的安全函数 literal_eval()
exec(source, globals=None, locals=None, /)	在参数 globals 和 locals 指定的上下文中执行 source 代码或者 compile() 函数编译得到的字节码对象
exit()	结束程序，退出当前 Python 环境
filter(function or None, iterable)	使用可调用对象 function 描述的规则对 iterable 中的元素进行过滤，返回 filter 对象（属于迭代器对象，具有惰性求值的特点）。filter 对象可生成可迭代对象 iterable 中使可调用对象 function 返回值等价于 True 的那些元素，第一个参数为 None 时返回的 filter 对象可生成 iterable 中所有等价于 True 的元素
float(x=0, /)	把整数或字符串 x 转换为浮点数
format(value, format_spec='', /)	把参数 value 按 format_spec 指定的格式转换为字符串。例如，format(5, '06d') 等价于'{:06d}'.format(5)，结果均为'000005'
frozenset() frozenset(iterable)	创建不可变、无序的集合对象
getattr(object, name[, default])	获取对象 object 的 name 属性，等价于 obj.name
globals()	返回当前作用域中能够访问的所有全局变量名与值组成的字典
hasattr(obj, name, /)	检查对象 obj 是否拥有 name 指定的属性，如 hasattr(int, '__call__') 的值为 True
hash(obj, /)	计算参数 obj 的哈希值，如果 obj 不可哈希则抛出异常。该函数常用于测试一个对象是否可哈希，但一般不需要关心具体的哈希值是什么。对于 Python 内置对象，可哈希与不可变是一个意思，不可哈希与可变是一个意思。从面向对象程序设计的角度来讲，可哈希对象是指同时实现了特殊方法__hash__()和__eq__()的类的对象
help(obj)	返回对象 obj 的帮助信息，例如，help(sum)可以查看内置函数 sum() 的使用说明，help('math')可以查看标准库 math 的使用说明，使用任意列表对象作为参数可以查看列表对象的使用说明，参数也可以是类、对象方法、标准库或扩展库函数等。直接调用 help() 函数不加参数时进入交互式帮助会话，输入字母 q 退出
hex(number, /)	返回整数 number 的十六进制形式的字符串，参数 number 可以是二进制、八进制、十进制或十六进制数
id(obj, /)	返回对象的内存地址
input(prompt=None, /)	输出参数 prompt 的内容作为提示信息，接收键盘输入的内容，回车表示输入结束，以字符串形式返回输入的内容（不包含最后的回车符）

续表

函　　数	功能简要说明
int([x]) int(x, base=10)	返回实数 x 的整数部分，或把字符串 x 看作 base 进制数并转换为十进制，base 默认为十进制，取值范围为 0 或 2～36 的整数
isinstance(obj, 　　　　class_or_tuple, /)	测试对象 obj 是否属于指定类型（如果有多个类型，需要放到元组中表示其中之一）的实例
issubclass(cls, 　　　　class_or_tuple, /)	检查参数 cls 是不是 class_or_tuple 或其（其中某个类的）子类
iter(iterable) iter(callable, sentinel)	第一种形式用来根据可迭代对象创建迭代器对象；第二种形式用来重复调用可调用对象，直到其返回参数 sentinel 指定的值
len(obj, /)	返回容器对象 obj 包含的元素个数，适用列表、元组、集合、字典、字符串及 range 对象，不适用具有惰性求值特点的生成器对象，以及 map、zip 等迭代器对象
list(iterable=(), /)	把对象 iterable 转换为列表，不带参数时返回空列表
locals()	返回包含当前作用域中局部变量名和值的字典
map(func, *iterables)	返回可生成若干函数值的 map 对象（属于迭代器对象，具有惰性求值的特点），函数 func 的参数分别来自 iterables 指定的一个或多个可迭代对象中对应位置的元素，直到最短的一个可迭代对象中的元素全部用完。形参前面加一个星号表示可以接收任意多个按位置传递的实参（见 5.3.3 节）
max(iterable, *[, default=obj, 　　key=func]) max(arg1, arg2, * args, *[, 　　key=func])	返回可迭代对象中所有元素或多个实参的最大值，允许使用参数 key 指定排序规则，使用参数 default 指定 iterable 为空时返回的默认值
memoryview(object)	创建引用指定对象的内存视图对象
min(iterable, *[, default=obj, 　　key=func]) min(arg1, arg2, *args, *[, 　　key=func])	返回可迭代对象中所有元素或多个实参的最小值，允许使用参数 key 指定排序规则，使用参数 default 指定 iterable 为空时返回的默认值
next(iterator[, default])	返回迭代器对象 iterator 中的下一个值，如果 iterator 为空则返回参数 default 的值；如果不指定 default 参数，当 iterator 为空时会抛出 StopIteration 异常
oct(number, /)	返回整数 number 的八进制形式的字符串，参数 number 可以是二进制、八进制、十进制或十六进制数
open(file, mode='r', buffering= 　　-1, 　　encoding=None, errors=None, 　　newline=None, closefd=True, 　　opener=None)	打开参数 file 指定的文件并返回文件对象，详见 7.1 节
ord(c, /)	返回单个字符 c 的 Unicode 编码
pow(base, exp, mod=None)	相当于 base**exp 或 (base**exp) % mod

续表

函　数	功能简要说明
print(value, ⋯, sep=' ', end='\n', file=sys.stdout, flush=False)	基本输出函数，可以输出一个或多个表达式的值，参数 sep 表示相邻数据之间的分隔符（默认为空格），参数 end 用于指定输出完所有值后的结束符（默认为换行符）
property(fget=None, fset=None, fdel=None, doc=None)	用于创建属性
quit(code=None)	结束程序，退出当前 Python 环境
range(stop) range(start, stop[, step])	返回 range 对象，其中包含区间[start, stop)内以 step 为步长的整数，start 默认值为 0，step 默认值为 1
reduce(function, sequence[, initial])	将双参数函数 function 以迭代的方式从左到右依次应用至可迭代对象 sequence 中的每个元素，并把中间计算结果作为下一次计算时函数 function 的第一个参数，最终返回单个值作为结果。在 Python 3.x 中 reduce() 不是内置函数，需要从标准库 functools 中导入再使用
repr(obj, /)	把对象 obj 转换为适合 Python 解释器内部识别的字符串形式，对于不包含反斜线的字符串和其他类型对象，repr(obj)与 str(obj)功能一样；对于包含反斜线的字符串，repr()会把单个反斜线转换为两个反斜线
reversed(sequence, /)	返回序列 sequence 中所有元素逆序的 reversed 对象（属于迭代器对象，具有惰性求值的特点）
round(number, ndigits=None)	对整数或实数 number 进行四舍五入，最多保留 ndigits 位小数，参数 ndigits 可以为负数。如果 number 本身的小数位数少于 ndigits，不再处理。例如，round(3.1, 3)的结果为 3.1，round(1234, -2)的结果为 1200
set() set(iterable)	创建空集合或把可迭代对象 iterable 转换为集合
setattr(obj, name, value, /)	设置对象属性，相当于 obj.name = value
slice(stop) slice(start, stop[, step])	创建切片对象，可以用作下标。例如，对于列表对象 data，那么 data[slice(start, stop, step)]等价于 data[start:stop:step]
sorted(iterable, /, *, key=None, reverse=False)	返回参数 iterable 中所有元素排序后组成的列表，参数 key 用于指定排序规则或依据，参数 reverse 用于指定升序或降序，默认值 False 表示升序。单个星号 * 作参数表示该位置后面的所有参数都必须为关键参数，星号本身不是参数（见 5.3 节）
staticmethod(function)	用于把普通成员方法转换为类的静态方法
str(object='') str(bytes_or_buffer[, encoding[, errors]])	把任意对象直接转换为字符串，或者把字节串使用参数 encoding 指定的编码格式转换为字符串，第二种用法相当于 bytes_or_buffer.decode(encoding)
sum(iterable, /, start=0)	返回参数 start 与可迭代对象 iterable 中所有元素之和
super() super(type) super(type, obj) super(type, type2)	返回基类

续表

函　　数	功能简要说明
tuple(iterable=(), /)	将可迭代对象 iterable 转换为元组，不加参数时返回空元组
type(object_or_name, bases, dict) type(object) type(name, bases, dict)	查看对象类型或创建新类型
vars([object])	不带参数时等价于 locals()，带参数时等价于 object.__dict__
zip(*iterables)	组合一个或多个可迭代对象中对应位置上的元素，返回 zip 对象（属于迭代器对象，具有惰性求值的特点），其中每个元素为 (iterables[0][i], iterables[1][i], iterables[2][i], …) 形式的元组，最终 zip 对象中可用的元素个数取决于所有参数可迭代对象中最短的那个

内置函数数量众多且功能强大，很难一下全部解释清楚，本书将在后面的章节(map()、reduce()、filter()这几个函数的介绍在 5.8 节，可提前学习)中根据内容组织的需要逐步展开并演示其用法。这里只通过几个例子来演示部分内置函数的使用，如果需要用到某个内置函数但还没有看到本书后面的讲解，可以通过内置函数 help()查看函数的使用帮助。

ord()用来返回单个字符的 Unicode 编码，chr()用来返回指定 Unicode 编码对应的字符，str()将任意类型参数转换为字符串。

```
>>> ord('A'), ord('董')
(65, 33891)
>>> chr(65), chr(20184)
('A', '付')
>>> chr(ord('A')+1)                          #Python 不支持字符与整数相加
'B'
>>> str(1234)                                #把任意对象直接转换为字符串
'1234'
>>> str([1, 2, 3])
'[1, 2, 3]'
>>> str((1, 2, 3))
'(1, 2, 3)'
>>> str({1, 2, 3})
'{1, 2, 3}'
```

max()函数

max()、min()、sum()这 3 个内置函数分别用于计算列表、元组或其他可迭代对象中所有元素的最大值、最小值以及所有元素之和，sum()默认只支持包含数值型元素的可迭代对象(除非指定 start 参数)，max()和 min()要求可迭代对象中的元素之间可比较大小。

```
>>> import random                            #导入模块，见 1.4.8 节
>>> a = [random.randint(1,100) for i in range(10)]   #列表推导式，见 2.1.9 节
>>> a
[72, 26, 80, 65, 34, 86, 19, 74, 52, 40]
>>> print(max(a), min(a), sum(a))
86 19 548
>>> max(['aa', 'b'], key=len)                #长度最大的字符串，key 参数的值应为可调用对象
'aa'
```

如果需要计算该列表中所有元素的平均值，可以使用下面的方法：

```
>>> a = [72, 26, 80, 65, 34, 86, 19, 74, 52, 40]
>>> sum(a) / len(a)
54.8
```

list()、tuple()、dict()、set()用于把其他类型的数据转换成列表、元组、字典和集合，或者创建空列表、空元组、空字典和空集合。

```
>>> list()                          #创建空列表，另外几个函数也有类似用法
[]
>>> list(range(5))                  #把 range 对象转换为列表
[0, 1, 2, 3, 4]
>>> tuple(range(5))                 #把 range 对象转换为元组
(0, 1, 2, 3, 4)
>>> dict(name='董付国', age=41)      #根据指定的"键"和"值"创建字典
{'name': '董付国', 'age': 41}
>>> set('Pythonnic')                #创建集合，自动删除重复，集合元素是无序的
{'o', 'i', 't', 'y', 'P', 'c', 'n', 'h'}
```

内置函数 eval()用于计算字符串或字节串内表达式的值，在有些场合也可以用来实现类型转换的功能，例如把字符串中的列表、元组、字典、集合还原为本来的类型。

```
>>> eval('3*5')                     #计算字符串内表达式的值
15
>>> eval('[1, 2, 3, 4]')            #字符串求值，还原为列表
[1, 2, 3, 4]
>>> list('[1, 2, 3, 4]')            #把字符串中所有元素都转换为列表中的元素
                                    #注意函数 list()和 eval()的区别
['[', '1', ',', ' ', '2', ',', ' ', '3', ',', ' ', '4', ']']
```

sorted()可以对可迭代对象中的元素进行排序并返回新列表，支持使用 key 参数指定排序规则，key 参数的值可以是函数、lambda 表达式(见 5.6 节)、类、方法(见第 6 章)等可调用对象(与 max()、min()函数的 key 参数相同)，不指定时表示直接按元素大小排列。还可以使用 reverse 参数指定是升序(False)排序还是降序(True)排序，默认为升序排序。

sorted()
函数

```
>>> x = list(range(11))             #创建列表
>>> import random
>>> random.shuffle(x)               #shuffle()用于随机打乱顺序
>>> x
[2, 4, 0, 6, 10, 7, 8, 3, 9, 1, 5]
>>> sorted(x)                       #按正常大小升序排序
[0, 1, 2, 3, 4, 5, 6, 7, 8, 9, 10]
>>> sorted(x, key=str)              #按转换成字符串后的大小升序排序
[0, 1, 10, 2, 3, 4, 5, 6, 7, 8, 9]
>>> x = ['aaaa', 'bc', 'd', 'b', 'ba']
>>> sorted(x, key=lambda item: (len(item), item))
                                    #先按长度排序，长度一样的按字符串大小排序
['b', 'd', 'ba', 'bc', 'aaaa']
>>> num = random.choices(range(1,10), k=5)  #5 个介于[1,10)区间的随机数
>>> num
[6, 3, 3, 1, 5]
```

```
>>> int(''.join(sorted(map(str, num), reverse=True)))
                              #几位数字能够组成的最大数,join()方法见4.1.2节
65331
```

range()函数的完整语法格式为range([start,] stop[, step]),有range(stop)、range(start, stop)和range(start, stop, step)3种用法。该函数返回的range对象中包含区间[start, stop)内以step为步长的整数。其中,参数start默认值为0,step默认值为1。

```
>>> range(5)                  #start默认值为0,step默认值为1
range(0, 5)
>>> list(range(5))            #把range对象转换为列表
[0, 1, 2, 3, 4]
>>> list(range(1, 10))        #指定start和stop参数
[1, 2, 3, 4, 5, 6, 7, 8, 9]
>>> list(range(1, 10, 2))     #同时指定start、stop和step参数
[1, 3, 5, 7, 9]
>>> list(range(9, 0, -2))     #步长为负数时,start应比stop大
[9, 7, 5, 3, 1]
>>> list(range(5, 10, -1))    #否则不会包含任何元素
[]
```

zip()函数把多个可迭代对象中对应位置上的元素分别组合到一起,返回一个zip对象,其中每个元素都是包含原来的多个可迭代对象对应位置上元素的元组,最终结果中包含的元素个数取决于所有参数可迭代对象中最短的那个。

```
>>> list(zip('abcdef', [1, 2, 3]))   #组合字符串和列表中对应位置上的元素
                                     #结果数量取决于最短的参数
[('a', 1), ('b', 2), ('c', 3)]
>>> list(zip('123', 'abc', ',.!'))   #可以处理任意多个可迭代对象
[('1', 'a', ','), ('2', 'b', '.'), ('3', 'c', '!')]
>>> x = zip('abcd', '1234')
>>> list(x)
[('a', '1'), ('b', '2'), ('c', '3'), ('d', '4')]
>>> list(x)                          #注意,zip对象中的每个元素只能使用一次
                                     #访问过的元素就不存在了
                                     #enumerate、filter、map对象以及生成器对象也
                                     #有这个特点
[]
```

对于初学者而言,dir()和help()这两个内置函数应该是最有用的。使用dir()函数可以查看指定模块中包含的所有成员或者指定对象类型所支持的操作,help()函数返回指定模块或函数的说明文档,这对于了解和学习新的模块与知识是非常重要的,能够熟练使用这两个函数也是学习能力的重要体现。

```
>>> import math
>>> dir(math)              #查看模块中的可用对象
>>> help(math.sqrt)        #查看指定函数的使用帮助
>>> help(math.sin)
>>> dir(3+4j)              #查看数字类型对象成员,参数可以是任意数字
>>> dir('')                #查看字符串类型成员,参数可以是任意字符串
```

1.4.7 基本输入输出

内置函数 input()用于接收用户的键盘输入，一般用法为

```
x = input('提示信息: ')
```

该函数会阻塞程序，输入结束按 Enter 键才会继续执行。不论输入什么，input()函数的返回值都是字符串，可以按需转换为相应的类型。

1.4.7

```
>>> x = input('Please input:')
Please input:3
>>> print(type(x))                    #可以使用 int()函数转换为整数
<class 'str'>
>>> x = input('Please input:')
Please input:[1, 2, 3]
>>> print(type(x))                    #如果要转换为列表，需要使用 eval()函数
<class 'str'>
```

基本输出函数 print()的语法如下：

```
print(value, …, sep=' ', end='\n', file=sys.stdout, flush=False)
```

其中，参数 sep 前面的一个或多个参数是要输出的值，参数 sep 用于指定分隔符，参数 end 用于指定结束符，参数 flush=True 时直接输出而不缓存。例如，下面的代码用来修改默认的分隔符：

```
>>>print(3, 5, 7, sep=':')
3:5:7
>>>print(3, 5, 7, sep=',')            #参数 sep 后面的参数必须是关键参数，见 5.3.2 节
3,5,7
```

下面的代码用于修改默认的结束符，每次输出之后以空格结束，不换行：

```
>>> for i in range(10, 20):           #循环结构见 3.3 节
    print(i, end=' ')
10 11 12 13 14 15 16 17 18 19
```

下面的程序综合演示了 print()函数几个参数的用法，把代码保存为程序文件 progress.py，然后在命令提示符或 Powershell 环境中执行命令 python progress.py，可以用来模拟一个进度条。

```
from time import sleep

for i in range(1, 101):
    print('\r', '.' * (i//5), i, '%', sep='', end='', flush=True)
    sleep(0.1)                        #暂停执行 0.1s，参数越小程序执行速度越快
```

1.4.8 模块导入与使用

Python 默认安装仅包含内置模块和标准库模块，但用户可以很方便地按需安装大量的其他扩展模块（见 1.3 节）。在 Python 启动时，也仅加载了很少的一部分模块（例如，内置模块__builtins__），在需要时再由程序员显式地加载（扩展库需要先安装）其他模块。这样可以减小程序运行的压力，仅加载真正需要的模块和对象，且具有很强的可扩展性，也可以减

小打包后可执行文件的大小。

1. import 模块名[as 别名]

使用这种方式导入模块以后,需要在要使用的对象之前加上前缀,即以"模块名.对象名"的方式访问。也可以为导入的模块设置一个别名,然后可以使用"别名.对象名"的方式使用其中的对象。

```
>>> import math
>>> math.sin(0.5)                #求 0.5 弧度的正弦值
0.479425538604203
>>> import random
>>> x = random.random()          #获得[0, 1)区间上的随机小数
>>> n = random.randint(1, 100)   #获得[1, 100]区间上的随机整数
>>> import numpy as np           #导入模块并设置别名,扩展库 NumPy 见 17.1 节
>>> a = np.array((1, 2, 3, 4))   #通过模块的别名来访问其中的对象
```

2. from 包名/模块名 import 模块名/对象名[as 别名]

使用这种方式仅导入明确指定的对象,并且可以为导入的对象起一个别名。这种导入方式可以减少查询次数,提高访问速度,同时也减少了程序员需要输入的代码量,不需要使用模块名作为前缀。例如:

```
>>> from math import sin
>>> sin(3)
0.1411200080598672
>>> from math import sin as f
>>> f(3)
0.1411200080598672
```

比较极端的情况是一次导入模块中的所有对象,例如:

```
from math import *
```

使用这种方式固然简单省事,但是并不推荐使用,一旦多个模块中有同名的对象,这种方式会导致混乱。

在测试自己编写的模块时,可能需要频繁地修改代码并重新导入模块,可以使用 imp 模块或 importlib 模块的 reload() 函数。不论使用哪种方法重新加载模块,都要求该模块已经被正确加载,第一次导入和加载模块时不能使用 reload() 函数。

在导入模块时,Python 首先在当前目录中查找需要导入的模块文件,如果没有找到,则从 sys 模块的 Path 变量所指定的目录中查找,如果仍没有找到模块文件,则提示模块不存在。可以使用 sys 模块的 Path 变量查看 Python 导入模块时搜索模块的路径,也可以使用 append() 方法向其中添加自定义的文件夹以扩展搜索路径。

在导入模块时,会优先导入相应的.pyc 文件,如果相应的.pyc 文件与.py 文件时间不相符或不存在对应的.pyc 文件,则导入.py 文件并重新将该模块文件编译为.pyc 文件。

在大的程序中可能会需要导入很多模块,此时应按照下面的顺序来依次导入模块。

(1)导入 Python 内置模块和标准库,如 math、os、sys、re。

(2)导入第三方扩展库,如 PIL、NumPy、SciPy。

(3)导入自己定义和开发的本地模块。

1.5 Python 代码编写规范

(1) 缩进。

Python 程序是依靠代码块的缩进来体现代码之间的逻辑关系的。对于 with 块(见 7.1 节)、类定义(见第 6 章)、函数定义(见第 5 章)、选择结构(见 3.1 节)、循环结构(见 3.2 节)以及异常处理结构(见第 8 章)来说,行尾的冒号以及下一行的缩进表示一个代码块的开始,缩进结束则表示一个代码块结束了。

一般以 4 个空格为基本缩进单位,在 IDLE 中可以使用下面的方式来修改基本缩进单位,单击菜单 Options→Configure IDLE,如图 1-7 所示。

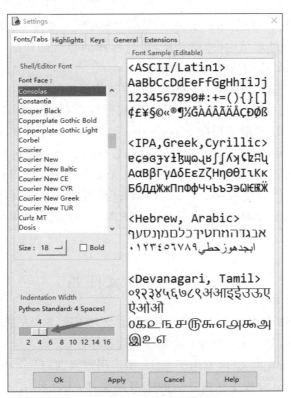

图 1-7　IDLE 中基本缩进量的设置

编写程序时,可以通过下面的菜单进行代码块的批量缩进和反缩进,需要提前将需要缩进或反缩进的代码块选中:

> Format→Indent Region/Dedent Region

也可以使用快捷键 `Ctrl+]` 或 `Tab` 键缩进,使用快捷键 `Ctrl+[` 反缩进。

(2) 注释。

注释对于程序理解和团队合作开发具有非常重要的意义。据统计,一个可维护性和可读性都很强的程序一般会包含 30% 以上的注释。Python 中常用的注释方式主要有两种。

① 以符号 # 开始,表示本行 # 之后的内容为注释,不会被当作代码进行解释和执行。

② 包含在一对三引号之间且不属于任何语句的内容也可以认为是注释,常用于文档字符串。

在 IDLE 中,可以选中代码块,然后使用下面的菜单快速注释/解除注释代码块:

Format→Comment Out Region/Uncomment Region

或者使用快捷键 Alt+3 和 Alt+4 分别进行代码块的批量注释和解除注释。

(3) 每个 import 语句只导入一个模块,尽量避免一次导入多个模块。

(4) 如果一行语句太长,可以在行尾使用续行符"\"来表示下面紧接的一行仍属于当前语句,也可以使用括号来包含多行内容。列表、元组、字典、集合、推导式、函数调用等可以写成多行。

(5) 使用必要的空格与空行增强代码的可读性。一般来说,运算符两侧(下标运算符和圆点运算符、圆括号除外)、逗号后面建议增加一个空格,正号和负号只在前面加一个空格;不同功能的代码块之间、不同的函数定义以及不同的类定义之间建议增加一个空行。

(6) 适当使用异常处理结构(见第 8 章)提高程序的容错性和健壮性,但不能过多依赖异常处理结构,适当的显式判断还是必要的。

(7) 软件应具有较强的可测试性,测试与开发齐头并进。

有很多成熟的工具可以检查 Python 代码的规范性,例如 pep8、flake8、pylint 等。可以使用 pip 安装 pep8 工具,然后使用命令 pep8 test.py 检查 test.py 文件中 Python 代码的规范性。pep8 常用的可选参数有--show-source、--first、--show-pep8 等。flake8 结合了 pyflakes 和 pep8 的特点,可以检查更多的内容,优先推荐使用。使用 pip install flake8 可以直接安装 flake8 工具,然后使用命令 flake8 test.py 检查 test.py 文件中代码的规范性。也可以使用 pip 安装 pylint 工具,然后使用命令行工具 pylint 或者可视化工具 pylint-gui 来检查程序的规范性。

1.6 Python 文件名

在 Python 中,不同扩展名的文件类型有不同的含义和用途,常见的扩展名主要有以下几种。

(1) py——Python 源文件,常用于控制台应用程序。

(2) pyw——Python 源文件,常用于 GUI(图形用户界面)程序文件。

(3) pyc——Python 源程序伪编译后的字节码文件,无法使用文本编辑器直接查看该类型文件内容,可用于隐藏 Python 源代码和提高加载速度。作为 Python 模块使用时,第一次被导入时将被编译成字节码的形式,并在以后再次导入时优先使用.pyc 文件,以提高模块的加载和运行速度。直接执行 Python 程序文件时并不生成.pyc 文件,但可以使用 py_compile 模块的 compile() 函数进行编译以提高加载和运行速度。另外,Python 还提供了 compileall 模块,其中包含 compile_dir()、compile_file() 和 compile_path() 等方法,用来支持批量 Python 源程序文件的伪编译。

(4) pyo——优化的 Python 字节码文件,同样无法使用文本编辑器直接查看其内容。可以使用 python -O -m py_compile file.py 或 python -OO -m py_compile file.py 进行优化编译。从 Python 3.5 开始不再支持.pyo 文件,而是使用.pyc 文件存储优化和非优化字节码。

(5) pyd——一般是由其他语言编写并编译生成的二进制文件,常用于实现某些软件工具的 Python 编程接口插件或 Python 动态链接库。也可以使用扩展库 easycython 把 Python 程序打包为 pyd 文件来保护源码。

在给自己的 Python 程序文件起名时,不能与标准库和扩展库文件名相同。

1.7　Python 程序的 __name__ 属性

每个 Python 程序在运行时都有一个 __name__ 属性。如果程序作为模块被导入,则其 __name__ 属性值被自动设置为模块名;如果程序独立运行,则其 __name__ 属性值被自动设置为'__main__';多进程程序设计(见 13.4 节)时子进程中 __name__ 的值为 '__mp_main__'。例如,假设文件 nametest.py 中只包含下面一行代码:

```
print(__name__)
```

直接运行该程序时,运行结果如下:

```
__main__
```

而将该文件作为模块导入时得到如下执行结果:

```
>>> import nametest
nametest
```

1.7

利用 __name__ 属性可以控制 Python 程序中代码的运行方式。例如,编写一个包含大量函数的模块,而不希望该模块直接运行,则可以在程序文件中添加以下代码:

```
if __name__ == '__main__':
    print('Please use me as a module.')
```

这样一来,程序直接执行时将会得到提示"Please use me as a module.",使用 import 语句将其作为模块导入后可以使用其中的类、方法、常量或其他成员。

*1.8　编写和使用自己的包

包是 Python 用来组织程序文件和命名空间的重要方式,可以看作包含大量 Python 程序模块的文件夹。在包的每个目录中都必须包含一个 __init__.py 文件,该文件可以是一个空文件,仅用于表示该文件夹是一个包。__init__.py 文件的主要用途是设置 __all__ 变量以及执行初始化包所需的代码,其中 __all__ 变量中定义的对象可以在使用"from…import *"时全部被正确导入。

假设有如下结构的包(或目录树),并且模块 moduleX 和 moduleZ 中分别定义了函数 funcX() 和 funcZ()。

```
package/
    __init__.py
    subpackage1/
        __init__.py
        moduleX.py
        moduleY.py
```

```
        subpackage2/
            __init__.py
            moduleZ.py
        moduleA.py
```

在程序中下面的几种导入方式都是正确的:

```
import package
import package.subpackage1
from package.subpackage1.moduleX import funcX
```

除上面的绝对导入方式外,在包内模块中还可以使用相对导入方式,使用同一个包里其他模块中的成员。相对导入以圆点开始,一个圆点表示当前模块所在的包(或目录),两个圆点表示上一级包(或父目录)。例如,在上面的模块 moduleY 中下面的几种导入方式都是正确的:

```
from .moduleX import func1
from . import moduleX
from ..subpackage2.moduleZ import funcZ
```

*1.9 Python 程序伪编译与打包

1. Python 程序伪编译

可以使用 py_compile 模块的 `compile()` 函数或 compileall 模块的 `compile_file()` 函数对 Python 源程序文件进行伪编译,得到扩展名为 pyc 的字节码以提高加载和运行速度,同时还可以隐藏源代码。假设有程序文件 Stack.py,并已导入 py_compile,那么可以使用语句 `py_compile.compile('Stack.py')` 把 Stack.py 伪编译为字节码,如果需要优化编译可以使用 `py_compile.compile('Stack.py', optimize=1)` 或 `py_compile.compile('Stack.py', optimize=2)` 生成不同优化级别的字节码文件。生成的字节码文件保存在 __pycache__ 文件夹中。

此外,Python 的 compileall 模块还提供了 `compile_dir()` 和 `compile_path()` 等方法,用于支持 Python 源程序文件的批量编译。

Python 程序伪编译的字节码并不能很好地保护源码和知识产权,Python 扩展库 uncompyle6 或某些网站都可以很容易地对字节码反编译得到源码。

2. Python 程序打包

保护源代码的更好方式是把 Python 程序转换为二进制可执行程序之后再发布,这样使得打包后的程序可以在没有安装 Python 环境和相应扩展库的系统中运行,极大地方便了用户。可以把 Python 程序打包为可执行程序的工具有 py2exe(仅适用于 Windows 平台)、pyinstaller、cx_Freeze、Nuitka、py2app(仅适用于 macOS)等。打包之前应该保证 Python 程序可以正常运行,并且本机已安装了所有需要的扩展模块和相关的动态链接库文件。

以 pyinstaller 为例,使用 pip 安装该工具之后在命令提示符环境中使用命令 pyinstaller -F -w kousuan.pyw 或者 python pyinstaller-script.py -F -w kousuan.pyw 即可将 Python

程序 kousuan.pyw 及其所有依赖包打包成为当前所用平台上的可执行文件,安装扩展库 tinyaes 并在打包时增加 -- key 选项可实现代码混淆,进一步保护代码。

如果使用 cx_Freeze 工具,假设有 Python 程序 hello.py,在命令提示符环境执行 `python cxfreeze hello.py` 即可快速创建可执行程序并自动搜集依赖包。除此之外,在 Windows 平台上 cx_Freeze 还支持制作 .msi 安装程序。

1.10 案例精选

本节通过几个例题(配套 PPT 中还补充了 9 个)快速了解如何使用 Python 解决实际的问题,map() 函数的用法见表 1-6 和 5.8 节。

例 1-1 用户输入一个三位自然数,计算并输出其百位、十位和个位上的数字。

```
x = input('请输入一个三位自然数: ')
print(*map(int, x))                         #查阅表 1-6 和 5.8 节内容并结合右侧微课进行理解
```

例 1-1

注:与大部分程序设计教材一样,本书中给出的很多代码一般都不是完整的代码,只是为了演示特定的功能用法,没有考虑过于细节的外围工作。例如,在本例中,完整的程序还应该检查用户输入是不是数字、是不是三位数等,可以使用选择结构在计算之前进行判断,也可以使用异常处理结构增加程序的健壮性和容错性,类似问题后面不再赘述。

例 1-2 已知三角形的两边长及夹角,求第三边长。这里用到了 math 模块中求平方根的函数 sqrt() 和余弦函数 cos(),本例给出的是比较传统的写法,参考前面的知识,相信你可以写出更加简洁的代码。

例 1-2

```
import math
x = input('输入两边长及夹角(度): ')                #输入时使用空格分隔 3 个值
a, b, theta = map(float, x.split())               #序列解包,split() 方法用于切分字符串
c = math.sqrt(a**2+b**2-2*a*b*math.cos(theta*math.pi/180))   #cos() 参数为弧度
print('c=', c)
```

在这段代码中使用了序列解包的知识,在 2.2.3 节会详细讲解,这里可以不必深究,用心体会 Python 的精妙和强大即可。

例 1-3 任意输入三个英文单词,按字典顺序输出。本例主要注意变量值的交换方法,这是 Python 序列解包的常用方法。

```
s = input('x,y,z=')                         #输入时三个英文单词之间使用逗号分隔
x, y, z = s.split(',')                      #切分得到三个英文单词
if x > y:                                   #选择结构见 3.2 节
    x, y = y, x                             #交换变量值
if x > z:
    x, z = z, x
if y > z:
    y, z = z, y
print(x, y, z)
```

也可以直接使用 Python 内置函数 sorted() 快速实现上述功能。

```
s = input('x,y,z=')
```

```
x, y, z = sorted(s.split(','))                          #序列解包
print(x, y, z)
```

例 1-4 计算两点间的曼哈顿距离。曼哈顿距离用于计算城市中两点之间的最短行走距离,类似于中国象棋中"车"行走的路线。图 1-8 描述了这个距离的计算方法:在行走时不能穿越表示楼房或障碍物的浅色方块,只能沿道路向目的地前进。在这样的约束下,1、2、3 这三条路径的总长度是一样的。

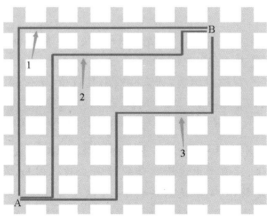

图 1-8 曼哈顿距离示意图

对于平面上的两个点(x_1,y_1)和(x_2,y_2),曼哈顿距离的定义为

$$|x_1-x_2|+|y_1-y_2|$$

对于高维空间中两个点$(x_1,x_2,\cdots,x_n)$和$(y_1,y_2,\cdots,y_n)$,曼哈顿距离的定义为

$$\sum_{k=1}^{n}|x_k-y_k|$$

```
def manhattanDistance(x, y):                            #函数定义与调用,见第 5 章
    return sum(map(lambda i, j: abs(i-j), x, y))

print(manhattanDistance([1, 2], [3, 4]))
print(manhattanDistance([1, 2, 3], [4, 5, 6]))
print(manhattanDistance([1, 2, 3, 4], [5, 6, 7, 8]))
```

例 1-5 统计一个字符串中所有字符在另一个字符串中出现的总次数。在使用机器学习算法进行垃圾邮件分类时,其中一个重要步骤是对邮件内容进行分词然后构造特征向量。然而,为了防止被分类为垃圾邮件,目前很多垃圾邮件中会在一些关键词中间插入类似【、】、*之类的字符来干扰分词。此时,可以反其道而行之,把这个特点作为一个判断依据,如果一封邮件中这样的干扰字符超过一定比例,则认为是垃圾邮件。

```
def check(s1, s2):
    return sum(map(lambda ch: s1.count(ch), s2))        #字符串方法 count()见 4.1.2 节

print(check('这是一个测*#试邮#件,内}含广'告', '[]*#/\\'))
```

*1.11　The Zen of Python

在本章最后，请大家静心阅读"Python 之禅"，不建议翻译下面这一段英文，这里也不打算过多地解读。只需要用心去体会，并在自己编写程序的时候多想想这段话，努力让自己编写的代码更加优雅、更符合 Python 的习惯。

```
>>> import this
The Zen of Python, by Tim Peters

Beautiful is better than ugly.
Explicit is better than implicit.
Simple is better than complex.
Complex is better than complicated.
Flat is better than nested.
Sparse is better than dense.
Readability counts.
Special cases aren't special enough to break the rules.
Although practicality beats purity.
Errors should never pass silently.
Unless explicitly silenced.
In the face of ambiguity, refuse the temptation to guess.
There should be one—and preferably only one—obvious way to do it.
Although that way may not be obvious at first unless you're Dutch.
Now is better than never.
Although never is often better than * right * now.
If the implementation is hard to explain, it's a bad idea.
If the implementation is easy to explain, it may be a good idea.
Namespaces are one honking great idea -- let's do more of those!
```

本 章 小 结

（1）选择 Python 版本时应首先充分了解自己的需求和可用的扩展库情况。

（2）pip 是 Python 扩展库管理的标准工具。

（3）在 Python 中一切都是对象。

（4）在 Python 中使用变量时不需要提前声明，直接为变量赋值即可创建一个变量。

（5）Python 采用基于值的内存管理方式。

（6）编程时优先考虑使用内置函数来实现自己的业务逻辑。

（7）在 Python 中，很多运算符具有多重含义。

（8）del 命令既可以删除一个变量，也可以删除列表、字典等可变序列中的部分元素。

（9）可以使用 import 语句导入模块中的对象，也可以为导入的模块或对象设置别名。

（10）一般建议每个 import 语句只导入一个模块。

（11）dir()和 help()是两个非常有用的内置函数，前者可以列出指定模块或对象中的成员，后者可以查看相应帮助文档和使用说明。

（12）Python 程序使用缩进来体现代码之间的逻辑关系，并且建议使用必要的空格、空行和注释来提高程序的可读性。

（13）Python 程序中的注释主要有两种形式。

① 以符号♯开头，表示本行该符号后的所有内容为注释。

② 放在一对三引号之间且不属于任何语句的内容也可以认为是注释。

（14）可以使用异常处理结构来提高程序的健壮性，但不建议过多依赖异常处理结构。

（15）可以通过 Python 程序的__name__属性控制程序的某些行为。

（16）Python 程序文件的扩展名为 py 或 pyw，Python 支持伪编译将程序转换为字节码，也可以打包为二进制可执行程序。

（17）Python 3.8 开始新增赋值运算符":="，可在选择结构和循环结构的条件表达式，以及函数调用和圆括号内的表达式中直接创建变量。

习　　题

1. _____ 是官方安装包自带的 Python 扩展库管理工具。

2. 运算符 % _____（可以、不可以）对浮点数进行求余数操作。

3. 一个数字 5 _____（是、不是）合法的 Python 表达式。

4. 在 Python 3.x 中 input()函数的返回值类型一律为 _____ 。

5. 简单说明如何选择正确的 Python 版本。

6. 为什么说 Python 采用的是基于值的内存管理模式？

7. 解释 Python 中的运算符 / 和 // 的区别。

8. 在 Python 中导入模块中的对象有哪几种方式？

9. 解释 Python 程序的__name__变量及其作用。

10. 编写程序，用户输入一个三位以上的整数，输出其百位以上的数字。例如，用户输入 1234，则程序输出 12（提示：使用整除运算）。

第 2 章　Python 可迭代对象

从面向对象程序设计的角度来讲,可迭代对象是指实现了特殊方法__iter__()的对象,可以简单理解为支持 for 循环从前向后逐个访问其中元素的对象。可迭代对象包括容器对象和迭代器(iterator)对象。容器对象包括列表、元组、字典、集合、字符串,其中列表为可变序列,元组和字符串为不可变序列,这三种对象都支持双向索引(0 表示第一个元素的下标,1 表示第二个元素的下标,2 表示第三个元素的下标……;-1 表示最后一个元素的下标,-2 表示倒数第二个元素的下标……)和切片操作。迭代器对象是指同时实现了特殊方法__iter__()和__next__()的类的实例,map 对象、zip 对象、filter 对象、enumerate 对象、生成器对象都属于迭代器对象。迭代器对象具有惰性求值的特点,只能从前往后逐个访问其中的元素,不支持下标和切片,并且每个元素只能使用一次。

本章通过大量案例介绍列表、元组、字典、集合等几种数据类型的用法,以及在实际应用中非常有用的切片操作、列表推导式、序列解包、生成器表达式等。在本章的最后,介绍了如何使用 Python 列表来实现栈、队列、树、图等较为复杂的数据结构并模拟其基本操作,其中用到的面向对象程序设计内容可参考第 6 章,函数相关内容见第 5 章。

2.1　列　　表

列表是 Python 内置的可变有序序列,是包含若干元素的有序连续内存空间。在形式上,列表的所有元素放在一对方括号"["和"]"中,相邻元素之间使用逗号分隔。当列表增加或删除元素时,列表对象自动进行内存的扩展或收缩,从而保证元素之间没有缝隙。Python 列表内存的自动管理可以大幅减少程序员的负担,但列表的这个特点会涉及列表中大量元素的移动,效率较低,并且对于某些操作可能会导致意外的错误结果。因此,除非确实有必要,否则应尽量从列表尾部进行元素的增加与删除操作,这会大幅提高列表处理速度。

同一个列表中元素的类型可以不相同,可以同时包含整数、实数、字符串等基本类型,也可以是列表、元组、字典、集合及其他自定义类型的对象。例如:

```
[10, 20, 30, 40]
['red flower', 'blue sky', 'green leaf']
['Python', 3.14, 5, [10, 20]]
[['file1', 200, 7], ['file2', 260, 9]]
```

都是合法的列表对象。

列表对象常用方法如表 2-1 所示。除此之外,Python 的很多内置函数和运算符也可以对列表和其他可迭代对象进行操作,后面将通过一些案例逐步进行介绍。

表 2-1 列表对象常用方法

方　　法	说　　明
append(object, /)	将任意对象 object 追加至当前列表的尾部
clear()	删除列表中的所有元素
copy()	返回当前列表对象的浅复制
count(value, /)	返回值为 value 的元素在当前列表中的出现次数
extend(iterable, /)	将有限长度的可迭代对象 iterable 中所有元素追加至当前列表的尾部
insert(index, object, /)	在当前列表的 index 位置前面插入对象 object
index(value, start=0, 　　　stop=9223372036854775807, /)	返回当前列表指定范围中第一个值为 value 的元素的索引,若不存在值为 value 的元素则抛出异常
pop(index=-1, /)	删除并返回当前列表中下标为 index 的元素
remove(value, /)	在当前列表中删除第一个值为 value 的元素
reverse()	对当前列表中的所有元素进行原地翻转,首尾交换
sort(*, key=None, 　　reverse=False)	对当前列表中的元素进行原地排序

2.1.1　列表的创建与删除

使用赋值分隔符"="直接将一个列表赋值给变量即可创建列表对象。

```
>>> a_list = ['a', 'b', 'mpilgrim', 'z', 'example']
>>> a_list = []                                        #创建空列表
```

也可以使用 list() 函数将元组、range 对象、字符串或其他类型有限长度的可迭代对象转换为列表。

```
>>> a_list = list((3,5,7,9,11))
>>> list(range(1,10,2))
[1, 3, 5, 7, 9]
>>> list('hello world')
['h', 'e', 'l', 'l', 'o', ' ', 'w', 'o', 'r', 'l', 'd']
>>> x = list()                                         #创建空列表
```

列表推导式也是一种常用的快速生成符合特定要求列表的方式,可参考 2.1.9 节。当不再使用时,使用 del 命令删除整个列表。

2.1.2　列表元素的增加

在实际应用中,列表元素的动态增加和删除是经常遇到的操作,Python 列表提供了多种不同的方法来实现这一功能。

（1）可以使用+运算符来实现将元素添加到列表中的功能。但这并不是真的为列表添加元素,而是创建一个新列表,并将原列表中的元素和新元素依次复制到新列表的内存空间。由于涉及大量元素的复制,该操作速度较慢,在涉及大量元素添加时不建议使用。

```
>>> aList = [3, 4, 5]
>>> aList = aList + [7]
>>> aList
[3, 4, 5, 7]
```

(2) 使用列表对象的 append() 方法,原地修改列表,不影响列表及原有元素的引用,速度较快,是真正意义上的在列表尾部添加元素,也是推荐使用的方法。

```
>>> aList.append(9)
>>> aList
[3, 4, 5, 7, 9]
```

Python 采用基于值的自动内存管理方式,当为对象修改值时,并不是直接修改变量的值,而是使变量指向新的值,这对所有类型的变量都是一样的。

对于列表、集合、字典等可变对象而言,情况稍微复杂一些。以列表为例,列表中包含的是元素值的引用,而不是直接包含元素值。如果是通过下标来修改列表中元素的值或通过列表对象自身提供的方法来增加和删除元素,列表对象在内存中的起始地址是不变的,仅仅是被改变值的元素地址发生变化。例如下面的代码:

```
>>> a = [1, 2, 3]              >>> a[0] = 5
>>> id(a)                      >>> a
2389572193096                  [5, 2, 4]
>>> a.append(4)                >>> id(a)
>>> a.remove(3)                2389572193096
```

(3) 使用列表对象的 extend() 方法可以将另一个可迭代对象的所有元素添加至该列表对象尾部。通过 extend() 方法来增加列表元素也不改变列表的内存首地址和已有的元素,属于原地操作。例如,继续上面的代码执行下面的代码:

```
>>> a.extend([7, 8, 9])         [3, 4, 5, 7, 9, 11, 13]
>>> a                           >>> aList.extend((15, 17))
[5, 2, 4, 7, 8, 9]              >>> aList
>>> aList.extend([11, 13])      [3, 4, 5, 7, 9, 11, 13, 15, 17]
>>> aList
```

(4) 使用列表对象的 insert() 方法将元素添加至列表的指定位置,也属于原地操作。

```
>>> aList.insert(3, 6)          #在下标 3 的位置插入元素 6,该位置之后的元素向后移动
>>> aList
[3, 4, 5, 6, 7, 9, 11, 13, 15, 17]
```

列表的 insert() 方法可以在列表的任意位置插入元素,但由于列表的自动内存管理功能,insert() 方法会涉及插入位置之后所有元素的移动,这会影响处理速度,并且导致插入位置后面元素的下标发生变化,类似的还有后面介绍的 remove() 方法以及使用 pop() 方法弹出列表非尾部元素和使用 del 命令删除列表非尾部元素的情况。因此,除非必要,应尽量避免在列表中间位置插入和删除元素的操作。

(5) 使用乘法来扩展列表对象,将列表与整数相乘,生成一个新列表,新列表是原列表中元素引用的重复。

```
>>> aList = [3, 5, 7]
>>> aList * 3                                    #得到新列表
[3, 5, 7, 3, 5, 7, 3, 5, 7]
```

从上面代码的运行结果可以看出,该操作实际上是创建了一个新列表(因为有返回值),而不是真的扩展了原列表。该操作同样适用于字符串和元组,并具有相同的特点。

需要注意的是,使用 * 运算符将包含列表的列表进行重复并创建新列表时,并不创建元素值的复制,而是创建已有元素的引用。当修改其中一个值时,相应的引用也会被修改。

```
>>> x = [[None] * 2] * 3
>>> x                                            #3个元素是同一个子列表的引用
[[None, None], [None, None], [None, None]]
>>> x[0][0] = 5                                  #通过其中一个子列表可以影响另外2个子列表
>>> x
[[5, None], [5, None], [5, None]]
```

2.1.3 列表元素的删除

(1) 使用 del 命令删除列表中指定位置上的元素,后面的元素自动向前移动。

```
>>> a_list = [3, 5, 7, 9, 11]
>>> del a_list[1]
>>> a_list                                       #后面元素的下标会发生变化
[3, 7, 9, 11]
```

(2) 使用列表的 pop() 方法删除并返回指定(默认为最后一个)位置上的元素,如果给定的索引超出列表的范围,则抛出异常。

```
>>> a_list = list((3, 5, 7, 9, 11))              >>> a_list.pop(1)
>>> a_list.pop()                                 5
11                                               >>> a_list        #元素下标发生变化
>>> a_list                                       [3, 7, 9]
[3, 5, 7, 9]
```

(3) 使用列表对象的 remove() 方法删除首次出现的指定元素,如果列表中不存在要删除的元素,则抛出异常。

```
>>> a_list = [3, 5, 7, 9, 7, 11]
>>> a_list.remove(7)                             #删除第一个7
>>> a_list                                       #元素下标发生变化
[3, 5, 9, 7, 11]
```

有时候可能需要删除列表中指定元素的所有重复,大家会很自然地想到使用"循环+remove()"的方法,但是具体操作时很有可能会出现意料之外的错误:代码运行没有出现错误,但结果却是错的;或者代码不稳定——对某些数据处理结果是正确的,对另外一些数据处理结果却是错误的。例如,下面的代码试图删除所有1,执行结果是完全正确的。

```
>>> x = [1, 2, 1, 2, 1, 2, 1, 2, 1]
>>> for i in x:
        if i == 1:
            x.remove(i)
```

```
>>> x
[2, 2, 2, 2]
```

然而，上面这段代码的逻辑是错误的。例如下面的代码，仅仅是所处理的数据发生了一点变化，循环结束后却发现并没有把所有的 1 都删除，只删除了一部分。

```
>>> x = [1, 2, 1, 2, 1, 1, 1]
>>> for i in x:
        if i == 1:
            x.remove(i)
>>> x
[2, 2, 1]
```

2.1.3

很容易看出，两组数据的本质区别在于，第一组数据中没有连续的 1，而第二组数据中存在连续的 1。出现这个问题的原因是列表的自动内存管理功能。前面已经提到，在删除列表元素时，Python 会自动对列表内存进行收缩并移动列表元素，以保证所有元素之间没有空隙；在增加列表元素时，也会自动扩展内存并对元素进行移动，以保证元素之间没有空隙。每当插入或删除一个元素之后，该元素位置后面所有元素的索引就都改变了。下面的代码很好地说明了这个问题：

```
>>> x = list(range(20))
>>> x
[0, 1, 2, 3, 4, 5, 6, 7, 8, 9, 10, 11, 12, 13, 14, 15, 16, 17, 18, 19]
>>> for i in range(len(x)):
        del x[0]
>>> x
[]
```

为了更清楚地解释这个问题带来的影响，对上面最开始给出的代码进行适当插桩，以便了解具体的执行过程。

```
>>> x = [1, 2, 1, 2, 1, 1, 1]
>>> for i in x:
        i
        if i == 1:
            x.remove(i)      #删除当前列表中的第一个 1，会影响循环
        x
1
[2, 1, 2, 1, 1, 1]
1
[2, 2, 1, 1, 1]
1
[2, 2, 1, 1]
1
[2, 2, 1]
```

上面这段代码的执行过程如图 2-1 所示。

x [1, 2, 1, 2, 1, 1, 1]
↑
x [2, 1, 2, 1, 1, 1]
↑
x [2, 2, 1, 1, 1]
↑
x [2, 2, 1, 1]
↑
x [2, 2, 1]

图 2-1　代码执行过程示意图（一）

```
>>> x = [1, 2, 1, 2, 1, 1, 1]
>>> for i in x[::]:              #遍历切片副本，删除原列表中的元素
        i
        if i == 1:
```

```
        x.remove(i)                               #不影响循环
    x
1
[2, 1, 2, 1, 1, 1]
2
1
[2, 2, 1, 1, 1]
2
1
[2, 2, 1, 1]
1
[2, 2, 1]
1
[2, 2]
```

上面这段代码的执行过程如图 2-2 所示。

```
>>> x = [1, 2, 1, 2, 1, 1, 1]
>>> for i in x[::-1]:              #和上一段代码等价
    i
    if i == 1:
        x.remove(i)
    x
1
[2, 1, 2, 1, 1, 1]
1
[2, 2, 1, 1, 1]
1
[2, 2, 1, 1]
2
1
[2, 2, 1]
2
1
[2, 2]
```

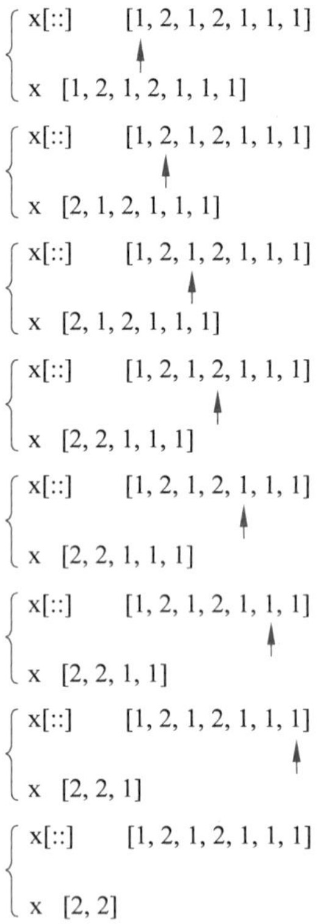

关于切片的讲解可以参考 2.1.6 节。另外，更建议按照从后向前的顺序删除列表中的重复元素，类似于视频处理时从一个视频中删除多段视频。例如下面的代码：

图 2-2 代码执行过程示意图（二）

```
>>> x = [1, 2, 1, 2, 1, 1, 1]
>>> for i in range(len(x)-1, -1, -1):
    i
    if x[i] == 1:
        del x[i]                                  #每次删除元素时不影响前面的元素
    x
6
[1, 2, 1, 2, 1, 1]
5
```

```
[1, 2, 1, 2, 1]
4
[1, 2, 1, 2]
3
2
[1, 2, 2]
1
0
[2, 2]
```

2.1.4 列表元素访问与计数

可以使用表示序号或位置的整数作下标直接访问列表中的元素。列表支持双向索引，有效下标范围为[-L,L-1]，L为列表长度。正向索引时从左向右并且下标从0开始递增，反向索引时从右向左并且下标从-1开始递减。如果指定的下标不存在，则抛出异常提示下标越界，例如：

```
>>> aList = [3, 4, 5, 6, 7, 9, 11, 13, 15, 17]
>>> aList[3]                          #访问元素
6
>>> aList[3] = 5.5                    #修改元素
>>> aList
[3, 4, 5, 5.5, 7, 9, 11, 13, 15, 17]
>>> aList[15]                         #下标不存在，出错
IndexError: list index out of range
```

使用列表对象的 index() 方法可以获取指定元素首次出现的下标，语法为 index(value, [start, [stop]])，其中，start 和 stop 用于指定搜索范围，start 默认值为 0，stop 默认值为列表长度。若列表对象中不存在指定元素，则抛出异常提示列表中不存在该值，例如：

```
>>> aList.index(7)
4
>>> aList.index(100)
ValueError: 100 is not in list
```

可以使用列表对象的 count() 方法统计指定元素在列表中出现的次数，例如下面的代码：

```
>>> aList
[3, 4, 5, 5.5, 7, 9, 11, 13, 15, 17]
>>> aList.count(7)
1
>>> aList.count(0)
0
```

该方法也可以用于元组、字符串及 range 对象。

2.1.5 元素存在性测试

如果需要判断列表中是否存在或包含指定的值，可以使用前面介绍的 count() 方法；如果存在指定的值，则返回大于 0 的数；如果返回 0，则表示不存在指定的值。或者，使用更加

简洁的 in 关键字来判断一个值是否存在于列表中,返回结果为 True 或 False。

```
>>> aList
[3, 4, 5, 5.5, 7, 9, 11, 13, 15, 17]
>>> 3 in aList
True
>>> 18 in aList
False
>>> bList =[[1], [2], [3]]
>>> 3 in bList
False
>>> 3 not in bList
True
>>> [3] in bList
True
```

关键字 in 和 not in 也可以用于其他可迭代对象,包括元组、字典、range 对象、字符串、集合等,常用在循环语句中对可迭代对象中的元素进行遍历。

2.1.6 切片操作

切片是 Python 中有序序列的重要操作之一,适用于列表、元组、字符串、range 对象等类型。切片使用两个冒号分隔的 3 个数字来完成:[start:stop:step],第一个数字表示切片开始位置(step>0 时,start 默认值为 0;step<0 时,start 默认值为 -1),第二个数字表示切片截止(但不包含)位置(step>0 时,stop 默认值为 L;step<0 时,stop 默认值为 -L-1,L 表示列表长度),第三个数字表示切片的步长(默认值为 1),当步长省略时可以顺便省略最后一个冒号。

与使用下标访问列表元素的方法不同,切片操作不会因为下标越界而抛出异常,而是简单地在列表尾部截断或者返回一个空列表,代码具有更强的健壮性。

```
>>> aList =[3, 4, 5, 6, 7, 9, 11, 13, 15, 17]
>>> aList[::]
[3, 4, 5, 6, 7, 9, 11, 13, 15, 17]
>>> aList[::-1]                   #步长为负数,从右往左切,得到反向副本
[17, 15, 13, 11, 9, 7, 6, 5, 4, 3]
>>> aList[::2]                    #从下标 0 开始隔一个元素取一个元素
[3, 5, 7, 11, 15]
>>> aList[1::2]                   #从下标 1 开始,隔一个元素取一个元素
[4, 6, 9, 13, 17]
>>> aList[3::]                    #下标 3 往后的所有元素
[6, 7, 9, 11, 13, 15, 17]
>>> aList[3:6]                    #下标范围介于[3,6)的元素
[6, 7, 9]
>>> aList[3:6:1]
[6, 7, 9]
>>> aList[-3:]                    #最后 3 个元素
[13, 15, 17]
>>> aList[:-6]                    #倒数第 6 个(不包含)之前的元素
[3, 4, 5, 6]
>>> aList[0:100:1]                #在尾部截断,不会抛出异常
```

```
[3, 4, 5, 6, 7, 9, 11, 13, 15, 17]
>>> aList[100:]
[]
```

切片作用于元组和字符串时只能用于访问其中的部分元素，作用于列表时可以实现很多目的，例如原地修改列表内容，列表元素的增、删、改、查等操作都可以通过切片来实现，并且不影响列表对象在内存中的起始地址。

```
>>> aList = [3, 5, 7]
>>> aList[len(aList):] = [9]          #在尾部追加元素
>>> aList
[3, 5, 7, 9]
>>> aList[:3] = [1, 2, 3]             #替换前 3 个元素
>>> aList
[1, 2, 3, 9]
>>> aList[:3] = []                    #删除前 3 个元素
>>> aList
[9]
>>> aList = list(range(10))
>>> aList
[0, 1, 2, 3, 4, 5, 6, 7, 8, 9]
>>> aList[::2] = [0] * (len(aList)//2)   #切片不连续，要求等号两侧元素数量一样
>>> aList
[0, 1, 0, 3, 0, 5, 0, 7, 0, 9]
>>> aList[::2] = [0, 0, 0]            #等号两侧元素数量不一样，代码出错抛出异常
ValueError: attempt to assign sequence of size 3 to extended slice of size 5
```

也可以结合使用 del 命令与切片操作来删除列表中的部分元素。

```
>>> aList = [3, 5, 7, 9, 11]
>>> del aList[:3]
>>> aList
[9, 11]
>>> aList = [3, 5, 7, 9, 11]
>>> del aList[::2]                    #切片可以不连续
>>> aList
[5, 9]
```

切片返回的是浅复制。所谓浅复制，是指生成一个新的列表，把原列表中所有元素的引用都复制到新列表中。如果原列表中只包含整数、实数、复数等基本类型或元组、字符串这样的不可变类型的数据是没有问题的，包含列表等可变对象时情况复杂一些。

```
>>> aList = [3, 5, 7]
>>> bList = aList[:]                  #切片，浅复制
>>> aList == bList                    #切片刚完成的瞬间，bList 和 aList 中包含同样的元素引用
True
>>> aList is bList                    #切片得到的 bList 与原来的 aList 不是同一个对象
False
>>> bList[1] = 8                      #列表中只包含可哈希对象，修改 bList 时不影响 aList
>>> bList
[3, 8, 7]
```

```
>>> aList
[3, 5, 7]
>>> aList = [3, [5], 7]                    #列表 aList 中包含可变的列表对象
>>> bList = aList[:]                       #切片
>>> bList[1].append(6)                     #调用子列表的 append()方法,这个方法是原地操作的
>>> bList
[3, [5, 6], 7]
>>> aList                                  #aList 受到影响
[3, [5, 6], 7]
>>> aList = [3, [5], 7]
>>> import copy
>>> bList = copy.deepcopy(aList)           #深复制,递归复制,直到遇到可哈希对象
                                           #aList 和 bList 完全独立,互相不影响
>>> aList == bList
True
>>> aList is bList
False
>>> bList[1].append(6)                     #修改 bList 不会影响 aList
>>> bList
[3, [5, 6], 7]
>>> aList
[3, [5], 7]
```

2.1.7 列表排序与逆序

2.1.7

使用列表对象自身提供的 sort()方法可以进行原地排序,没有返回值,或者说返回空值 None(见第 5 章)。

```
>>> aList = [3, 4, 5, 6, 7, 9, 11, 13, 15, 17]
>>> import random
>>> random.shuffle(aList)                            #随机打乱顺序
>>> aList
[3, 4, 15, 11, 9, 17, 13, 6, 7, 5]
>>> aList.sort()                                     #默认为升序排列
>>> aList
[3, 4, 5, 6, 7, 9, 11, 13, 15, 17]
>>> aList.sort(reverse=True)                         #降序排列
>>> aList
[17, 15, 13, 11, 9, 7, 6, 5, 4, 3]
>>> aList.sort(key=lambda x: len(str(x)))            #按转换为字符串后的长度排序
>>> aList
[9, 7, 6, 5, 4, 3, 17, 15, 13, 11]
>>> aList.sort(key=lambda num: sum(map(int,str(num))))
                                                     #按各位数字之和升序排列,稳定排序
                                                     #各位数字之和相等的保持原来的相对顺序
>>> aList
[11, 3, 4, 13, 5, 6, 15, 7, 17, 9]
```

也可以使用内置函数 sorted()对列表进行排序,与列表对象的 sort()方法不同,内置函数 sorted()返回新列表,并不对原列表进行任何修改。

```
>>> sorted(aList)
```

```
[3, 4, 5, 6, 7, 9, 11, 13, 15, 17]
>>> sorted(aList, reverse=True)              #降序排列
[17, 15, 13, 11, 9, 7, 6, 5, 4, 3]
```

列表对象的 reverse() 方法用于将所有元素原地逆序：

```
>>> import random
>>> aList = [random.randint(50, 100) for i in range(10)]
>>> aList
[87, 79, 52, 96, 56, 59, 74, 80, 53, 79]
>>> aList.reverse()                           #原地逆序，首尾交换
>>> aList
[79, 53, 80, 74, 59, 56, 96, 52, 79, 87]
```

内置函数 reversed() 支持对列表元素进行逆序（注意，不是降序）排列，与列表对象的 reverse() 方法不同，内置函数 reversed() 不对原列表做任何修改，而是返回一个逆序排列后的迭代器对象。例如：

```
>>> aList = [3, 4, 5, 6, 7, 9, 11, 13, 15, 17]
>>> newList = reversed(aList)                 #返回可迭代的 reversed 对象
>>> newList
<listreverseiterator object at 0x0000000003624198>
>>> list(newList)                             #可以把 reversed 对象转换为列表
[17, 15, 13, 11, 9, 7, 6, 5, 4, 3]
>>> for i in newList:                         #可以使用 for 循环遍历
    print(i, end=' ')
```

上面代码中最后的 for 循环没有输出任何内容，因为在之前的 list() 函数执行时，reversed 对象已遍历结束，需要重新创建可迭代的 reversed 对象才能再次访问其内容，即

```
>>> newList = reversed(aList)
>>> for i in newList:
    print(i, end=' ')
17 15 13 11 9 7 6 5 4 3
```

2.1.8 用于序列操作的常用内置函数

很多内置函数可以作用于列表，本节简单介绍其中一部分，map()、reduce() 和 filter() 函数对列表操作的讲解可参考 5.8 节。

（1）all() 和 any()：all() 函数用于测试列表、元组等序列对象以及 map 对象、zip 对象等类似对象中是否所有元素都等价于 True（作为参数传递给内置函数 bool() 时返回值为 True），any() 函数用于测试序列或可迭代对象中是否存在等价于 True 的元素。例如：

```
>>>all([1, 2, 3])
True
>>>all([0, 1, 2, 3])
False
>>>any([0, 1, 2, 3])
True
>>>any([0])
False
```

（2）len()：返回列表中的元素个数，也适用于元组、字典、集合、字符串、range 对象等。

（3）max()、min()：返回列表中的最大或最小元素，同样适用于元组、字符串、集合、range 对象、字典和迭代器对象等，要求所有元素之间可以进行大小比较。这两个函数支持使用 key 参数指定排序规则。

（4）sum()：对数值型列表的元素进行求和运算，对非数值型列表运算则需要指定第二个参数，适用于元组、集合、range 对象、字典、map 对象及 filter 对象等。

```
>>> a = {1:1, 2:5, 3:8}
>>> sum(a)                              #对字典的"键"进行求和
6
>>> sum(a.values())
14
>>> sum([[1], [2]], [])                 #元素不是数值时需要指定第二个参数
[1, 2]
```

（5）zip()：将多个可迭代对象对应位置的元素组合为元组，并返回包含这些元组的 zip 对象。

zip()函数

```
>>> aList = [1, 2, 3]
>>> bList = [4, 5, 6]
>>> cList = zip(aList, bList)           #返回可迭代的 zip 对象
>>> cList
<zip object at 0x0000000003728908>
>>> list(cList)                         #可以转换为列表或使用 for 循环遍历
[(1, 4), (2, 5), (3, 6)]
```

（6）enumerate()：枚举列表、字典、元组或字符串中的元素，返回枚举对象，枚举对象中的每个元素是包含下标和元素值的元组。

enumerate()
函数

```
>>> for index, num in enumerate([1,2,3,4]):     #序列解包，见 2.2.3 节
        print((index,num), end=',')
(0, 1),(1, 2),(2, 3),(3, 4),
>>> for index, (num,ch) in enumerate(zip([1,2,3],'ab')):
        print(index, num, ch)
0 1 a
1 2 b
```

2.1.9 列表推导式

列表推导式(list comprehension，也称列表解析式)可以使用非常简洁的形式对列表或其他可迭代对象的元素进行遍历、过滤或再次计算，快速生成满足特定需求的列表。列表推导式的完整语法形式：

```
[expression for element1 in iterable1 if condition1
            for element2 in iterable2 if condition2
            for element3 in iterable3 if condition3
            ...
            for elementN in iterableN if conditionN]
```

列表推导式在逻辑上等价于一个循环结构，第一个循环相当于最外层的循环，最后一个循环相当于最内层的循环。实际使用时一般建议不超过两层循环，否则应使用循环结构

改写。

列表推导式使用非常简洁的方式快速生成满足特定需求的列表,例如:

```
>>> aList = [x*x for x in range(10)]
```

相当于

```
>>> aList = []
>>> for x in range(10):
        aList.append(x*x)
```

而

```
>>> freshfruit = [' banana', ' loganberry ', 'passion fruit ']
>>> freshfruit = [w.strip() for w in freshfruit]
```

等价于下面的代码:

```
>>> freshfruit = [' banana', ' loganberry ', 'passion fruit ']
>>> for i, v in enumerate(freshfruit):
        freshfruit[i] = v.strip()
```

也等价于

```
>>> freshfruit = [' banana', ' loganberry ', 'passion fruit ']
>>> freshfruit = list(map(str.strip, freshfruit))
```

但是不等价于下面的代码,因为修改后的内容没有放回列表:

```
>>> freshfruit = [' banana', ' loganberry ', 'passion fruit ']
>>> for i in freshfruit:
        i = i.strip()
```

接下来再通过几个示例来进一步展示列表推导式的强大功能。

(1) 使用列表推导式实现嵌套列表的平铺。

```
>>> vec = [[1, 2, 3], [4, 5, 6], [7, 8, 9]]
>>> [num for elem in vec for num in elem]
[1, 2, 3, 4, 5, 6, 7, 8, 9]
```

在配套 PPT 中补充了更多平铺的方法。

(2) 过滤不符合条件的元素。

在列表推导式中可以使用 if 子句来筛选,只在结果列表中保留符合条件的元素。例如,下面的代码可以列出当前文件夹下所有 Python 源文件,os 模块见 7.4 节。

```
>>> import os
>>> [filename for filename in os.listdir() if filename.endswith(('.py','.pyw'))]
```

下面的代码用于从列表中选择符合条件的元素组成新的列表:

```
>>> aList = [-1, -4, 6, 7.5, -2.3, 9, -11]
>>> [i for i in aList if i>0]
[6, 7.5, 9]
```

再如,已知有一个包含一些同学成绩的字典,计算成绩的最高分、最低分和平均分,并查找所有最高分同学，代码如下：

```
>>> scores = {"Zhang San": 45, "Li Si": 78, "Wang Wu": 40, "Zhou Liu": 96, "Zhao Qi": 65,
              "Sun Ba": 90, "Zheng Jiu": 78, "Wu Shi": 99, "Dong Shiyi": 60}
>>> highest = max(scores.values())
>>> lowest = min(scores.values())
>>> average = sum(scores.values()) / len(scores)
>>> highestPerson = [name for name, score in scores.items() if score==highest]
```

（3）生成笛卡儿积。

```
>>> [(x, y) for x in range(3) for y in range(3)]
[(0, 0), (0, 1), (0, 2), (1, 0), (1, 1), (1, 2), (2, 0), (2, 1), (2, 2)]
>>> from itertools import product
>>> list(product(range(3), range(3)))              #等价代码
[(0, 0), (0, 1), (0, 2), (1, 0), (1, 1), (1, 2), (2, 0), (2, 1), (2, 2)]
>>> [(x, y) for x in [1, 2, 3] for y in [3, 1, 4] if x != y]
[(1, 3), (1, 4), (2, 3), (2, 1), (2, 4), (3, 1), (3, 4)]
```

（4）使用列表推导式实现矩阵转置。

```
>>> matrix = [[1, 2, 3, 4], [5, 6, 7, 8], [9, 10, 11, 12]]
>>> [[row[i] for row in matrix] for i in range(4)]
[[1, 5, 9], [2, 6, 10], [3, 7, 11], [4, 8, 12]]
```

也可以使用内置函数来实现矩阵转置，使用扩展库 NumPy 实现的方法见 17.1 节。

```
>>> list(map(list, zip(*matrix)))
[[1, 5, 9], [2, 6, 10], [3, 7, 11], [4, 8, 12]]
```

（5）列表推导式中可以使用函数或复杂表达式。

```
>>> def f(v):
    if v%2 == 0:
        v = v ** 2
    else:
        v = v +1
    return v
>>> print([f(v) for v in [2, 3, 4, -1] if v>0])
[4, 4, 16]
>>> print([v**2 if v%2 == 0 else v+1 for v in [2, 3, 4, -1] if v>0])
[4, 4, 16]
```

（6）列表推导式支持文件对象迭代。

```
>>> fp = open('C:\install.log', 'r')
>>> print([line for line in fp])                #为节约篇幅，这里没有给出代码运行结果
>>> fp.close()
```

（7）使用列表推导式生成 100 以内的所有素数。

```
>>> [p for p in range(2, 100) if 0 not in [p%d for d in range(2, int(sqrt(p))+1)]]
[2, 3, 5, 7, 11, 13, 17, 19, 23, 29, 31, 37, 41, 43, 47, 53, 59, 61, 67, 71, 73, 79, 83, 89, 97]
```

*2.1.10 使用列表实现向量运算

向量运算经常涉及这样的操作，例如，所有分量同时加、减、乘、除同一个数，或者向量之

间的加、减、乘运算,Python 列表对象本身不直接支持这样的操作。下面的代码演示了如何使用列表结合内置函数和 operator 库提供的函数实现向量运算,关于 NumPy 中数组与矩阵运算的知识可参考第 17 章。

```
>>> import random
>>> x = [random.randint(1, 100) for i in range(10)]    #生成 10 个[1, 100]区间内的随机数
>>> list(map(lambda i: i+5, x))                        #所有元素同时加 5
>>> x = [random.randint(1, 10) for i in range(10)]
>>> y = [random.randint(1, 10) for i in range(10)]
>>> import operator
>>> sum(map(operator.mul, x, y))                       #向量内积,函数式编程,见 5.8 节
>>> sum((i*j for i, j in zip(x, y)))                   #向量内积,生成器表达式
>>> list(map(operator.add, x, y))                      #两个等长的向量对应元素相加
```

2.2 元　　组

元组可以看作轻量级列表,功能比列表简单很多,属于有序不可变序列。元组的形式与列表相似,区别在于元组的所有元素放在一对圆括号中,而不是方括号中。

2.2.1 元组的创建与删除

使用"="将一个元组赋值给变量即可创建一个元组变量。

```
>>> a_tuple = ('Python', C++', 'PHP', 'Java')
>>> x = ()                                    #空元组
```

如果要创建只包含一个元素的元组,只把元素放在圆括号里是不行的,还需要在元素后面加一个逗号,创建包含多个元素的元组没有这个限制。

```
>>> a = 3                              >>> a = 3,         #有逗号时可以没有圆括号
>>> a                                  >>> a
3                                      (3,)
>>> a = (3)        #这里的圆括号没有用   >>> a = 1, 2       #创建元组
>>> a                                  >>> a
3                                      (1, 2)
```

还可以使用 tuple() 函数将列表、字符串、字典、集合、map 对象等其他类型有限长度的可迭代对象转换为元组。

```
>>> print(tuple('abcdefg'))
('a', 'b', 'c', 'd', 'e', 'f', 'g')
>>> s = tuple()                               #空元组
```

可以使用 del 命令删除整个元组对象,不能只删除元组中的部分元素,因为元组属于不可变序列。

2.2.2 元组与列表的区别

列表属于可变序列(不可哈希),可以随意地修改列表中的元素值,以及增加和删除列表元素;元组属于不可变序列(可哈希),元组一旦定义就不允许通过任何方式更改其中元素的数量

和引用。元组没有提供 append()、extend() 和 insert() 等方法,无法向元组中添加元素;同样,元组也没有 remove() 和 pop() 方法,不支持对元组元素进行 del 操作,不能从元组中删除元素,只能使用 del 命令删除整个元组。元组也支持切片操作,但是只能通过切片来访问元组中的元素,不支持使用切片修改元组中元素的值,也不支持使用切片操作为元组增加或删除元素。

元组元素的访问速度比列表略快,占用内存也略少。如果定义了一系列常量值,主要用途仅是对它们进行遍历或其他类似操作,不需要对其元素进行任何修改,那么一般建议使用元组而不用列表。可以认为元组对不需要修改的数据进行了"写保护",从内部实现上不允许修改其元素值,从而使得代码更加安全。

另外,作为不可变序列,与整数、字符串一样,元组可用作字典的"键"和集合的元素;而列表则永远都不能这样使用,因为列表不是不可变的。

最后,虽然元组属于不可变序列,其元素的数量和引用都是不可改变的,但是如果元组中包含可变序列,情况略有不同,例如下面的代码:

```
>>> x = ([1, 2], 3)
>>> x[0][0] = 5
>>> x
([5, 2], 3)
>>> x[0].append(8)
>>> x
([5, 2, 8], 3)
>>> x[0] = x[0] + [10]                      #不允许修改元组中元素的引用
TypeError: 'tuple' object does not support item assignment
>>> x
([5, 2, 8], 3)
```

如果元组中包含列表、字典、集合或其他可变类型的对象,这样的元组不能作为字典的"键"或集合的元素。

2.2.3 序列解包

在实际开发中,序列解包是非常重要和常用的一个语法特性,大幅提高了代码的可读性,并且减少了程序员的代码输入量。例如,可以使用序列解包对多个变量同时进行赋值,要求等号左侧变量数量与右侧值的数量相等:

2.2.3

```
>>> x, y, z = 1, 2, 3                       #等号两侧实际都是元组
>>> v_tuple = (False, 3.5, 'exp')
>>> (x, y, z) = v_tuple
>>> x, y, z = v_tuple                       #与上一行代码等价
```

序列解包也可以用于列表、字典、集合、字符串及迭代器对象,但对字典使用时,默认是对字典"键"操作;如果需要对"键:值"元素操作,则需要使用字典的 items() 方法说明;如果需要对字典"值"操作,则需要使用字典的 values() 方法明确指定。

```
>>> a = [1, 2, 3]                           >>> b, c, d = s          #对"键"进行解包
>>> b, c, d = a                             >>> b, c, d = s.values()
>>> s = {'a':1, 'b':2, 'c':3}               >>> a, b = b, a          #交换两个变量的值
>>> b, c, d = s.items()
```

使用序列解包可以很方便地同时遍历多个序列。

```
>>> keys = ['a', 'b', 'c', 'd']
>>> values = [1, 2, 3, 4]
>>> for k, v in zip(keys, values):        #循环变量数量与 zip 对象中每个元素长度相等
    print(k, v)
a 1
b 2
c 3
d 4
```

在调用函数时，在实参前面加上一个或两个星号也表示序列解包，将可迭代对象中的元素依次传递给相同数量的形参，详见 5.3.4 节。

序列解包还支持下面代码演示的用法，可结合第 5 章内容理解。

```
>>> print(*[1], *[2], 3, *[4, 5])
1 2 3 4 5
>>> def demo(a, b, c, d):
    print(a, b, c, d)
>>> demo(**{'a': 1, 'c': 3}, **{'b': 2, 'd': 4})    #调用函数时参数序列解包
1 2 3 4
>>> *range(4), 4                                     #创建元组
(0, 1, 2, 3, 4)
>>> [*range(4), 4]                                   #创建列表
[0, 1, 2, 3, 4]
>>> {*range(4), 4, *(5, 6, 7)}                       #创建集合
{0, 1, 2, 3, 4, 5, 6, 7}
>>> {'x': 1, **{'y': 2}}                             #创建字典
{'x': 1, 'y': 2}
>>> data = [(1,2,3,4), (1,2,3,4,5), (1,2,3,4,5,6)]
>>> for i, j, *k in data:                            #变量 k 用于收集第 3 个往后的所有元素
    print(k)
[3, 4]
[3, 4, 5]
[3, 4, 5, 6]
```

2.2.4 生成器表达式

2.2.4

从形式上看，生成器表达式与列表推导式非常接近，只是生成器表达式使用圆括号，而不是列表推导式所使用的方括号。与列表推导式本质上不同的是，生成器表达式的结果是一个生成器对象，属于迭代器对象。使用生成器对象的元素时，可以根据需要将其转换为列表、集合或元组，也可以使用内置函数 next()、生成器对象的 __next__() 方法获取下一个元素，或者直接使用 for 循环遍历。但是不管用哪种方法访问其元素，生成器对象中的每个元素只能使用一次，并且只能从前往后顺序访问。

```
>>> g = ((i+2)**2 for i in range(10))
>>> g
<generator object <genexpr> at 0x02B15C60>
>>> tuple(g)                                         #转换为元组
(4, 9, 16, 25, 36, 49, 64, 81, 100, 121)
>>> tuple(g)                                         #元素遍历结束,生成器对象已空
```

```
()
>>> g = ((i+2)**2 for i in range(10))          #重新创建生成器对象
>>> list(g)                                     #转换为列表
[4, 9, 16, 25, 36, 49, 64, 81, 100, 121]
>>> g = ((i+2)**2 for i in range(10))
>>> g.__next__()
4
>>> next(g)                                     #与上一行代码功能等价
9
>>> next(g)
16
>>> g = ((i+2)**2 for i in range(10))
>>> for i in g:                                 #使用循环遍历生成器对象中的元素
    print(i, end=' ')
4 9 16 25 36 49 64 81 100 121
```

2.3 字　　典

字典是包含若干"键:值"对的可变容器对象,字典中的每个元素包含两部分:"键"和"值"。定义字典时,每个元素的"键"和"值"用冒号分隔,相邻元素之间用逗号分隔,所有的元素都放在一对花括号"{}"中。

字典中的"键"可以是任意不可变或可哈希数据,如整数、实数、复数、字符串、元组等,但不能使用列表、集合、字典作为字典的"键",包含列表、集合、字典的元组也不能作为字典的"键"。另外,字典中的"键"不允许重复,"值"可以重复。

2.3.1 字典的创建与删除

使用"="将一个字典赋值给一个变量即可创建一个字典变量。

```
>>> a_dict = {'server': 'db.diveintopython3.org', 'database': 'mysql'}
>>> x = {}                                      #空字典
```

可以使用内置函数 dict() 通过已有数据快速创建字典:

```
>>> keys = ['a', 'b', 'c', 'd']
>>> values = [1, 2, 3, 4]
>>> dictionary = dict(zip(keys, values))
>>> x = dict()                                  #空字典
```

或者使用内置函数 dict() 根据给定的"键:值"对来创建字典:

```
>>> d = dict(name='Dong', age=37)               #关键参数的语法见 5.3.2 节
```

还可以以给定内容为"键",创建"值"为空或特定值(通过 fromkeys() 方法的第 2 个参数指定)的字典:

```
>>> adict = dict.fromkeys(['name', 'age', 'sex'])
>>> adict
{'name': None, 'age': None, 'sex': None}
>>> data = dict.fromkeys(['a', 'b', 'c'], [])
```

```
>>> data                                          #所有"值"引用同一个列表
{'a': [], 'b': [], 'c': []}
>>> data['a'].append(97)                          #影响另外两个"值"
>>> data
{'a': [97], 'b': [97], 'c': [97]}
>>> data['b'] = 98                                #修改了"值"的引用,不影响另外两个"值"
>>> data
{'a': [97], 'b': 98, 'c': [97]}
```

当不再需要某个字典时,可以使用 del 命令删除整个字典,也可以使用 del 命令删除字典中指定的元素,可参考 2.3.3 节的内容。

2.3.2 字典元素的访问

可以使用字典的"键"作为下标来访问字典元素的"值",若指定的"键"不存在则抛出异常。

```
>>> aDict = {'name':'Dong', 'sex':'male', 'age':37}
>>> aDict['name']
'Dong'
>>> aDict['tel']                                  #略去了详细错误信息
KeyError: 'tel'
```

比较推荐的也是更加安全的字典元素访问方式是字典对象的 get() 方法。使用字典对象的 get() 方法可以获取指定"键"对应的"值",并且可以在指定"键"不存在的时候返回指定值,如果不指定则默认返回 None。

```
>>> print(aDict.get('address'))
None
>>> print(aDict.get('address', 'SDIBT'))
SDIBT
```

另外,使用字典对象的 items() 方法可以返回字典的"键:值"对,使用字典对象的 keys() 方法可以返回字典的"键",使用字典对象的 values() 方法可以返回字典的"值"。

```
>>> aDict = {'name':'Dong', 'sex':'male', 'age':37}
>>> aDict.keys()                                  #返回值支持集合运算
dict_keys(['name', 'sex', 'age'])
>>> aDict.values()                                #返回值不支持集合运算
dict_values(['Dong', 'male', 37])
>>> aDict.items()                                 #返回值支持集合运算
dict_items([('name', 'Dong'), ('sex', 'male'), ('age', 37)])
>>> for item in aDict.items():                    #遍历每个元素
    print(item)
('name', 'Dong')
('sex', 'male')
('age', 37)
>>> for key, value in aDict.items():              #序列解包,遍历每个元素的"键"和"值"
    print(key, value, sep=':')
name:Dong
sex:male
age:37
```

2.3.3 字典元素的添加与修改

当以指定"键"为下标为字典元素赋值时,若该"键"存在,则表示修改该"键"的值;若不存在,则表示添加一个新的"键:值"对,也就是添加一个新元素。

```
>>> aDict = {'name':'Dong', 'sex':'male', 'age':37}
>>> aDict
{'name':'Dong', 'sex':'male', 'age':37}
>>> aDict['age'] = 40                         #修改元素的"值"
>>> aDict['address'] = 'Yantai'               #添加新元素
>>> aDict
{'name':'Dong', 'sex':'male', 'age':40, 'address':'Yantai'}
```

使用字典对象的 `update()` 方法将另一个字典的"键:值"元素全部添加到当前字典,如果两个字典中存在相同的"键",则以另一个字典中的"值"为准对当前字典进行更新。

当需要删除字典元素时,可以根据具体需要使用 del 命令删除字典中指定"键"对应的元素,或者也可以使用字典对象的 `clear()` 方法删除字典中的所有元素,还可以使用字典对象的 `pop()` 方法删除并返回指定"键"的元素,或者使用字典对象的 `popitem()` 方法删除并返回字典中的一个元素,读者可以自行练习这些用法。

2.3.4

2.3.4 字典应用案例

下面的代码首先生成包含 1000 个随机字符的字符串,然后统计每个字符的出现次数。其中用到的字符串方法 join() 参考 4.1.2 节。

```
from random import choices
from string import ascii_letters, digits

z = ''.join(choices(ascii_letters+digits, k=1000))
d = dict()                                    #字典中的"键"表示字符,"值"表示出现的次数
for ch in z:
    d[ch] = d.get(ch,0) +1                    #更新字符出现次数,或添加新元素
print(d)                                      #此处省略结果
```

也可以使用 collections 模块的 defaultdict 类来实现该功能。

```
>>> from collections import defaultdict
>>> frequences = defaultdict(int)             #每个元素的"值"默认为整数 0
>>> frequences
defaultdict(<type 'int'>, {})
>>> for item in z:
    frequences[item] += 1
>>> frequences.items()
```

使用 collections 模块的 Counter 类可以快速实现这个功能,并且能够满足其他需要,如查找出现次数最多的元素。下面的代码演示了 Counter 类的用法:

```
>>> from collections import Counter
>>> frequences = Counter(z)
>>> frequences.items()                        #略去输出结果
```

```
>>> frequences.most_common(1)                    #出现次数最多的 1 个字符和次数
>>> frequences.most_common(3)                    #出现次数最多的 3 个字符和次数
```

类似列表推导式，Python 也支持字典推导式快速生成符合特定条件的字典。

```
>>> {i:str(i) for i in range(1, 5)}
{1:'1', 2:'2', 3:'3', 4:'4'}
>>> x = ['A', 'B', 'C', 'D']
>>> y = ['a', 'b', 'b', 'd']
>>> {i:j for i,j in zip(x,y)}
{'A':'a', 'B':'b', 'C':'b', 'D':'d'}
```

2.4 集　　合

集合是无序可变的容器对象，与字典一样使用一对花括号作为界定符，同一个集合的元素之间不允许重复，集合中每个元素都是唯一的。

2.4.1 集合的创建与常用操作

直接将集合赋值给变量即可创建一个集合对象。

```
>>> a = {3, 5}
```

也可以使用 set() 函数将列表、元组等其他长度有限的可迭代对象转换为集合，如果原来的数据中存在重复元素，在转换为集合的时候只保留一个。

```
>>> a_set = set(range(8, 14))
>>> a_set
{8, 9, 10, 11, 12, 13}
>>> b_set = set([0, 1, 2, 3, 0, 1, 2, 3, 7, 8])
>>> b_set
{0, 1, 2, 3, 7, 8}
>>> x = set()                                    #空集合
```

可以使用集合对象的 add() 方法增加元素（元素已存在时自动忽略）。另外，可以使用集合对象的 pop() 方法弹出并删除其中一个元素，或者使用集合对象的 remove() 或 discard() 方法删除指定元素，以及使用集合对象的 clear() 方法清空集合删除所有元素。可以使用 dir(set()) 查看全部方法。当不再使用某个集合时，可以使用 del 命令删除整个集合。

```
>>> a = {1, 4, 2, 3}
>>> a.pop()                                      #删除并返回一个元素
1
>>> a.remove(3)                                  #删除指定元素，不存在时抛出异常
>>> a
{2, 4}
>>> a.pop(2)                                     #pop()方法不接收参数
TypeError: pop() takes no arguments (1 given)
```

2.4.2 集合运算

Python 集合支持交集、并集、差集及子集测试等运算。

```
>>> a_set = set([8, 9, 10, 11, 12, 13])
>>> b_set = set([0, 1, 2, 3, 7, 8])
>>> a_set | b_set                       #并集
{0, 1, 2, 3, 7, 8, 9, 10, 11, 12, 13}
>>> a_set & b_set                       #交集
{8}
>>> a_set - b_set                       #差集
{9, 10, 11, 12, 13}
>>> a_set ^ b_set                       #对称差集
{0, 1, 2, 3, 7, 9, 10, 11, 12, 13}
>>> x = {1, 2, 3}
>>> y = {1, 2, 5}
>>> z = {1, 2, 3, 4}
>>> x < y                               #比较集合大小
False
>>> x < z                               #x 是 z 的真子集，返回 True
True
>>> y < z                               #y 不是 z 的真子集，返回 False
False
>>> max({1}, {2}, {3}, {4})             #先假设第一个集合最大，然后检查有没有更大的
                                        #如果没有更大的集合，返回第一个
{1}
>>> min({1}, {2}, {3}, {4})
{1}
```

2.4.3 集合运用案例

Python 集合的内部实现保证了元素不重复，并做了大量优化。下面代码使用 3 种方法生成不重复随机数，大量实验数据表明，使用集合可以获得最高的执行效率。由于篇幅限制，在配套 PPT 中补充了另几个案例，更多案例可以关注微信公众号"Python 小屋"进行学习。

例 2-1　生成不重复随机数的效率比较，不允许使用 random 模块的 sample() 函数。

例 2-1

```python
import random
import time

def randomNumbers(number, start, end):
    '''使用列表来生成 number 个介于 start~end 的不重复随机数'''
    data = []
    n = 0
    while True:
        element = random.randint(start, end)
        if element not in data:
            data.append(element)
            n += 1
            if n == number - 1:
                break
```

```
        return data

def randomNumbers1(number, start, end):
    '''使用列表来生成 number 个介于 start~end 的不重复随机数'''
    data = []
    while True:
        element = random.randint(start, end)
        if element not in data:
            data.append(element)
            if len(data) == number:
                break
    return data

def randomNumbers2(number, start, end):
    '''使用集合来生成 number 个介于 start~end 的不重复随机数'''
    data = set()
    while True:
        data.add(random.randint(start, end))
        if len(data) == number:
            break
    return data

start = time.time()
for i in range(10000):
    randomNumbers(50, 1, 100)
print('Time used:', time.time()-start)
```

说明：另两个函数的测试代码见配套源代码，或根据上面的代码自行编写。改变函数参数，增加不重复数的个数要求会发现，序列越长，使用集合的效率相对来说越高。

例 2-2 电影推荐问题。假设已有若干用户名及其喜欢的电影清单，现有某用户，已看过并喜欢一些电影，现在想找个新电影看看，又不知道看什么好，特来寻求推荐。根据已有数据，查找与该用户爱好最相似的用户，也就是看过并喜欢的电影与该用户最接近，然后从那个用户喜欢的电影中选取一个当前用户还没看过的电影，进行推荐。

```
from random import randrange

#其他用户喜欢看的电影清单
data = {'user'+str(i):{'film'+str(randrange(1, 10)) for j in range(randrange(15))}
        for i in range(10)}
#待测用户曾经看过并感觉不错的电影
user = {'film1', 'film2', 'film3'}
#查找与待测用户最相似的用户和他喜欢看的电影,忽略与待测用户完全一样的用户
similarUser, films = max(data.items(),
                        key=lambda item: (item[1]!=user, len(item[1]&user)))
print('历史数据: ')
for u, f in data.items():
    print(u, f, sep=':')
print('和您最相似的用户是: ', similarUser)
print('他最喜欢看的电影是: ', films)
print('他看过的电影中您还没看过的有: ', films-user)
```

例 2-3 过滤无效书评。如果一条评论中重复字超过一定比例就认为是无效的。

```python
comments = ['这是一本非常好的书,作者用心了',
            '作者大大辛苦了',
            '好书,感谢作者提供了这么多的好案例',
            '书在运输的路上破损了,我好悲伤。。。 ',
            '啊啊啊啊啊啊,我怎么才发现这么好的书啊,相见恨晚',
            '好好好好好好好好好好好',
            '好难啊看不懂好难啊看不懂好难啊看不懂',
            '书的内容很充实',
            '你的书上好多代码啊,不过想想也是,编程的书嘛,肯定代码多一些',
            '书很不错!!一级棒!!买书就上当当,正版,价格又实惠,让人放心!!! ',
            '无意中来到你小铺就淘到心仪的宝贝,心情不错! ',
            '送给朋友的、很不错',
            '这是一本好书,讲解内容深入浅出又清晰明了,推荐给所有喜欢阅读的朋友同好们。']
result = filter(rule, lambda s:len(set(s))/len(s)>0.5)
print('原始书评:')
for comment in comments:
    print(comment)
print('=' * 30)
print('过滤后的书评:')
for comment in result:
    print(comment)
```

最后,除了前面几节介绍的列表推导式、生成器表达式、字典推导式,Python也支持集合推导式。

```
>>>{x.strip() for x in ('he', 'she', 'I')}
{'I', 'she', 'he'}
>>>import random
>>>x = {random.randint(1,500) for i in range(100)}    #生成随机数,自动删除重复元素
>>>len(x)                                             #一般输出结果会小于100
```

2.5 再谈内置函数 sorted()

2.5

列表对象提供了sort()方法支持原地排序,而内置函数sorted()返回新的列表,不对原列表做任何修改。sorted()函数还可以对元组、字典、集合、字符串等有限长度的可迭代对象排序,借助key参数可以实现更复杂的排序。

```
>>> persons =[{'name':'Dong', 'age':37}, {'name':'Zhang', 'age':40},
              {'name':'Li', 'age':50}, {'name':'Dong', 'age':43}]
#使用key来指定排序依据,先按姓名升序排序,姓名相同的按年龄降序排序
>>> print(sorted(persons, key=lambda x:(x['name'], -x['age'])))
[{'name': 'Dong', 'age': 43}, {'name': 'Dong', 'age': 37}, {'name': 'Li', 'age': 50},
 {'name': 'Zhang', 'age': 40}]
>>> phonebook = {'Linda':'7750', 'Bob':'9345', 'Carol':'5834'}
>>> from operator import itemgetter
>>> sorted(phonebook.items(), key=itemgetter(1))      #按字典中元素值排序
[('Carol', '5834'), ('Linda', '7750'), ('Bob', '9345')]
>>> sorted(phonebook.items(), key=itemgetter(0))      #按字典中元素的键排序
```

```
[('Bob', '9345'), ('Carol', '5834'), ('Linda', '7750')]
>>> gameresult =[['Bob', 95.0, 'A'], ['Alan', 86.0, 'C'], ['Mandy', 83.5, 'A'],
                 ['Rob', 89.3, 'E']]
>>> sorted(gameresult, key=itemgetter(0, 1))    #按姓名升序排序,姓名相同的按分数升序排序
[['Alan', 86.0, 'C'], ['Bob', 95.0, 'A'], ['Mandy', 83.5, 'A'], ['Rob', 89.3, 'E']]
>>> sorted(gameresult, key=itemgetter(1, 0))    #按分数升序排序,分数相同的按姓名升序排序
[['Mandy', 83.5, 'A'], ['Alan', 86.0, 'C'], ['Rob', 89.3, 'E'], ['Bob', 95.0, 'A']]
>>> sorted(gameresult, key=itemgetter(2, 0))    #按等级升序排序,等级相同的按姓名升序排序
[['Bob', 95.0, 'A'], ['Mandy', 83.5, 'A'], ['Alan', 86.0, 'C'], ['Rob', 89.3, 'E']]
>>> gameresult =[{'name':'Bob', 'wins':10, 'losses':3, 'rating':75.0},
                 {'name':'David', 'wins':3, 'losses':5, 'rating':57.0},
                 {'name':'Carol', 'wins':4, 'losses':5, 'rating':57.0},
                 {'name':'Patty', 'wins':9, 'losses':3, 'rating':72.8}]
#按'wins'升序排序,该值相同的按'name'升序排序,略去结果
>>> sorted(gameresult, key=itemgetter('wins', 'name'))
```

下面的代码演示如何根据另一个列表的值来对当前列表元素排序。

```
>>> list1 = ["what", "I'm", "sorting", "by"]
>>> list2 = ["something", "else", "to", "sort"]
>>> pairs = zip(list1, list2)
>>> pairs = sorted(pairs)              #根据list1中的元素排序list2中的元素
>>> pairs
[("I'm", 'else'), ('by', 'sort'), ('sorting', 'to'), ('what', 'something')]
>>> result = [x[1] for x in pairs]
>>> result
['else', 'sort', 'to', 'something']
>>> sorted(list2,                       #这个用法仅限于list2中元素各不相同的场合
          key=lambda item:list1[list2.index(item)])
['else', 'sort', 'to', 'something']
```

下面的代码用于把包含若干整数的列表中所有奇数都放到前面,偶数都放到后面。

```
>>> from random import randint
>>> x = [randint(1,100) for i in range(20)]    #可以使用random.choices()函数改写
>>> x
[19, 32, 76, 82, 23, 63, 38, 50, 20, 30, 39, 14, 19, 50, 81, 27, 77, 12, 55, 29]
>>> sorted(x, key=lambda item:item%2==0)       #lambda表达式见5.6节
[19, 23, 63, 39, 19, 81, 27, 77, 55, 29, 32, 76, 82, 38, 50, 20, 30, 14, 50, 12]
```

*2.6 复杂数据结构

在应用开发中,除了 Python 内置数据类型外,还经常需要使用其他一些复杂的数据结构,如堆、队列、栈、树、图等。其中,有些结构 Python 本身已经提供了,有些则需要自己利用 Python 基本数据类型进行集成和二次开发来实现。

2.6.1 堆

堆是一种重要的数据结构,在进行排序时使用较多,优先队列也是堆结构的一个重要应用。堆是一棵二叉树,其中每个父节点的值都小于或等于其所有子节点的值。使用列表来实现堆时,对于所有的 k(下标,从 0 开始)都满足 heap[k] <= heap[2 * k +1]和 heap[k] <= heap[2 * k +2],

并且整个堆中最小的元素总是位于二叉树的根节点。Python 在 heapq 模块中提供了对堆的支持。下面的代码演示了堆的原理及 heapq 模块的用法，同时也请注意 random 模块的用法。另外，当堆中没有元素时，进行 heappop()操作将会抛出异常。

```
>>> import heapq
>>> import random
>>> data = list(range(10))
>>> random.shuffle(data)                    #随机打乱列表中元素的顺序
>>> data
[6, 1, 3, 4, 9, 0, 5, 2, 8, 7]
>>> heap = []
>>> for n in data:                          #建堆
    heapq.heappush(heap, n)
>>> heap
[0, 2, 1, 4, 7, 3, 5, 6, 8, 9]
>>> heapq.heappush(heap, 0.5)               #新数据入堆,堆会自动重建
>>> heap
[0, 0.5, 1, 4, 2, 3, 5, 6, 8, 9, 7]
>>> heapq.heappop(heap)                     #弹出最小的元素,堆会自动重建
0
>>> heapq.heappop(heap)                     #可以增加代码查看堆的内容和结构
0.5
>>> heapq.heappop(heap)
1
>>> myheap = [1, 2, 3, 5, 7, 8, 9, 4, 10, 333]
>>> heapq.heapify(myheap)                   #将列表转换为堆
>>> myheap
[1, 2, 3, 4, 7, 8, 9, 5, 10, 333]
>>> heapq.heapreplace(myheap, 6)            #替换堆中的最小元素值,自动重新构建堆
1
>>> myheap
[2, 4, 3, 5, 7, 8, 9, 6, 10, 333]
>>> heapq.nlargest(3, myheap)               #返回前3个最大的元素,myheap 不必须是堆
[333, 10, 9]
>>> heapq.nsmallest(3, myheap)              #返回前3个最小的元素,还支持 key 参数
[2, 3, 4]
>>>heapq.nlargest(3, (1,2,3,4,5,6,7,8),
                 key=lambda num: num%3)     #除以3的余数最大的前3个元素
[2, 5, 8]
>>>heapq.nlargest(2, [111,22,3], key= str)  #转换为字符串后最大的两个元素
[3, 22]
>>> heapq.nlargest(1, [111,22,3],           #各位数字之和最大的一个元素
                 key=lambda num: sum(map(int,str(num))))
[22]
```

2.6.2 队列

　　队列的特点是先进先出（First In First Out，FIFO）和后进后出（Last In Last Out，LILO），在某些应用中有着重要的作用，如多线程编程、作业处理等。Python 提供了 queue 模块和 collections.deque 对象支持队列的操作，也可以使用 Python 列表进行二次开发来实现自定义的队列结构。

```
>>> import queue
>>> q = queue.Queue()
>>> q.put(0)                                    #元素入队,添加到队列尾部
>>> q.put(1)
>>> q.put(2)
>>> q.queue
deque([0, 1, 2])
>>> q.get()                                     #队列头元素出队
0
>>> q.queue
deque([1, 2])
>>> q.get()
1
>>> q.queue
deque([2])
```

另外,queue 模块还提供了后进先出队列和优先级队列:

```
>>> import queue
>>> LiFoQueue = queue.LifoQueue(5)              #后进先出队列
>>> LiFoQueue.put(1)
>>> LiFoQueue.put(2)
>>> LiFoQueue.put(3)
>>> LiFoQueue.get()
3
>>> LiFoQueue.get()
2
>>> LiFoQueue.get()
1
>>> PriQueue = queue.PriorityQueue(5)           #优先级队列
>>> PriQueue.put(3)
>>> PriQueue.put(5)
>>> PriQueue.put(1)
>>> PriQueue.put(8)
>>> PriQueue.queue
[1, 5, 3, 8]
>>> PriQueue.get()
1
>>> PriQueue.get()
3
>>> PriQueue.get()
5
>>> PriQueue.get()
8
```

下面的代码使用列表模拟队列结构,考虑了入队、出队、判断队列是否为空、是否已满及改变队列大小等基本操作。面向对象程序设计的知识可参考第 6 章。

例 2-4　自定义队列结构。

```
class MyQueue:
    #构造方法,默认队列大小为 10
    def __init__(self, size=10):
        self.__content = []
```

```python
        self.__size = size
        self.__current = 0

    def setSize(self, size):
        if size < self.__current:
            #如果缩小队列,应删除后面的元素
            del self.__content[size:self.current]
            self.__current = size
        self.__size = size

    def put(self, v):
        if self.__current < self.__size:
            self.__content.append(v)
            self.__current = self.__current + 1
        else:
            print('The queue is full')

    def get(self):
        if self.__content:
            self.__current = self.__current - 1
            return self.__content.pop(0)
        else:
            print('The queue is empty')

    def show(self):
        if self.__content:
            print(self.__content)
        else:
            print('The queue is empty')

    def empty(self):
        self.__content, self.__current = [], 0

    def isEmpty(self):
        return not self.__content

    def isFull(self):
        return self.__current == self.__size

if __name__ == '__main__':
    print('Please use me as a module.')
```

将上面的代码保存为 myQueue.py 文件,下面的代码演示了自定义队列类的用法。

```
>>> import myQueue
>>> q = myQueue.MyQueue()
>>> q.get()
The queue is empty
>>> q.put(5)
>>> q.put(7)
>>> q.isFull()
False
>>> q.put('a')
>>> q.put(3)
>>> q.show()
[5, 7, 'a', 3]
>>> q.setSize(3)
>>> q.show()
[5, 7, 'a']
>>> q.put(10)
The queue is full
>>> q.setSize(5)
>>> q.put(10)
>>> q.show()
[5, 7, 'a', 10]
```

2.6.3 栈

栈是一种后进先出（Last In First Out，LIFO）或先进后出（First In Last Out，FILO）的数据结构，Python 列表本身就可以实现栈结构的基本操作。例如，列表对象的 append() 方法是在列表尾部追加元素，类似于入栈操作；pop() 方法默认是弹出并返回列表的最后一个元素，类似于出栈操作。但是直接使用 Python 列表对象模拟栈操作并不是很方便，当列表为空时，若再执行 pop() 出栈操作，则会抛出一个异常；另外，也无法限制栈的大小。例如下面的代码：

```
>>> myStack = []                        7
>>> myStack.append(3)                   >>> myStack.pop()
>>> myStack.append(5)                   5
>>> myStack.append(7)                   >>> myStack.pop()
>>> myStack                             3
[3, 5, 7]                               >>> myStack.pop()
>>> myStack.pop()                       IndexError: pop from empty list
```

下面的代码使用列表模拟栈结构的用法，实现了入栈、出栈、判断栈是否为空、是否已满及改变栈大小等操作。

例 2-5 自定义栈结构。

```
class Stack:
    def __init__(self, size=10):
        self.__content = []               #使用列表存放栈的元素
        self.__size = size                #初始栈大小
        self.__current = 0                #栈中元素个数初始化为 0

    def empty(self):
        self.__content = []
        self.__current = 0

    def isEmpty(self):
        return not self.__content:

    def setSize(self, size):
        #如果缩小栈空间,则删除指定大小之后的已有元素
        if size < self.__current:
            del self.__content[size:self.__current]
            self.__current = size
        self.__size = size

    def isFull(self):
        return self.__current == self.__size

    def push(self, v):
        if len(self.__content) < self.__size
            self.__content.append(v)
            self.__current = self.__current +1        #栈中元素个数加 1
        else:
```

```
                print('Stack Full!')

    def pop(self):
        if self.__content:
            self.__current = self.__current - 1        #栈中元素个数减1
            return self.__content.pop()
        else:
            print('Stack is empty!')

    def show(self):
        print(self.__content)

    def showRemainderSpace(self):
        print('Stack can still PUSH', self.__size-self.__current, 'elements.')

if __name__ == '__main__':
    print('Please use me as a module.')
```

将代码保存为 Stack.py 文件，下面的代码演示了自定义栈结构的用法。

```
>>> import Stack                          >>> s.showRemainderSpace()
>>> s = Stack.Stack()                     Stack can still PUSH 6 elements.
>>> s.isEmpty()                           >>> s.setSize(3)
True                                      >>> s.isFull()
>>> s.isFull()                            True
False                                     >>> s.show()
>>> s.push(5)                             [5, 8, 'b']
>>> s.push(8)                             >>> s.setSize(5)
>>> s.push('a')                           >>> s.push('d')
>>> s.pop()                               >>> s.push('dddd')
'a'                                       >>> s.push(3)
>>> s.push('b')                           Stack Full!
>>> s.push('c')                           >>> s.show()
>>> s.show()                              [5, 8, 'b', 'd', 'dddd']
[5, 8, 'b', 'c']
```

2.6.4 链表

可以直接使用列表及其基本操作来实现链表的功能，很方便地完成链表创建以及节点的插入和删除操作；也可以对列表进行封装或全新定义节点和指针来实现自定义的链表结构，完成特殊功能或更加完美的外围检查工作。

```
>>> linkTable = []
>>> linkTable.append(3)                   #在链表尾部追加节点
>>> linkTable.append(5)
>>> linkTable
[3, 5]
>>> linkTable.insert(1, 4)                #在链表中间插入节点
>>> linkTable
[3, 4, 5]
>>> del linkTable[1]                      #删除节点
>>> linkTable
[3, 5]
```

如前所述，使用列表直接模拟链表结构时，同样存在一些问题，例如，链表为空或删除的元素不存在时会抛出异常，插入和删除元素会有额外开销。可以对列表进行封装来实现完整的链表操作，可以参考队列与栈，以及 2.6.5 节二叉树的代码，此处不再赘述。

2.6.5 二叉树

下面的代码实现了二叉树结构，即二叉树创建、插入子节点，以及前序遍历、后序遍历和中序遍历等遍历方式，同时还支持二叉树中任意子树的节点遍历。

例 2-6　自定义二叉树结构。

```python
class BinaryTree:
    def __init__(self, value):
        self.__left = None                          #左子节点指针
        self.__right = None                         #右子节点指针
        self.__data = value                         #当前节点数据

    def insertLeftChild(self, value):               #创建左子树
        if self.__left:
            print('left child tree already exists.')
        else:
            self.__left = BinaryTree(value)
            return self.__left

    def insertRightChild(self, value):              #创建右子树
        if self.__right:
            print('Right child tree already exists.')
        else:
            self.__right = BinaryTree(value)
            return self.__right

    def show(self):
        print(self.__data)                          #显示当前节点的数据

    def preOrder(self):                             #前序遍历
        print(self.__data)                          #输出根节点的值
        if self.__left:
            self.__left.preOrder()                  #遍历左子树，函数递归调用见第 5 章
        if self.__right:
            self.__right.preOrder()                 #遍历右子树

    def postOrder(self):                            #后序遍历
        if self.__left:
            self.__left.postOrder()
        if self.__right:
            self.__right.postOrder()
        print(self.__data)

    def inOrder(self):                              #中序遍历
        if self.__left:
            self.__left.inOrder()
```

```
            print(self.__data)
        if self.__right:
            self.__right.inOrder()

if __name__ == '__main__':
    print('Please use me as a module.')
```

把上面的代码保存为 BinaryTree.py 文件，下面的代码创建了图 2-3 所示的二叉树，并对该树进行遍历。

```
>>> import BinaryTree
>>> root = BinaryTree.BinaryTree('root')
>>> b = root.insertRightChild('B')
>>> a = root.insertLeftChild('A')
>>> c = a.insertLeftChild('C')
>>> d = c.insertRightChild('D')
>>> e = b.insertRightChild('E')
>>> f = e.insertLeftChild('F')
>>> root.inOrder()
C  D  A  root  B  F  E
>>> root.postOrder()
D  C  A  F  E  B  root
>>> b.inOrder()
B  F  E
```

图 2-3 二叉树

2.6.6 有向图

下面的代码模拟了有向图的创建和路径搜索功能。有向图由若干节点和边组成，其中每条边都有明确方向，从一个节点指向另一个节点。若有向图中两个节点之间存在一条或多条有向边，则表示从起点可以到达终点，认为存在一条路径。

例 2-7 自定义有向图结构。

例 2-7

```
def __generatePath(graph, path, end, results):
    current = path[-1]
    if current == end:
        results.append(path)
    else:
        for n in graph[current]:
            if n not in path:
                __generatePath(graph, path+[n], end, results)

def searchPath(graph, start, end):
    results = []
    __generatePath(graph, [start], end, results)
    results.sort(key=len)                #按所有路径的长度排序
    return results

def showPath(results):
    print('The path from', results[0][0], 'to', results[0][-1], 'is:')
    for path in results:
```

```
            print(path)

if __name__ == '__main__':
    graph = {'A':['B', 'C', 'D'],
             'B':['E'],
             'C':['D', 'F'],
             'D':['B', 'E', 'G'],
             'E':['D'],
             'F':['D', 'G'],
             'G':['E']}
    r1 = searchPath(graph, 'A', 'D')
    showPath(r1)
```

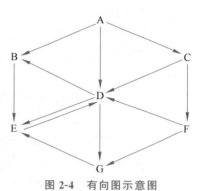

图 2-4 有向图示意图

代码中用到的有向图如图 2-4 所示,程序运行结果为

```
The path from   A   to   D   is:
['A', 'D']
['A', 'C', 'D']
['A', 'B', 'E', 'D']
['A', 'C', 'F', 'D']
['A', 'C', 'F', 'G', 'E', 'D']
```

本 章 小 结

(1) 列表、字符串、元组属于有序序列,支持双向索引和切片;支持使用负整数作为下标来访问其中的元素,-1 表示最后一个元素位置,-2 表示倒数第二个元素位置,以此类推。

(2) 同一个列表中元素的数据类型可以各不相同,即同时分别为整数、实数、字符串等基本类型,也可以是列表、元组、字典、集合及其他自定义类型的对象,并且支持复杂数据类型对象的嵌套。

(3) 集合不支持使用下标的方式来访问其中的元素,可以使用字典的"键"作为下标来访问字典中的"值"。

(4) 如果要创建只包含一个元素的元组,只把元素放在圆括号里是不可以的,还需要在元素后面加一个逗号。

(5) 将列表、元组或字符串对象与一个整数进行 * 运算,表示将对象中的元素进行重复并返回一个新的同类型对象。

(6) 虽然+运算符可以连接两个列表对象,但并不是原地修改列表,而是返回一个新列表,不对原列表对象做任何修改。该运算符涉及大量的元素复制操作,效率较低,建议优先考虑使用列表对象的 append()或 extend()方法。

(7) 推荐使用字典对象的 get()方法来访问其中元素的值。

(8) 列表、字典、集合属于可变对象,元组、字符串属于不可变对象。

(9) 虽然列表支持在中间任意位置插入和删除元素,但一般建议尽量从尾部进行元素的增加与删除,这样可以获得更快的速度。

(10) 切片操作不仅可以用来返回列表、元组、字符串中的部分元素,还可以对列表中的元素值进行修改,以及增加或删除元素。

（11）关键字 in 可以用于列表及其他可迭代对象,包括元组、字典、range 对象、字符串、集合及迭代器对象,常用在 for 循环语句中对可迭代对象中的元素进行遍历。

（12）列表推导式可以使用简洁的形式来生成满足特定需要的列表。

（13）序列解包在多个场合具有重要的应用,是 Python 的基本操作之一。

（14）字典中的"键"可以是任意不可变数据,如整数、实数、复数、字符串、元组等,但不能使用列表、集合、字典作为字典的"键",包含它们的元组也不能,因为这些类型的对象是可变的。

（15）字典中的"键"不允许重复,"值"可以重复。

（16）集合中的所有元素不允许重复,可以使用集合快速提取其他序列中的唯一元素。

（17）内置函数 len()可以用来返回列表中元素的个数,同样适用于元组、字典、集合、字符串、range 对象。

（18）内置函数 zip()可以将多个可迭代对象中对应位置的元素组合为元组,并返回"包含"这些元组的 zip 对象。

（19）内置函数 enumerate()可以用来枚举列表、元组或其他可迭代对象的元素,返回枚举对象,枚举对象中每个元素是包含下标和元素值的元组。

（20）列表、元组、字典、集合、字符串等实际包含元素的对象称为容器对象,与具有惰性求值特点的迭代器对象统称为可迭代对象。

习　题

1. Python 3.x 的 range()函数返回一个_____。

2. 表达式[3] in [1,2,3,4]的值为_____。

3. 列表对象的 sort()方法用于对列表元素进行原地排序,该函数返回值为_____。

4. 列表对象的_____方法删除首次出现的指定元素,如果列表中不存在要删除的元素,则抛出异常。

5. 假设列表对象 aList 的值为[3,4,5,6,7,9,11,13,15,17],那么切片 aList[3:7]得到的值是_____。

6. 字典和集合都是用一对_____作为界定符,字典的每个元素由两部分组成,即_____和_____,其中_____不允许重复。

7. 使用字典对象的_____方法可以返回字典的"键:值"对,使用字典对象的_____方法可以返回字典的"键",使用字典对象的_____方法可以返回字典的"值"。

8. 假设有列表 a=['name','age','sex']和 b=['Dong',38,'Male'],使用一条语句将这两个列表的内容转换为字典,并且以列表 a 中的元素为"键",以列表 b 中的元素为"值",这个语句可以写为_____。

9. 假设有一个列表 a,现要求从列表 a 中每 3 个元素取 1 个,并且将取到的元素组成新的列表 b,可以使用语句_____。

10. 使用列表推导式生成包含 10 个数字 5 的列表,语句可以写为_____。

11. _____(可以、不可以)使用 del 命令来删除元组中的部分元素。

12. 已知列表 data 中有若干字符串,要求编写程序,对 data 中的字符串进行过滤,只输出重复字符不超过一半的字符串。

13. 编写程序,使用筛选法查找并输出小于 1000 的所有素数。

14. 为什么应尽量从列表的尾部进行元素的增加与删除操作?

15. 编写程序,生成包含 1000 个 0~100 的随机整数,并统计每个元素的出现次数。

16. 编写程序,用户输入一个列表和两个整数作为下标,然后输出列表中介于两个下标闭区间之间的元素组成的子列表。例如,用户输入[1,2,3,4,5,6]和 2,5,程序输出[3,4,5,6]。

17. 设计一个字典,并编写程序,用户输入内容作为"键",然后输出字典中对应的"值",如果用户输入的"键"不存在,则输出"您输入的键不存在!"。

18. 编写程序,生成包含 20 个随机数的列表,然后将前 10 个元素升序排列,后 10 个元素降序排列,并输出结果。

第 3 章 选择与循环

在传统的面向过程程序设计中有 3 种经典的控制结构,即顺序结构、选择结构和循环结构。即使是在面向对象程序设计语言中以及事件驱动或消息驱动应用开发中,也无法脱离这 3 种基本的程序结构。可以说,不管使用哪种程序设计语言,在实际开发中,为了实现特定的业务"逻辑或"算法,都不可避免地要用到大量的选择结构和循环结构,并且经常需要将选择结构和循环结构嵌套使用,如果把选择结构、循环结构、异常处理结构(见第 8 章)、函数定义、类定义、with 块等分别看作一个整体,那么它们之间又属于顺序结构。本章首先介绍条件表达式和 Python 中选择结构与循环结构的语法,然后通过几个示例来理解其用法。

3.1 条件表达式

在选择结构和循环结构中,都要使用条件表达式来确定下一步的执行流程。在 Python 中,单个常量、变量或者任意合法表达式都可以作为条件表达式。

在选择和循环结构中,条件表达式的值只要不是 False、0(或 0.0、0j 等)、空值 None、空列表、空元组、空集合、空字典、空字符串、空 range 对象或其他空的容器对象,Python 解释器均认为与 True 等价。map、zip 等所有迭代器对象都等价于 True。一个对象等价于 True 是指作为内置函数 bool() 的参数可返回 True。

关于运算符和表达式的详细内容在 1.4.5 节中已有介绍,此处不再赘述,只简单介绍一下条件表达式中比较常用的几个运算符。首先是关系运算符,与很多语言不同的是,在 Python 中的关系运算符可以连续使用,例如:

```
>>> print(1<2<3)                              #等价于 print(1<2 and 2<3)
True
>>> print(1<2>3)
False
>>> print(1<3>2)                              #其中的表达式等价于 1<3 and 3>2
True
```

比较常用的运算符还有逻辑运算符 and 和 or,这两个运算符具有短路求值或惰性求值的特点,简单地说,就是只计算必须计算的表达式的值。在设计条件表达式时,在表示复杂条件时如果能够巧妙利用逻辑运算符 and 和 or 的短路求值或惰性求值特性来组织多个子表达式的顺序,可以大幅提高程序的运行效率,减少不必要的计算与判断。

以 and 为例,对于表达式"表达式 1 and 表达式 2"而言,如果"表达式 1"的值为 False 或其他等价值时,不论"表达式 2"的值是什么,整个表达式的值都等价于 False,此时"表达式 2"的值无论是什么都不影响整个表达式的值,因此将不会被计算,从而减少不必要的计算和判断。"逻辑或"运算符 or 也具有类似的特点,读者可以自行分析。

在 Python 中,条件表达式中不允许使用"=",避免了某些语言中误将关系运算符"=="写作"="带来的麻烦,在条件表达式中使用"="将抛出异常,提示语法错误,比逻辑错误更容易发现和修改。在 Python 3.8 及之后的版本中,可以在表达式中使用海象运算符":="。

```
>>> if a = 3:
SyntaxError: invalid syntax
>>> if a := 3:                              #会输出字符串 ok,同时创建变量 a
    print('ok')
```

3.2 选择结构

选择结构通过判断某些特定条件是否满足来决定下一步的执行流程。常见的有单分支选择结构、双分支选择结构、嵌套的选择结构、多分支选择结构,形式比较灵活多变,具体使用哪种最终取决于所要实现的业务逻辑。后面章节中讲到的循环结构和异常处理结构中也可以带有 else 子句,可以看作选择结构的变形。

3.2.1 单分支选择结构

单分支选择结构是最简单的一种形式,其语法如下,其中表达式后面的冒号是不可缺少的,表示一个语句块的开始,后面几种其他形式的选择结构和循环结构中的冒号也是必须有的。注意,语句块必须相对于 if 向右缩进,一般以 4 个空格为缩进单位。

```
if 表达式:
    语句块
```

当表达式值为 True 或其他等价值时,表示条件满足,语句块将被执行,否则该语句块将不被执行。

```
x = input('Input two numbers:')             #输入时两个数字之间使用空格分隔
a, b = map(int, x.split())                  #序列解包,见 2.2.3 节和例 1-2
if a > b:
    a, b = b, a                             #交换两个变量的值
print(a, b)
```

3.2.2 双分支选择结构

双分支选择结构的语法为

```
if 表达式:
    语句块 1
else:
    语句块 2
```

当表达式值为 True 或其他等价值时,执行语句块 1,否则执行语句块 2。下面的代码演示了双分支选择结构的用法:

```
>>> chTest = ['1', '2', '3', '4', '5']
>>> if chTest:
    print(chTest)                           #前面的 3 个大于号可以理解为不占位置
else:
    print('Empty')                          #else虽然顶格,但逻辑上和 if 是对齐的
['1', '2', '3', '4', '5']
```

Python 还支持如下形式的表达式：

```
value1 if condition else value2
```

当条件表达式 condition 的值与 True 等价时，表达式的值为 value1，否则为 value2。另外，在 value1 和 value2 中还可以使用复杂表达式，包括函数调用。下面的代码演示了上面的表达式的用法，可以看出，这个结构的表达式也具有惰性求值的特点。

```
>>> x = math.sqrt(9) if 5>3 else random.randint(1,100)   #此时还没有导入 math 模块
NameError: name 'math' is not defined
>>> import math
#此时还没有导入 random 模块，但由于条件表达式 5 >3 的值为 True，所以可以正常运行
>>> x = math.sqrt(9) if 5>3 else random.randint(1,100)
#此时还没有导入 random 模块，由于条件表达式 2 >3 的值为 False，需要计算第二个表达式的值
#因此出错
>>> x = math.sqrt(9) if 2>3 else random.randint(1,100)
NameError: name 'random' is not defined
>>> import random
>>> x = math.sqrt(9) if 2>3 else random.randint(1,100)
```

3.2.3 嵌套的选择结构

嵌套的选择结构为用户提供了更多的选择，可以实现复杂的业务逻辑，一种语法形式为

```
if 表达式 1:
    语句块 1
elif 表达式 2:
    语句块 2
elif 表达式 3:
    语句块 3
    ⋮
else:
    语句块 n
```

其中，关键字 elif 是 else if 的缩写。下面的代码演示了利用该语法将成绩从百分制变换到等级制的实现方法，函数定义的语法见 5.1 节。

```python
def func(score):
    if score >100:
        return 'wrong score. must<=100.'
    elif score >= 90:
        return 'A'
    elif score >= 80:
        return 'B'
    elif score >= 70:
        return 'C'
    elif score >= 60:
        return 'D'
    elif score >= 0:
        return 'F'
    else:
        return 'wrong score. must>0.'
```

另一种嵌套选择结构的语法形式如下：

```
if 表达式 1:
    语句块 1
    if 表达式 2:
        语句块 2
    else:
        语句块 3
else:
    if 表达式 4:
        语句块 4
```

使用该结构时，一定要严格控制好不同级别代码块的缩进量，因为这决定了不同代码块的从属关系以及业务逻辑是否被正确地实现、是否能够被 Python 正确理解和执行。例如，百分制转等级制的示例，作为一种编程技巧，还可以尝试下面的写法：

```
def func(score):
    degree = 'DCBAAF'                    #第一个 A 对应[90,100)区间，第二个 A 对应 100 分
    if score>100 or score<0:
        return 'wrong score. must between 0 and 100.'
    else:
        index = (score-60) // 10
        if index >= 0:
            return degree[index]
        else:
            return degree[-1]
```

3.2.4 多分支选择结构

在 Python 3.9 及之前的版本中，没有提供真正意义上的多分支选择结构，如果确实需要，可以通过字典构造跳转表来实现。例如下面的代码：

```
status = {200:'ok', 201:'Created', 202:'Accepted',
          203:'Non-Authoritative Information', 204:'No Content'}
s = int(input('请输入状态码: '))
print('对应的状态为: ', status.get(s, 'unknown'))
funcs = {'1': lambda x: x**1, '2': lambda x: x**2,
         '3': lambda x: x**3, '4': lambda x: x**4}
k = input('请输入一个数字: ')
func = funcs.get(k)
if func:
    print(func(int(k)))
else:
    print('输入错误。')
```

Python 3.10 新增了软关键字（只在特定场合作为关键字，普通场合也可作为变量名）match 和 case，实现了真正意义上的多分支选择结构。

下面代码使用多分支选择结构实现了 HTTP 状态码到含义的转换。

```
code = int(input('请输入 HTTP 状态码: '))
match code:
    case 200:
        print('ok')
```

```
        case 201:
            print('Created')
        case 202:
            print('Accepted')
        case 203:
            print('Non-Authoritative Information')
        case 204:
            print('No Content')
        case 401 | 403 | 404:
            #同时匹配401、403、404这3种情况
            print('Not allowed')
        case _:
            #通配符,表示任意内容,如果前面的都不匹配,就执行这里的代码
            print('Sorry, please try later.')
```

下面的代码演示了下画线在元组中表示任意内容的用法:

```
while (point:=input('表示三维空间坐标的元组,0表示结束: ')) != '0':
    point = eval(point)
    match point:
        case (0, 0, 0):
            print('坐标原点')
        case (0, _, _):
            print('YOZ平面上的点')
        case (_, 0, _):
            print('XOZ平面上的点')
        case (_, _, 0):
            print('XOY平面上的点')
```

在match…case…多分支选择结构中,下画线除了上面的用法之外,还可以和星号组合使用,例如下面的代码中"*_"表示从当前位置往后还有0到任意多项:

```
while (content:=eval(input('请输入列表: '))) != 0:
    match content:
        case [1, 2, 3, 4, *_]:
            print('前4项匹配成功')
        case [1, 2, 3, *_]:
            print('前3项匹配成功')
        case [1, 2, *_]:
            print('前2项匹配成功')
        case [1, *_]:
            print('前1项匹配成功')
        case [*_]:
            print('匹配失败')
        case _:
            print('格式不对')
```

在下面的代码中,使用if对当前匹配项进行约束,如果条件不成立就继续检查下一项是否匹配,其中的x和y可以是任意变量名。

```
match (3, 5):
    case (x, y) if x<y:
        print('<')
    case (x, y) if x==y:
```

```
        print('==')
    case (x, y) if x>y:
        print('>')
```

3.2.5 选择结构应用案例

例 3-1 面试资格确认。

```
age = 24
subject = '计算机'
college = '非重点'
if (age>25 and subject=='电子信息工程') or (college=='重点' and subject=='电子信息工程')\
    or (age<=28 and subject=='计算机'):
    print('恭喜,您已获得我公司的面试机会!')
else:
    print('抱歉,您未达到面试要求')
```

例 3-2 用户输入若干分数,每输入一个分数后询问是否继续输入下一个分数,回答 yes 就继续输入下一个分数,回答 no 就停止输入分数,然后计算所有分数的平均分。

```
numbers = []                                    #使用列表存放临时数据
while True:                                     #循环结构,参考 3.3 节
    x = input('请输入一个成绩:')
    try:                                        #异常处理结构,参考第 8 章
        numbers.append(float(x))
    except:
        print('不是合法成绩')
    while True:
        flag = input('继续输入吗?(yes/no)').lower()
        if flag not in ('yes', 'no'):           #限定用户输入内容必须为 yes 或 no
            print('只能输入 yes 或 no')
        else:
            break
    if flag == 'no':
        break

print(sum(numbers)/len(numbers))
```

例 3-2

例 3-3 编写程序,判断某个日期是该年第几天。

```
import time

def demo(year, month, day):                     #函数定义,参考第 5 章
    day_month = [31, 28, 31, 30, 31, 30, 31, 31, 30, 31, 30, 31]    #每个月的天数
    if year%400==0 or (year%4==0 and year%100!=0):                   #判断是否为闰年
        day_month[1] = 29                                            #闰年 2 月为 29 天
    if month == 1:
        return day
    else:
        return sum(day_month[:month-1]) + day

date = time.localtime()                         #获取当前的日期时间
```

例 3-3

```
year, month, day = date[:3]
print(demo(year, month, day))
```

标准库 calendar 提供了 isleap() 函数用于判断是否闰年。另外,标准库 datetime 提供了 date、datetime、time 和 timedelta 对象,可以很方便地计算指定年、月、日、时、分、秒之前或之后的日期、时间,前两个对象还提供了返回结果中包含"今天是今年第几天""今天是本周第几天"等属性的 timetuple() 方法及其他属性和方法。

```
>>> import datetime
>>> today = datetime.date.today()
>>> today
datetime.date(2023, 9, 2)
>>> today- datetime.date(Today.year, 1, 1) +datetime.timedelta(days=1)
datetime.timedelta(245)
>>> today.timetuple().tm_yday              #今天是今年的第几天
245
>>> today.replace(year=2018)               #替换日期中的年
datetime.date(2018, 9, 2)
>>> today.replace(month=1)                 #替换日期中的月
datetime.date(2023, 1, 2)
>>> now = datetime.datetime.now()
>>> now
datetime.datetime(2023, 12, 6, 16, 1, 6, 313898)
>>> now.replace(second=30)                 #替换日期时间中的秒
datetime.datetime(2023, 12, 6, 16, 1, 30, 313898)
>>> now +datetime.timedelta(days=5)        #计算 5 天后的日期时间
datetime.datetime(2023, 12, 11, 16, 1, 6, 313898)
>>> now +datetime.timedelta(weeks=-5)      #计算 5 周前的日期时间
datetime.datetime(2023, 11, 1, 16, 1, 6, 313898)
```

3.3 循环结构

3.3.1 while 循环与 for 循环

Python 提供了两种基本的循环结构:while 循环和 for 循环。其中,while 循环一般用于循环次数难以提前确定的情况,也可以用于循环次数确定的情况;for 循环一般用于循环次数可以提前确定的情况,尤其适用于遍历可迭代对象中元素的场合。循环结构之间可以互相嵌套,也可以与选择结构嵌套使用,用于实现更为复杂的逻辑。

while 循环和 for 循环常见的用法为

```
while 条件表达式:
    循环体
```

和

```
for 循环变量 in 可迭代对象:
    循环体
```

另外,while 循环和 for 循环都可以带 else 子句,如果循环因为条件表达式不成立而自然结束(不是因为执行了 break 而结束循环),继续执行 else 结构中的语句;如果循环因为执

行了 break 语句而导致循环提前结束，则不执行 else 中的语句。其语法形式为

```
while 条件表达式：
    循环体
else:
    else 子句代码块
```

和

```
for 变量 in 可迭代对象：
    循环体
else:
    else 子句代码块
```

3.3.2 循环结构的优化

为了优化程序以获得更高的效率和运行速度，在编写循环语句时，应尽量减少循环内部不必要的计算，将与循环变量无关的代码尽可能地提取到循环之外。对于使用多重循环嵌套的情况，应尽量减少内层循环中不必要的计算，尽可能地向外提。例如下面的代码，第二段明显比第一段的运行效率要高。

```python
import time
digits = (1, 2, 3, 4)

start = time.time()
for i in range(1000):
    result = []
    for i in digits:
        for j in digits:
            for k in digits:
                result.append(i*100+j*10+k)      #这一行的乘法重复了很多次
print(time.time()-start)
print(result)

start = time.time()
for i in range(1000):
    result = []
    for i in digits:
        i = i * 100                              #乘法计算尽量往外提
        for j in digits:
            j = j * 10
            for k in digits:
                result.append(i+j+k)
print(time.time()-start)
print(result)
```

在循环中应尽量使用局部变量，因为局部变量的查询和访问速度比全局变量略快，可参考 5.5 节的介绍。另外，在使用模块中的对象时，可以通过将其直接导入来减少查询次数和提高运行速度，可参考 1.4.8 节的介绍。

3.4　break 和 continue 语句

break 语句和 continue 语句在 while 循环和 for 循环中都可以使用,并且一般常与选择结构结合使用,以达到在特定条件得到满足时改变代码执行流程的目的。一旦 break 语句被执行,将使得逻辑上距离最近的循环提前结束。continue 语句的作用是终止本次循环,并忽略 continue 之后的所有语句,直接回到循环的顶端,提前进入下一次循环。

continue 的用法在后面章节会进行演示,下面的代码用来计算小于 100 的最大素数,请注意 break 语句和 else 子句的用法。

3.4

```
>>> for n in range(100, 1, -1):
        for i in range(2, n):                #循环范围可缩小为 n 的平方根
            if n%i == 0:
                break
        else:
            print(n)
            break
97
```

删除上面代码中最后一个 break 语句,可以用于输出 100 以内的所有素数。

3.5　案 例 精 选

例 3-4　计算表达式 1+2+3+…+100 的值。

```
s = 0
for i in range(1, 101):
    s = s + i
print('1+2+3+…+100=', s)
print('1+2+3+…+100=', sum(range(1, 101)))   #直接使用内置函数来实现题目的要求
```

例 3-5　枚举并输出序列中的元素。

```
a_list = ['a', 'b', 'mpilgrim', 'z', 'example']
for i, v in enumerate(a_list):
    print('列表的第', i+1, '个元素是: ', v)   #可以使用 enumerate()函数的参数 start 来简化
```

对于类似元素遍历的问题,同样也可以使用 while 循环来解决,但是代码要麻烦一些,可读性也较差,例如:

```
a_list = ['a', 'b', 'mpilgrim', 'z', 'example']
i = 0
number = len(a_list)
while i < number:
    print('列表的第', i+1, '个元素是: ', a_list[i])
    i += 1
```

例 3-6　求 1~100 能被 7 整除,但不能同时被 5 整除的所有整数。

```
for i in range(1, 101):
```

例 3-6

```
        if i%7==0 and i%5!=0:
            print(i)
```

例 3-7　输出 3 位"水仙花数"。所谓 n 位水仙花数是指一个 n 位的十进制数,其各位数字的 n 次方之和恰好等于该数本身。例如,153 是水仙花数,因为 $153 = 1^3 + 5^3 + 3^3$。

例 3-7

```
for i in range(10**2, 10**3):           #遍历所有 3 位自然数
    bai, shi, ge = map(int, str(i))     #获取百位数、十位数、个位数
    if bai**3 + shi**3 + ge**3 == i:
        print(i)
```

例 3-8　统计考试成绩中优秀、良、中、及格、不及格的人数。

方法一:

```
scores = [89, 70, 49, 87, 92, 84, 73, 71, 78, 81, 90, 37,
          77, 82, 81, 79, 80, 82, 75, 90, 54, 80, 70, 68, 61]
groups = {'优秀':0, '良':0, '中':0, '及格':0, '不及格':0}
for score in scores:
    if score >= 90:
        groups['优秀'] = groups['优秀'] +1
    elif score >= 80:
        groups['良'] = groups['良'] +1
    elif score >= 70:
        groups['中'] = groups['中'] +1
    elif score >= 60:
        groups['及格'] = groups['及格'] +1
    else:
        groups['不及格'] = groups['不及格'] +1
print(groups)
```

方法二:

```
from itertools import groupby

scores = [89, 70, 49, 87, 92, 84, 73, 71, 78, 81, 90, 37,
          77, 82, 81, 79, 80, 82, 75, 90, 54, 80, 70, 68, 61]
def classify(score):
    if score >= 90:
        return '优秀'
    elif score >= 80:
        return '良'
    elif score >= 70:
        return '中'
    elif score >= 60:
        return '及格'
    else:
        return '不及格'

groups = {category:len(tuple(score)) for category, score in groupby(sorted(scores), classify)}
print(groups)
```

方法三:

```
from collections import Counter
```

```
from pandas import cut                           #需要先安装扩展库 Pandas 才能使用
scores = [89, 70, 49, 87, 92, 84, 73, 71, 78, 81, 90, 37,
          77, 82, 81, 79, 80, 82, 75, 90, 54, 80, 70, 68, 61]
groups = Counter(cut(scores, [0, 60, 70, 80, 90, 101],
                     labels=['不及格', '及格', '中', '良', '优秀'],
                     right=False))               #设置各分数段为左闭右开区间
print(groups)
```

例 3-9 打印九九乘法表。

例 3-9

```
for i in range(1, 10):
    for j in range(1, i+1):
        print('{0}*{1}={2}'.format(i, j, i*j).ljust(6), end=' ')
    print()
```

例 3-10 求 200 以内能被 17 整除的最大正整数。

```
for i in range(200, 0, -1):                      #从大到小遍历
    if i%17 == 0:
        print(i)
        break                                    #遇到第一个符合条件的数就结束
```

例 3-11 判断一个正整数是否为素数。

```
import math

n = int(input('Input an integer:'))
m = math.ceil(math.sqrt(n)+1)
for i in range(2, m):
    if n%i==0 and i<n:                           #有因数,不是素数
        print('No')
        break
else:
    print('Yes')                                 #没有因数,是素数
```

math 是用于数学计算的标准库,除了平方根函数 sqrt()和向上取整函数 ceil(),还提供了最大公约数函数 gcd()、最小公倍数函数 lcm(),sin()、asin()等三角函数与反三角函数,弧度与角度转换函数 degrees()、radians(),误差函数 erf()、剩余误差函数 erfc()、伽马函数 gamma(),对数函数 log()、log2()、log10(),阶乘函数 factorial(),连乘函数 prod(),组合数函数 comb()和排列数函数 perm(),多维欧氏距离函数 hypot(),常数 pi 和 e,等等。

例 3-12 鸡兔同笼问题。假设共有鸡、兔 30 只,脚 90 只,求鸡、兔各有多少只。

例 3-12

```
for ji in range(0, 31):
    if 2*ji+(30-ji)*4 == 90:
        print('ji:', ji, ' tu:', 30-ji)
        break                                    #如果有解,肯定唯一
```

例 3-13 编写程序,输出由 1、2、3、4 这 4 个数字组成的每位数都不相同的所有 3 位数。

例 3-13

```
digits = (1, 2, 3, 4)
for i in digits:
```

```
        for j in digits:
            for k in digits:
                if i!=j and j!=k and i!=k:
                    print(i*100+j*10+k)
```

从代码优化的角度来讲,上面这段代码并不是很好,其中有些判断完全可以在外层循环来做,从而提高运行效率,可以扫描二维码观看优化方法以及更多实现。

例 3-14 有一箱苹果,4 个 4 个地数最后余下 1 个,5 个 5 个地数最后余下 2 个,9 个 9 个地数最后余下 7 个。编写程序计算这箱苹果至少有多少个。

先确定除以 9 余 7 的最小正整数,对这个数字重复加 9(确保总是除以 9 余 7),如果得到的数字除以 5 余 2 就停止;然后对得到的数字重复加 45(5 和 9 的最小公倍数),如果得到的数字除以 4 余 1 就停止。这时得到的数字就是题目的答案。

```
from itertools import count

for num in count(16, 9):                    #遍历从 16 开始且以 9 为步长的所有自然数
    if num%5 == 2:
        break
for result in count(num, 45):
    if result%4 == 1:
        break
print(result)
```

例 3-14

例 3-15 编写程序,计算组合数 $C(n,i)$,即从 n 个元素中任选 i 个,有多少种选法。Python 3.8 开始在标准库 math 中新增了组合数函数 `comb()`,本例主要是演示更多编程技巧。

根据组合数定义,需要计算 3 个数的阶乘,在很多编程语言中都很难直接使用整型变量表示大数的阶乘结果,虽然 Python 并不存在这个问题,但是计算大数的阶乘仍需要相当多的时间。本例提供另一种计算方法:以 cni(8,3) 为例,按定义式展开为 cni(8,3)=8!/3!/(8-3)!=(8×7×6×5×4×3×2×1)/(3×2×1)/(5×4×3×2×1)=(8×7×6)/(3×2×1),对于 (5,8] 区间的数,分子上出现一次而分母上没出现;[3,5] 区间的数在分子、分母上各出现一次,可以约掉;[1,3] 区间的数分子上出现一次而分母上出现两次,约分后只有分母上有。

```
def cni1(n, i):
    if not (isinstance(n, int) and isinstance(i, int) and n>=i):
        print('n and i must be integers and n must be larger than or equal to i.')
        return
    result = 1
    Min, Max = sorted((i, n-i))
    for i in range(n, 0, -1):
        if i > Max:
            result *= i                     #一般更建议写成 result = result * i,后同
        elif i <= Min:
            result //= i
    return result
print(cni1(6, 2))
```

例 3-15

也可以使用 math 库中的阶乘函数直接按组合数定义实现。

```
>>> def cni2(n, i):
    import math
    return int(math.factorial(n)/math.factorial(i)/math.factorial(n-i))
>>> cni2(6, 2)
15
```

还可以借助于 Python 标准库 itertools 提供的组合函数计算组合数，下面的代码占用内存较多，n 和 i 太大时会卡顿甚至崩溃。

```
>>> import itertools
>>> len(tuple(itertools.combinations(range(60), 2)))
1770
```

除了组合函数 combinations()，itertools 模块还提供了排列函数 permutations()、用于循环遍历可迭代对象元素的函数 cycle()、根据一个序列的值对另一个序列进行过滤的函数 compress()、根据函数返回值对序列进行分组的函数 groupby()、用于计数的函数 count()等大量函数。

```
>>> import itertools
>>> x = 'Private Key'
>>> y = itertools.cycle(x)                          #循环遍历序列中的元素
>>> for i in range(20):
    print(next(y), end=',')
P, r, i, v, a, t, e, , K, e, y, P, r, i, v, a, t, e, , K,
>>> for i in range(5):
    print(next(y), end=',')
e, y, P, r, i,
>>> x = range(1, 20)
>>> y = (1, 0)*9 +(1,)
>>> y
(1, 0, 1, 0, 1, 0, 1, 0, 1, 0, 1, 0, 1, 0, 1, 0, 1, 0, 1)
>>> list(itertools.compress(x, y))                  #根据一个序列的值对另一个序列进行过滤
[1, 3, 5, 7, 9, 11, 13, 15, 17, 19]
>>> def group(v):                                   #用于确定分组的自定义函数
    if v >10:
        return 'greater than 10'
    elif v < 5:
        return 'less than 5'
    else:
        return 'between 5 and 10'
>>> x = range(20)                                   #待分组的数据,要求已按序排列
>>> y = itertools.groupby(x, group)                 #根据函数返回值对元素进行分组
>>> for k, v in y:
    print(k, ':', list(v))
less than 5 : [0, 1, 2, 3, 4]
between 5 and 10 : [5, 6, 7, 8, 9, 10]
greater than 10 : [11, 12, 13, 14, 15, 16, 17, 18, 19]
>>> list(itertools.permutations([1, 2, 3, 4], 3))   #从 4 个元素中任选 3 个的所有排列
```

```
>>> x = itertools.permutations([1, 2, 3, 4], 4)    #输出结果(略)
                                                    #4个元素全排列
>>> next(x)                                         #获取下一个排列
(1, 2, 3, 4)
>>> next(x)
(1, 2, 4, 3)
>>> next(x)
(1, 3, 2, 4)
```

例 3-16 编写程序,计算理财产品收益。此处假设利息和本金一起滚动。

```
def licai(base, rate, days):
    #初始投资金额
    result = base
    #整除,用于计算一年可以滚动多少期
    times = 365 // days
    for i in range(times):
        result = result + result*rate/365*days
    return result
#14天理财,利率为 0.0385,投资 10 万元
print(licai(100000, 0.0385, 14))
```

例 3-17 计算前 n 个自然数的阶乘之和 $1!+2!+3!+\cdots+n!$ 的值。

```
def factorialBefore(n):
    result, t = 1, 1
    for i in range(2, n+1):
        t *= i                #利用相邻两项的关系进行加速
        result += t
    return result
print(factorialBefore(6))
```

例 3-17

例 3-18 验证 6174 猜想。1955 年,卡普耶卡(D.R.Kaprekar)对 4 位数字进行了研究,发现一个规律:对任意各位数字不相同的 4 位数,使用各位数字能组成的最大数减去能组成的最小数,对得到的差重复这个操作,最终会得到 6174 这个数字,并且这个操作最多不会超过 7 次。下面的代码可以结束并且没有任何输出,说明 6174 猜想是正确的。

例 3-18

```
from string import digits
from itertools import combinations

for item in combinations(digits, 4):
    times = 0
    while True:
        #当前选择的 4 个数字能够组成的最大数和最小数
        big = int(''.join(sorted(item, reverse=True)))
        little = int(''.join(sorted(item)))
        difference = big - little
        times = times +1
        #如果最大数和最小数相减得到 6174 就退出
        #否则就对得到的差重复这个操作
        #最多 7 次,总能得到 6174
```

```
            if difference == 6174:
                if times >7:
                    print(times)
                break
            else:
                item = str(difference)
```

例 3-19 检测序列中的元素是否满足严格升序关系。

例 3-19

```
def lessThan1(seq):
    for index, value in enumerate(seq[:-1]):
        if value >= seq[index+1]:
            return False
    return True

def lessThan2(seq):
    func = lambda x, y: x < y
    return all(map(func, seq[:-1], seq[1:]))

tests = ('abcdeff', [1, 2, 3, 5, 4], (3, 5, 7, 9))
for test in tests:
    print(lessThan1(test), lessThan2(test))
```

本 章 小 结

（1）所有合法的 Python 表达式都可以作为选择结构和循环结构中的条件表达式。

（2）Python 的关系运算符可以连续使用，例如，3<4<5>2 的值为 True。

（3）数字 0、0.0、0j、逻辑假 False、空列表[]、空集合 set()或空字典{}、空元组()、空字符串''、空值 None，以及任意与这些值等价的值作为条件表达式时均被认为条件不成立，否则认为条件表达式成立。

（4）逻辑运算符 and 和 or 具有短路求值或惰性求值的特点，只计算必须计算的表达式的值。充分利用这个特点可以提高程序运行速度。

（5）选择结构和循环结构往往会互相嵌套使用来实现复杂的业务逻辑。

（6）关键字 elif 表示 else if 的意思。

（7）在 for 和 while 循环中，应优先考虑使用 for 循环，尤其是列表、元组、字典、集合或其他 Python 可迭代对象中元素遍历的场合。

（8）编写循环语句时，应尽量减少内循环中的无关计算，对循环进行必要的优化。

（9）for 循环和 while 循环都可以带 else 子句，如果循环因为条件表达式不满足而自然结束时，执行完循环中的代码后继续执行 else 子句中的代码；如果循环因为执行了 break 语句而提前结束，则不执行 else 子句中的代码。

（10）break 语句用于提前结束其所在循环，continue 语句用于提前结束本次循环并进入下一次循环。

（11）除非 break 和 continue 语句可以让代码变得更简单或更清晰，否则不要轻易使用。

（12）Python 3.8 开始标准库 math 提供了组合数函数 comb()和排列数函数 perm()，

Python 3.9 开始在标准库 math 中新增了最小公倍数函数 lcm(),并且增强了最大公约数函数 gcd()的功能。

习　　题

1. Python 提供了两种基本的循环结构：＿＿＿＿和＿＿＿＿。
2. 分析逻辑运算符 or 的短路求值特性。
3. 编写程序,运行后用户输入 4 位整数作为年份,判断其是否为闰年。如果年份能被 400 整除,则为闰年；如果年份能被 4 整除但不能被 100 整除也为闰年。
4. 编写程序,生成一个包含 50 个随机整数的列表,然后删除其中所有奇数(提示：从后向前删)。
5. 编写程序,生成一个包含 20 个随机整数的列表,然后对其中偶数下标的元素进行降序排列,奇数下标的元素不变(提示：使用切片)。
6. 编写程序,用户从键盘输入小于 1000 的整数,对其进行因式分解。例如,$10=2\times 5$,$60=2\times 2\times 3\times 5$。
7. 编写程序,至少使用两种不同的方法计算 100 以内所有奇数的和。
8. 编写程序,输出所有由 1、2、3、4 这 4 个数字组成的素数,并且在每个素数中每个数字只使用一次。
9. 编写程序,实现分段函数计算,如表 3-1 所示。

表 3-1　分段函数计算

x	y
$x<0$	0
$0\leqslant x<5$	x
$5\leqslant x<10$	$3x-5$
$10\leqslant x<20$	$0.5x-2$
$20\leqslant x$	0

10. 查阅资料编写程序,模拟蒙蒂霍尔悖论游戏。
11. 查阅资料编写程序,模拟尼姆游戏。
12. 查阅资料编写程序,输出小于 100 的所有丑数。

第 4 章 字符串与正则表达式

最早的字符串编码是美国信息交换标准码(ASCII),仅对 10 个数字、26 个大写英文字母、26 个小写英文字母及一些标点符号和控制符号进行了编码。ASCII 采用 1 字节来对字符进行编码,表示能力非常有限。

随着信息技术的发展和信息交换的需要,各国的文字都需要进行编码,于是分别设计了不同的编码格式,并且编码格式之间有较大的区别,其中我国常用的编码有 UTF-8、GB 2312、GBK、CP936 等。采用不同的编码格式意味着不同的表示和存储形式,把同一字符存入不同编码格式的文本文件时,写入的内容可能会不同,在理解文件内容时必须了解编码规则并进行正确的解码。其中,UTF-8 编码是国际通用的编码,以 1 字节表示英语字符(兼容 ASCII),以 3 字节表示常见汉字,对全世界所有国家需要用到的字符进行了编码。

GB 2312 是我国制定的中文编码标准,使用 1 字节兼容 ASCII,2 字节表示中文;GBK 是 GB 2312 的扩充;CP936 是微软公司在 GBK 基础上开发的编码方式。GB 2312、GBK 和 CP936 都是使用 2 字节表示中文,三者互相兼容,能表示的字符比 UTF-8 少。

Python 3.x 程序默认使用 UTF-8 编码,无论是一个数字、英文字母、标点符号,还是一个汉字,都按一个字符对待和处理。在 Python 3.x 中可以使用中文作为变量名。

```
>>> s = '中国山东烟台'
>>> len(s)      #统计字符数量
6
>>> s = 'Python 小屋'
>>> len(s)
5
>>> s = '中国山东烟台 SDIBT'
>>> len(s)
11
>>> 姓名 = '张三'
>>> 年龄 = 40
>>> print(姓名)
张三
>>> print(年龄)
40
```

4.1 字　符　串

在 Python 中,字符串使用单引号、双引号、三单引号或三双引号作为界定符,不同的界定符之间可以互相嵌套。除了支持序列通用方法(包括比较大小、计算长度、元素访问、切片等操作)以外,字符串类型还支持一些特有的操作,例如,编码、格式化、字符串查找、字符串替换等。字符串属于不可变序列,不能对字符串对象进行元素增加、修改与删除等操作。字符串对象提供的 replace()、translate() 以及其他类似方法并不是对原字符串直接修改和替换,而是返回一个修改替换后的结果字符串。

4.1.1 字符串格式化

字符串格式化用于把整数、实数、列表等对象转换为特定格式的字符串或与其他字符拼接为复杂字符串。使用%运算符进行字符串格式化的格式如图 4-1 所示,%运算符之前的字符串为格式字符串,之后的部分为需要进行格式化的内容。

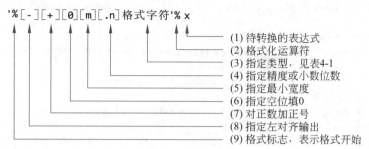

图 4-1 字符串格式化

Python 支持大量的格式字符,常见的格式字符如表 4-1 所示。

表 4-1 常见的格式字符

格式字符	说明	格式字符	说明
%s	字符串（采用 str() 的显示）	%e	指数（基底写为 e）
%r	字符串（采用 repr() 的显示）	%E	指数（基底写为 E）
%c	单个字符	%f,%F	浮点数
%d	十进制整数	%g	指数(e)或浮点数（根据显示长度确定）
%i	十进制整数	%G	指数(E)或浮点数（根据显示长度确定）
%o	八进制整数	%%	字符"%"
%x	十六进制整数		

下面的代码简单演示了字符串格式化的用法:

```
>>> x = 1235
>>> '%o' % x
'2323'
>>> '%#x' % x
'0x4d3'
>>> '%e' % x
'1.235000e+03'
>>> '%d, %c' %(65, 65)          #如果要求格式化的对象多于一个,要放在元组中
'65, A'
>>> '%d' % '555'                #试图将字符串按整数进行转换,抛出异常
TypeError: %d format: a number is required, not str
>>> '%s' % 65                   #类似于 str()
'65'
>>> '%s' % [1, 2, 3]
'[1, 2, 3]'
```

使用 format() 方法进行格式化更加灵活,不仅可以使用位置进行格式化,还支持使用与位置无关的参数名（关键参数,见 5.3.2 节）进行格式化,并且支持序列解包格式化字符串,为程序员提供了非常大的方便。例如:

```
>>> print('The number {0:,} in hex is: {0:#x}, the number {1} in oct is {1:#o}'.format
(5555, 55))
```

```
The number 5555 in hex is: 0x15b3, the number 55 in oct is 0o67
>>> print('The number {1:,} in hex is: {1:#x}, the number {0} in oct is {0:#o}'.format
(5555, 55))
The number 55 in hex is: 0x37, the number 5555 in oct is 0o12663
>>> print('my name is {name}, my age is {age}, and my QQ is {qq}'.format(name='Dong Fuguo',
qq='306467***', age=37))
my name is Dong Fuguo, my age is 37, and my QQ is 306467***
>>> position = (5, 8, 13)
>>> print("X:{0[0]};Y:{0[1]};Z:{0[2]}".format(position))
X:5;Y:8;Z:13
>>> '{0:<8d},{0:^8d},{0:>8d}'.format(65)            #设置对齐方式
'65      ,   65   ,      65'
>>> '{0:+<8d},{0:-^8d},{0:=>8d}'.format(65)
'65++++++,---65---,======65'
>>> print('{0:.4f}'.format(10/3))                   #指定小数位数
3.3333
>>> print('{0:.2%}'.format(1/3))                    #格式化为百分数
33.33%
>>> print('{0:>10.2%}'.format(1/3))                 #符号>表示右对齐,<表示左对齐,^表示居中对齐
    33.33%
>>> weather = [('Monday', 'rainy'), ('Tuesday', 'sunny'),
               ('Wednesday', 'sunny'), ('Thursday', 'rainy'),
               ('Friday', 'Cloudy')]
>>> formatter = "Weather of '{0[0]}' is '{0[1]}'".format
>>> for item in map(formatter, weather):
    print(item)
Weather of 'Monday' is 'rainy'
Weather of 'Tuesday' is 'sunny'
Weather of 'Wednesday' is 'sunny'
Weather of 'Thursday' is 'rainy'
Weather of 'Friday' is 'Cloudy'
```

4.1.1

关于内置函数 map() 的介绍可以参考 5.8 节的介绍。上面最后一段代码也可以改为下面的写法：

```
>>> for item in weather:
    print(formatter(item))
```

Python 3.6 开始支持在数字常量的中间位置使用单个下画线作为分隔符来提高可读性，相应地，字符串格式化方法 format() 也提供了对下画线的支持。

```
>>> print('{0:_},{0:#_x}'.format(10000000))
10_000_000,0x98_9680
```

Python 3.6 开始支持一种新的字符串格式化方式，官方叫作 Formatted String Literals，简称 f-字符串，其含义与字符串对象的 format() 方法类似，但形式更加简洁。其中，花括号里面的变量名表示占位符，在进行格式化时，使用已定义的同名变量的值对格式字符串中的占位符进行替换。

```
>>>width, height = 8, 6
>>>print(f'Rectangle of {width}*{height}\nArea:{width*height}')
Rectangle of 8*6
Area:48
```

```
>>> f'{width= },{height= },{width * height= }'        #Python 3.8开始支持等号
'width= 8,height= 6,width * height=48'
>>> f'result:{"\n".join(["a","b","c"])}'              #Python 3.12开始支持反斜线
'result:a\nb\nc'
```

4.1.2 字符串常用方法

可以使用 dir("") 查看字符串操作所有方法列表,使用内置函数 help() 查看每个方法的帮助。字符串也是 Python 序列的一种,除了本节介绍的字符串处理方法,很多 Python 内置函数也支持对字符串的操作,例如,用于计算序列长度的 len() 函数,求最大值的 max() 函数,见 1.4.6 节。字符串编码方法 encode() 和字节串解码方法 decode() 见 10.4 节。

1. find()、rfind()、index()、rindex()、count()

find() 和 rfind() 方法分别用于查找一个字符串在当前字符串指定范围(默认是整个字符串)中首次和最后一次出现的位置,如果不存在则返回 -1;index() 和 rindex() 方法分别用于返回一个字符串在当前字符串指定范围中首次和最后一次出现的位置,如果不存在则抛出异常;count() 方法用于返回一个字符串在当前字符串中出现的次数,如果不存在则返回 0。

```
>>> s = 'apple,peach,banana,peach,pear'
>>> s.find('peach')                    #返回第一次出现的位置
6
>>> s.find('peach', 7)                 #从下标 7 开始查找
19
>>> s.find('peach', 7, 20)             #在下标[7,20)的范围中查找
-1
>>> s.rfind('p')                       #从字符串尾部向前查找
25
>>> s.index('p')                       #返回首次出现的位置
1
>>> s.index('pe')
6
>>> s.index('pear')
25
>>> s.index('ppp')                     #指定子字符串不存在时抛出异常
ValueError: substring not found
>>> s.count('p')                       #统计子字符串出现的次数
5
```

2. split()、rsplit()、partition()、rpartition()

split() 和 rsplit() 方法分别用于以指定字符为分隔符从字符串左端和右端开始将其分隔为多个字符串,返回包含分隔结果的列表;partition() 和 rpartition() 用于以指定字符串为分隔符将原字符串分隔为 3 部分,分别为分隔符前的字符串、分隔符字符串、分隔符后的字符串,如果指定的分隔符不在原字符串中,则返回原字符串和两个空字符串组成的元组。

split()方法

```
>>> s = 'apple,peach,banana,pear'
>>> li = s.split(',')                  #使用逗号分隔
>>> li
```

```
['apple', 'peach', 'banana', 'pear']
>>> s.partition(',')                              #只分隔1次
('apple', ',', 'peach,banana,pear')
>>> s.rpartition(',')
('apple,peach,banana', ',', 'pear')
>>> s.rpartition('banana')
('apple,peach,', 'banana', ',pear')
```

对于 split() 和 rsplit() 方法, 如果不指定分隔符, 则字符串中的任何空白符号 (包括空格、换行符、制表符等) 都将被认为是分隔符, 返回包含最终分隔结果的列表并丢弃所有空字符串。

```
>>> s = '\n\nhello\t\t world \n\n\n My name\t is Dong    '
>>> s.split()                                      #自动删除分隔结果中的空字符串
['hello', 'world', 'My', 'name', 'is', 'Dong']
```

然而, 明确传递参数指定 split() 使用的分隔符时, 情况是不一样的, 紧邻的分隔符之间会得到空字符串, 并且这些空字符串不会被丢弃。

```
>>> 'a,,,bb,,ccc'.split(',')        #每个逗号都被作为独立的分隔符,不删除空字符串
['a', '', '', 'bb', '', 'ccc']
>>> 'a\t\tbb\t\tccc'.split('\t')    #每个制表符都被作为独立的分隔符,不删除空字符串
['a', '', '', 'bb', '', 'ccc']
>>> 'a\t\tbb\t\tccc'.split()        #不指定分隔符时自动删除分隔得到的空字符串
['a', 'bb', 'ccc']
```

split() 和 rsplit() 方法允许指定最大分隔次数, 例如:

```
>>> s = '\n\nhello\t\t world \n\n\n My name is Dong    '
>>> s.split(None, 2)                   #也可以使用s.split(maxsplit=2)这样的写法
['hello', 'world', 'My name is Dong    ']
>>> s.rsplit(None, 2)
['\n\nhello\t\t world \n\n\n My name', 'is', 'Dong']
>>> s.split(None, 6)                   #指定的最大分隔次数大于实际可分隔次数时自动忽略
['hello', 'world', 'My', 'name', 'is', 'Dong']
```

3. join()

join() 方法用于将可迭代对象 (要求只包含字符串) 中的全部字符串进行连接, 并在相邻两个字符串之间插入指定字符串。

```
>>> li = ['apple', 'peach', 'banana', 'pear']
>>> sep = ','
>>> s = sep.join(li)                  #连接列表中的字符串,在相邻字符串之间插入逗号
>>> s
'apple,peach,banana,pear'
```

使用+运算符也可以连接字符串, 但效率较低, 应优先使用 join() 方法。下面的代码演示了两者之间速度的差异, 重点体会测试代码运行时间的方法。

```
import timeit

strlist = ['This is a long string that will not keep in memory.' for n in range(100)]
```

```python
def use_join():
    return ''.join(strlist)

def use_plus():
    result = ''
    for strtemp in strlist:
        result = result + strtemp
    return result

if __name__ == '__main__':
    times = 1000
    jointimer = timeit.Timer('use_join()', 'from __main__ import use_join')
    print('time for join:', jointimer.timeit(number=times))
    plustimer = timeit.Timer('use_plus()', 'from __main__ import use_plus')
    print('time for plus:', plustimer.timeit(number=times))
```

上面代码使用 timeit 模块的 Timer 类对代码运行时间进行测试。另外，该模块还支持下面代码演示的用法。

```
>>> import timeit
>>> timeit.timeit('"-".join(str(n) for n in range(100))', number=10000)
0.6054277848162267
>>> timeit.timeit('"-".join([str(n) for n in range(100)])', number=10000)
0.5314926897133567
>>> timeit.timeit('"-".join(map(str, range(100)))', number=10000)
0.33093395948368
```

4. lower()、upper()、capitalize()、title()、swapcase()

lower()、upper()、capitalize()、title()和 swapcase()方法分别用于将字符串中的英文字母转换为小写、大写字符串，将字符串首字母变为大写（如果第一个字符是英文字母），将每个单词的首字母变为大写以及大小写互换。

```
>>> s = 'What is Your Name?'
>>> s2 = s.lower()                 #英文字母全部小写
>>> s2
'what is your name?'
>>> s.upper()                       #英文字母全部大写
'WHAT IS YOUR NAME?'
>>> s2.capitalize()                 #第一个字符是英文字母就变为大写，其余全部小写
'What is your name?'
>>> s.title()                        #每个连续英文字母子串的第一个字母大写，其余小写
'What Is Your Name?'
>>> s.swapcase()
'wHAT IS yOUR nAME?'
```

5. replace()

replace()方法用于替换字符串中指定字符或子字符串的所有重复出现（类似于 Word 的全部替换功能），每次只能替换一个字符或一个子字符串，返回处理后的新字符串，不修改原字符串。

```
>>>words = ('测试', '非法', '暴力', '话')
>>>text = '这句话里含有非法内容'
```

```
>>> for word in words:
        if word in text:
            text = text.replace(word, '***')
>>> text
'这句***里含有***内容'
```

6. maketrans()、translate()

maketrans()方法用于生成字符映射表，translate()方法按字符映射表中的对应关系转换字符串并替换其中的字符，使用这两个方法的组合可以同时处理多个字符，replace()方法无法满足这一要求。下面的代码演示了这两个方法的用法。

maketrans()方法

```
#将字符'abcdef123'一一对应地转换为'uvwxyz@#$'
>>> table = ''.maketrans('abcdef123', 'uvwxyz@#$')
>>> s = 'Python is a great programming language. I like it!'
>>> s.translate(table)
'Python is u gryut progrumming lunguugy. I liky it!'
>>> table = ''.maketrans('0123456789', '零一二三四五六七八九')
>>> '2023 年 12 月 31 日'.translate(table)
'二零二三年一二月三一日'
```

下面的代码实现了凯撒加密算法，把文本中每个英文字母替换为其在字母表中后面的第 k 个字母，其中 k 为密钥，对于同一段文本，使用不同的密钥 k 可以得到不同的密文。

```
from string import ascii_letters, ascii_lowercase, ascii_uppercase

def kaisaEncrypt(text, k):
    #凯撒加密
    lower = ascii_lowercase[k:] +ascii_lowercase[:k]
    upper = ascii_uppercase[k:] +ascii_uppercase[:k]
    table = ''.maketrans(ascii_letters, lower+upper)
    return text.translate(table)

s = 'Python is a great programming language. I like it!'
print(kaisaEncrypt(s, 3))
```

输出结果为

```
'Sbwkrq lv d juhdw surjudpplqj odqjxdjh. L olnh lw!'
```

7. strip()、rstrip()、lstrip()

strip()、rstrip()、lstrip()方法分别用于删除当前字符串两端、右端或左端的空白字符或指定字符。

```
>>> s = ' abc   '
>>> s.strip()                               #删除两端的空白字符
'abc'
>>> '\n\nhello world    \n\n'.strip()       #删除两端的空白字符
'hello world'
>>> 'aaaassddf'.strip('a')                  #删除两端的指定字符
'ssddf'
>>> 'aaaassddf'.strip('af')
'ssdd'
```

```
>>> 'aaaassddfaaa'.rstrip('a')          #删除字符串右端的指定字符
'aaaassddf'
>>> 'aaaassddfaaa'.lstrip('a')          #删除字符串左端的指定字符
'ssddfaaa'
```

这 3 个方法的参数指定的字符串并不作为一个整体对待,而是在原字符串的两端、右端、左端删除参数字符串中包含的所有字符,一层一层地从外往里扒。

```
>>> 'aabbccddeeeffg'.strip('af')        #字母 f 不在字符串两端,所以不删除
'bbccddeeeffg'
>>> 'aabbccddeeeffg'.strip('gaf')       #右端删除字母 g 之后继续删 f
'bbccddeee'
>>> 'aabbccddeeeffg'.strip('gbaefcd')
''
```

8. eval()

内置函数 eval()尝试把任意字符串转换为 Python 表达式并求值。

```
>>> eval('3+4')
7
>>> a = 3
>>> b = 5
>>> eval('a+b')
8
```

使用 eval()时要注意的一个问题是,它可以计算任意合法表达式的值,如果用户巧妙地构造输入的字符串,可以执行任意外部程序,有一定风险,建议改用 ast 模块的 literal_eval()函数。例如,下面的代码运行后可以启动记事本程序:

```
>>> a = input('Please input:')
Please input:__import__('os').startfile(r'C:\Windows\notepad.exe')
>>> eval(a)
```

再执行下面的代码试试,然后看看当前工作目录中多了什么。还可以调用命令来删除这个文件夹或其他文件,或者精心构造其他字符串来达到其他特殊目的。

```
>>> eval("__import__('os').system('md testtest')")
```

9. 关键字 in

与列表、元组、字典、集合和迭代器对象一样,也可以使用关键字 in 和 not in 来判断一个字符串是否出现在另一个字符串中,返回 True 或 False。

```
>>> 'ab' in 'abcde'
True
>>> 'j' in 'abcde'
False
```

10. startswith()、endswith()

startswith()、endswith()方法分别用于判断字符串是否以指定字符串开始或结束。这两个方法可以接收两个整数参数来限定字符串的检测范围,例如:

```
>>> s = 'Beautiful is better than ugly.'
```

```
>>> s.startswith('Be')
True
>>> s.startswith('Be', 5)                    #从下标 5 开始检测
False
>>> s.startswith('Be', 0, 5)                 #在下标[0,5)内检测
True
```

另外,这两个方法还可以接收一个包含若干字符串的元组作为参数来表示前缀或后缀。例如,下面的代码可以列出 D 盘根目录下所有扩展名为 bmp、jpg 或 gif 的图片。

```
>>> import os
>>> [filename for filename in os.listdir(r'D:\\')
     if filename.endswith(('.bmp', '.jpg', '.gif'))]
```

11. isalnum()、isalpha()、isdigit()、isspace()、isupper()、islower()

isalnum()、isalpha()、isdigit()、isspace()、isupper()、islower()方法分别用于测试字符串是否全部为数字字符或英文字母、是否全部为英文字母、是否全部为数字字符、是否全部为空白字符、英文字母是否全部为大写及英文字母是否全部为小写。

```
>>> '1234abcd'.isalnum()           >>> 'abcd'.isalpha()
True                               True
>>> '1234abcd'.isalpha()           >>> '1234.0'.isdigit()
False                              False
>>> '1234abcd'.isdigit()           >>> '1234'.isdigit()
False                              True
```

12. center()、ljust()、rjust()

center()、ljust()、rjust()方法返回指定宽度的新字符串,原字符串分别以居中、左对齐或右对齐方式出现在新字符串中,如果指定的宽度大于原字符串长度,则使用指定的字符(默认为空格)填充。

```
>>> 'Hello world!'.center(20)           #原来的内容居中,两端填充空格
'    Hello world!    '
>>> 'Hello world!'.center(20, '=')      #两端填充等号
'====Hello world!===='
>>> 'Hello world!'.ljust(20, '=')       #原来的内容居左,右端填充等号
'Hello world!========'
>>> 'Hello world!'.rjust(20, '=')       #原来的内容居右,右端填充等号
'========Hello world!'
```

4.1.3 字符串常量

在 string 模块中定义了多个字符串常量,包括数字字符、标点符号、英文字母、大写英文字母、小写英文字母等。下面的代码演示了 8 位长度随机密码生成算法的原理。

```
>>> import string
>>> x = string.digits + string.ascii_letters + string.punctuation
>>> import random
>>> ''.join([random.choice(x) for i in range(8)])
'(CrZ[44M'
```

```
>>> ''.join(random.sample(x, 8))                    #不重复的 8 个字符
'o_?[M>iF'
```

random 是与随机数有关的 Python 标准库,除了用于从序列中任意选择一个元素的函数 choice(),还提供了用于生成指定二进制位数的随机整数的函数 getrandbits()、生成指定范围内随机数的函数 randrange() 和 randint()、列表原地乱序函数 shuffle()、从序列中随机选择指定数量不重复元素的函数 sample()、返回 [0,1] 区间内符合 beta 分布的随机数函数 betavariate()、符合伽玛分布的随机数函数 gammavariate()、符合高斯分布的随机数函数 gauss()、从指定分布中选取 k 个允许重复的元素的函数 choices() 等,同时还提供了 SystemRandom 类支持生成加密级别要求的不可再现伪随机数序列。

```
>>> import random
>>> random.getrandbits(17)
6233
>>> random.getrandbits(17)
55217
>>> x = list(range(20))
>>> random.shuffle(x)
>>> x
[2, 16, 15, 14, 18, 3, 10, 8, 5, 12, 6, 19, 1, 9, 13, 7, 0, 4, 17, 11]
>>> random.sample(x, 3)
[19, 4, 14]
```

下面的代码使用 Python 标准库 random 中的函数模拟了发红包算法。

4.1.3

```
import random

def hongbao(total, num):
    #total 表示拟发红包总金额,单位为分;num 表示拟发红包数量
    each = []                                       #每个人抢到的红包大小
    already = 0                                     #已发红包总金额

    for i in range(1, num):
        #为当前抢红包的人随机分配金额,至少给剩下的每人留一分钱
        t = random.randint(1, (total-already)-(num-i))
        each.append(t)
        already = already + t
    each.append(total-already)                      #剩余所有钱发给最后一个人
    return each

if __name__ == '__main__':
    total, num = 500, 5
    for i in range(30):                             #模拟 30 次
        each = hongbao(total, num)
        print(each)
```

上面程序中的函数 hongbao() 也可以使用下面的实现方式,其中演示了标准库 random 中 choices() 函数的用法和红包金额修正的思路。

```
import random

def hongbao(total, num):
```

```
    if total < num:
        return '钱太少了不够发。'
each = random.choices(range(1,50), k=num)
sum_ = sum(each)
#计算每个人的金额
each = [int(n/sum_ * total) for n in each]
#修正最后一个人的金额,确保正好发完总金额
each[-1] = total - sum(each[:-1])
#修正所有人金额,不得出现金额为 0 的情况
while (zero:=each.count(0)) > 0:
    max_ = max(each)
    each[each.index(max_)] = max_ - zero
    for i, v in enumerate(each):
        if zero == 0:
            break
        elif v == 0:
            each[i] = 1
            zero = zero - 1
return each
```

*4.1.4 可变字符串

在 Python 中,字符串属于不可变对象,不支持原地修改,如果需要修改其中的值,只能重新创建一个新的字符串对象。如果确实需要一个支持原地修改的 Unicode 数据对象,可以使用 io.StringIO 对象或 array 模块。

```
>>> import io                                  'Hello, there!'
>>> s = 'Hello, world'                         >>> import array
>>> sio = io.StringIO(s)                       >>> a = array.array('u', s)
>>> sio.getvalue()                             >>> print(a)
'Hello, world'                                 array('u', 'Hello, world')
>>> sio.seek(7)                                >>> a[0] = 'y'
7                                              >>> print(a)
>>> sio.write('there!')                        array('u', 'yello, world')
6                                              >>> a.tounicode()
>>> sio.getvalue()                             'yello, world'
```

4.1.5 中文分词与拼音处理

Python 扩展库 jieba 提供了分词功能,用法如下:

```
>>> import jieba                               #导入 jieba 模块
>>> x = '分词的准确度直接影响了后续文本处理和挖掘算法的最终效果。'
>>> jieba.cut(x)                               #使用默认词库进行分词,返回生成器对象
<generator object Tokenizer.cut at 0x000000000342C990>
>>> list(_)                                    #把上面得到的生成器对象转换为列表
['分词', '的', '准确度', '直接', '影响', '了', '后续', '文本处理', '和', '挖掘', '算法', '的', '最终', '效果', '。']
>>> jieba.lcut('Python可以这样学,Python 程序设计开发宝典') # 直接给出列表
Dumping model to file cache C:\Users\d\AppData\Local\Temp\jieba.cache
```

```
['Python', '可以', '这样', '学', ',', 'Python', '程序设计', '开发', '宝典']
>>>list(jieba.cut('花纸杯'))
['花', '纸杯']
>>>jieba.add_word('花纸杯')                          #增加词条
>>>list(jieba.cut('花纸杯'))                         #使用新词库进行分词
['花纸杯']
```

Python 扩展库 pypinyin 提供了中文拼音处理功能,用法如下:

```
>>>from pypinyin import lazy_pinyin, pinyin
>>>lazy_pinyin('董付国')                             #返回拼音
['dong', 'fu', 'guo']
>>>lazy_pinyin('董付国', 1)                          #带声调的拼音
['dǒng', 'fù', 'guó']
>>>lazy_pinyin('董付国', 2)                          #另一种拼音形式,数字表示前面字母的声调
['do3ng', 'fu4', 'guo2']
>>>lazy_pinyin('董付国', 3)                          #只返回拼音首字母,也就是声母
['d', 'f', 'g']
>>>lazy_pinyin('重要', 1)                            #能够根据词组智能识别多音字
['zhòng', 'yào']
>>>lazy_pinyin('重阳', 1)
['chóng', 'yáng']
>>>pinyin('重阳')                                    #返回拼音
[['chóng'], ['yáng']]
>>>pinyin('重阳节', heteronym=True)                  #返回当前上下文中每个多音字的所有读音
[['chóng'], ['yáng'], ['jié', 'jiē']]
>>>x = '中英文混合 test123'
>>>lazy_pinyin(x)
['zhong', 'ying', 'wen', 'hun', 'he', 'test123']
>>>x = '山东烟台的大樱桃真好吃啊'
>>>sorted(x, key=lazy_pinyin)                        #按拼音对汉字进行排序
['啊', '吃', '大', '的', '东', '好', '山', '台', '桃', '烟', '樱', '真']
>>>''.join(map(lambda word:''.join(sorted(word, key=pinyin)) if len(word)==2 else word, jieba.cut(x)))
'东山台烟的大樱桃真吃好啊'
```

4.1.6 字符串应用案例精选

例 4-1 编写函数实现字符串加密和解密,循环使用指定密钥,采用简单的异或算法。

例 4-1

```
from itertools import cycle

def crypt(source, key):
    result = ''
    temp = cycle(key)
    for ch in source:
        result = result + chr(ord(ch) ^ ord(next(temp)))
    return result

source = 'Shandong Institute of Business and Technology'
key = 'Dong Fuguo'
print('Before Encrypted:'+source)
```

```
encrypted = crypt(source, key)
print('After Encrypted:'+encrypted)
decrypted = crypt(encrypted, key)
print('After Decrypted:'+decrypted)
```

输出结果如图 4-2 所示。

```
Before Encrypted:Shandong Institute of Business and Technology
After Encrypted:|●✿     D)← U&*-β13 ┐U "O,↓S/¬-dʃ └ ┤┘  +└  Y
After Decrypted:Shandong Institute of Business and Technology
```

图 4-2　字符串加密与解密结果

如果使用函数式编程模式，上面的函数 crypt() 可以写作下面的形式：

```
def crypt(source, key):
    func = lambda x, y: chr(ord(x)^ord(y))
    return ''.join(map(func, source, cycle(key)))
```

例 4-2　编写程序，生成大量随机信息。

本例代码演示了如何使用 Python 标准库 random 来生成随机数据，这在需要获取大量数据来测试或演示软件功能的时候非常有用，不仅能展示软件功能或算法，还可以避免泄露真实数据或者引起不必要的争议。请自行运行程序并查看输出结果。

```
import random
import string

#常用汉字 Unicode 编码表，可以自行搜索，也可以直接使用汉字
StringBase = '\u7684\u4e00\u4e86\u662f\u6211\u4e0d\u5728\u4eba\u4eec'\
             '\u6709\u6765\u4ed6\u8fd9\u4e0a\u7740\u4e2a\u5730\u5230'\
             '\u5927\u91cc\u8bf4\u5c31\u53bb\u5b50\u5f97\u4e5f\u548c'

def getEmail():
    #常见域名后缀，可以随意扩展该列表
    suffix = ['.com', '.org', '.net', '.cn']
    characters = string.ascii_letters + string.digits + '_'
    username = ''.join(random.choices(characters, k=random.randrange(6,12)))
    domain = ''.join(random.choices(characters, k=random.randrange(3,7)))
    return username +'@' +domain +random.choice(suffix)

def getTelNo():
    return ''.join(random.choices(string.digits, k=11))

def getNameOrAddress(flag):
    '''flag=1 表示返回随机姓名,flag=0 表示返回随机地址'''
    if flag == 1:
        #大部分中国人姓名为 2~4 个汉字
        rangestart, rangeend = 2, 5
    elif flag == 0:
        #假设地址为 10~30 个汉字
        rangestart, rangeend = 10, 31
    else:
        print('flag must be 1 or 0')
```

```python
        return ''
    #生成并返回随机信息
    return ''.join(random.choices(StringBase,
                                  k=random.randrange(rangestart, rangeend)))

def getSex():
    return random.choice('男女')

def getAge():
    return str(random.randint(18,100))

def main():
    print('Name,Sex,Age,TelNO,Address,Email')
    #生成200个人的随机信息
    for i in range(200):
        name, sex, age, tel = getNameOrAddress(1), getSex(), getAge(), getTelNo()
        address, email = getNameOrAddress(0), getEmail()
        line = ','.join([name,sex,age,tel,address,email])
        print(line)

if __name__ == '__main__':
    main()
```

例 4-3 检查并判断密码字符串的安全强度。

```python
import string

def check(pwd):
    #密码必须为字符串且至少包含6个字符
    if not isinstance(pwd,str) or len(pwd)<6:
        return 'not suitable for password'
    #密码强度等级与包含字符种类的对应关系
    d = {1:'weak', 2:'below middle', 3:'above middle', 4:'strong'}
    #分别用来标记pwd是否有数字、小写字母、大写字母和指定的标点符号
    r = [False] * 4
    for ch in pwd:
        #检查当前字符是否为数字
        if not r[0] and ch in string.digits:
            r[0] = True
        #检查当前字符是否为小写字母
        elif not r[1] and ch in string.ascii_lowercase:
            r[1] = True
        #检查当前字符是否为大写字母
        elif not r[2] and ch in string.ascii_uppercase:
            r[2] = True
        #检查当前字符是否为指定的标点符号
        elif not r[3] and ch in ',.!;?<>':
            r[3] = True
    #统计包含的字符种类,返回密码强度
    return d.get(r.count(True), 'error')

print(check('a2Cd,'))
```

4.2 正则表达式

正则表达式是字符串处理的有力工具和技术,使用预定义的特定模式匹配一类具有共同特征的字符串,可以快速、准确地完成复杂的查找、替换、删除等处理任务,主要用于文本处理和网络爬虫。

Python 标准库 re 提供了正则表达式操作所需要的功能。本节首先介绍正则表达式的基础知识,然后介绍 re 模块提供的正则表达式函数与相关对象的用法。

4.2.1 正则表达式语法

正则表达式由元字符及其不同组合来构成,通过巧妙地构造正则表达式可以匹配任意字符串。常用的正则表达式元字符如表 4-2 所示。

表 4-2 常用的正则表达式元字符

元字符	含 义
.	默认匹配除换行符以外的任意单个字符,使用标志位 re.S 声明为单行模式时也可以匹配换行符。如果要匹配字符串中的圆点字符,需要在前面加反斜线,即'\.';圆点在表示范围的方括号中是普通字符,只匹配圆点本身
*	匹配*前面字符或子模式的 0 次或多次重复
+	匹配+前面字符或子模式的 1 次或多次重复
-	在方括号中用于表示范围(如'[0-9]'可以匹配任意单个数字字符),在其他位置表示普通减号字符
|	匹配位于|之前或之后的模式,匹配其中任意一个,可以连用表示多选一
^	① 匹配以^后面的字符或模式开头的字符串;② 在方括号中开始处表示不匹配方括号里的字符
$	匹配以$前面的字符或模式结束的字符串
?	① 表示问号之前的字符或子模式可有可无;② 当问号紧随*、+、?、{n}、{n,}、{,m}、{n,m}之后时,表示匹配模式是"非贪心的"。"非贪心的"模式匹配搜索到的、尽可能短的字符串,而默认的"贪心的"模式匹配搜索到的、尽可能长的字符串。例如,re.findall('abc{,3}?', 'abccc')返回['ab'],re.findall('abc{,3}', 'abccc')返回['abccc']
\num	① 使用原始字符串或 num 前面有两个反斜线时表示前面子模式的编号,num 按十进制数理解。例如,r'(.)\1'匹配两个连续的相同字符,\1 表示当前正则表达式中编号为 1 的子模式内容在这里又出现了一次。整个正则表达式编号为 0,肉眼可见的第一对圆括号是编号为 1 的子模式,肉眼可见的第二对圆括号是编号为 2 的子模式,以此类推。② 不使用原始字符串且 num 前面只有一个反斜线时表示转义字符,num 按八进制数理解。例如,转义字符'\101'匹配字符'A','\141'匹配字符'a','\60'和'\060'匹配字符'0'
\f	匹配一个换页符
\n	匹配一个换行符
\r	匹配一个回车符
\b	匹配单词头或单词尾

续表

元字符	含 义
\B	与'\b'含义相反,匹配单词内部
\d	匹配任意单个数字字符,'\d'等价于'[0-9]'
\D	与'\d'含义相反,'\D'相当于'[^0-9]',匹配除数字之外的任意单个字符
\s	匹配单个任意空白字符,包括空格、制表符、换页符、换行符,'\s'等价于'[\f\n\r\t\v]'
\S	与'\s'含义相反,匹配除空白字符之外的任意单个字符
\w	匹配任意单个字母、汉字、数字以及下画线
\W	与'\w'含义相反
()	将位于()内的内容作为一个整体来对待,称为一个子模式
{m,n} {m,} {,n} {m}	按花括号中指定的次数进行匹配,{m,n}表示前面的字符或子模式重复 m～n 次,{m,}表示前面的字符或子模式重复 m 至任意多次,{,n}表示前面的字符或子模式重复 0～n 次,{m}表示前面的字符或子模式恰好出现 m 次,注意花括号内任何位置都不要有空格。例如,{3,8}表示前面的字符或子模式重复 3～8 次
[]	表示范围,匹配位于[]中的任意单个字符,如果[]内以^开始则表示不匹配[]内的字符。例如,'[a-zA-Z0-9]'可以匹配单个任意大小写字母或数字,'[^a-zA-Z0-9]'可以匹配除英文字母和数字字符之外的任意单个字符

如果以\开头的元字符与转义字符形式相同但含义不同,则需要使用\\或者原始字符串,在字符串前加上字符 r 或 R。原始字符串可以减少用户的输入,主要用于正则表达式和文件路径字符串,如果字符串以一个\结束,则需要多写一个\,以\\结束。

下面给出一些常见的正则表达式,不用刻意记忆,用到的时候再查阅作为参考即可。

(1) 最简单的正则表达式是普通字符串,只能匹配自身。

(2) '[pjc]ython'可以匹配'python'、'jython'、'cython'。

(3) '[a-zA-Z0-9]'可以匹配一个任意大小写字母或数字。

(4) '[^abc]'可以匹配一个除'a'、'b'、'c'之外的任意字符。

(5) 'python|perl'或'p(ython|erl)'都可以匹配'python'或'perl'。

(6) 子模式后面加?表示可选。r'(http://)?(www\.)?python\.org'只能匹配'http://www.python.org'、'http://python.org'、'www.python.org'和'python.org'。

(7) '^http'只能匹配所有以'http'开头的字符串。

(8) '(pattern)*': 允许模式重复 0 次或多次。

(9) '(pattern)+': 允许模式重复 1 次或多次。

(10) '(pattern){m,n}': 允许模式重复 m～n 次,在书写时注意逗号后面不能有空格。

(11) '(pattern){,n}': 允许模式最多重复 n 次,最少 0 次。

(12) '(pattern){m,}': 允许模式最少重复 m 次,不限制最多重复次数。

(13) '(a|b)*c': 匹配多个(包含 0 个)a 或 b,后面紧跟一个字母 c。

(14) 'ab{1,}': 等价于'ab+',匹配以字母 a 开头后面带一个或多个字母 b 的字符串。

(15) '^[a-zA-Z]{1}([a-zA-Z0-9._]){4,19}$': 匹配长度为 5～20 的字符串,必须以字母开头并且可带字母、数字、_、.的字符串。

(16) `'^(\w){6,20}$'`：匹配长度为 6~20 的字符串，可以包含字母、数字、下画线。

(17) `'^\d{1,3}\.\d{1,3}\.\d{1,3}\.\d{1,3}$'`：检查给定字符串是否为合法 IP 地址格式，圆点前要加反斜线，否则可以匹配任意单个字符。

(18) `'^(13[0-9]|15[012356789]|17[678]|18[0-9]|14[57])[0-9]{8}$'`：检查给定字符串是否为手机号码格式，只检查格式，不保证手机号一定有效。

(19) `'^[a-zA-Z]+$'`：检查给定字符串是否只包含英文字母大小写。

(20) `'^\w+@(\w+\.)+\w+$'`：检查给定字符串是否为合法电子邮件地址格式。

(21) `r'(\w)(?!.*\1)'`：查找字符串中每个字符的最后一次出现。子模式扩展语法见表 4-4。

(22) `r'(\w)(?=.*\1)'`：查找字符串中所有出现次数大于 1 的字符。

(23) `'^(\-)?\d+(\.\d{1,2})?$'`：检查给定字符串是否为最多带有两位小数的正数或负数。

(24) `'[\u4e00-\u9fa5]'`：匹配给定字符串中的常见汉字。

(25) `'^\d{18}|\d{15}|\d{17}X$'`：检查给定字符串是否为合法身份证格式。

(26) `'\d{4}-\d{1,2}-\d{1,2}'`：匹配指定格式的日期，如 2024-1-31。

(27) `'^(?=.*[a-z])(?=.*[A-Z])(?=.*\d)(?=.*[,._]).{8,}$'`：检查给定字符串是否为强密码，必须同时包含英文大写字母、英文小写字母、数字或特殊符号（如英文逗号、英文句号、下画线），并且长度必须至少 8 位。

(28) `'(?!.*[\'\"\/;=%?]).+'`：如果给定字符串中包含 '、"、/、;、=、%、? 则匹配失败。

(29) `'(.)\\1+'`：匹配任意字符的两次或多次重复出现。

(30) `'((?P<f>\b\w+\b)\s+(?P=f))'`：匹配连续出现两次的单词。

(31) `'((?P<f>.)(?P=f)(?P<g>.)(?P=g))'`：匹配 AABB 形式的成语或字母组合。

(32) `'\d+(?=[a-z]+)'`：匹配连续的数字，并且最后一个数字后面跟着小写字母。

(33) `'\d+(?![a-z]+)'`：匹配连续的数字，并且最后一个数字后面不能跟小写字母。

(34) `'(?<=[a-z])\d+'`：匹配连续的数字，并且第一个数字的前面是小写字母。

(35) `'(?<![a-z])\d+'`：匹配连续的数字，并且第一个数字的前面不能是小写字母。

(36) `'\d{3}(?!\d)'`：匹配三位数字，而且这三位数字的后面不能是数字。

(37) `'\b((?!abc)\w)+\b'`：匹配不包含连续字符串 'abc' 的单词。

(38) `'(?<![a-z])\d{7}'`：匹配前面不是小写字母的 7 位数字。

(39) `r'(?<=<(\w{4})>)(.*)(?=<\/\1>)'`：匹配 "\<span\>hello world\</span\>" 中的 span 和 hello world。

4.2.2 re 模块主要函数

在 Python 中，主要使用 re 模块来实现正则表达式的操作。re 模块的常用函数如表 4-3 所示，具体使用时，既可以直接使用 re 模块的函数进行字符串处理，也可以将模式编译为正则表达式对象，然后使用正则表达式对象的方法来操作字符串。

表 4-3 re 模块的常用函数

函　　数	功　能　说　明
compile(pattern[, flags])	创建正则表达式对象
search(pattern, string[, flags])	在整个字符串中寻找模式,返回 Match 对象或 None
match(pattern, string[, flags])	从字符串的开始处匹配模式,返回 Match 对象或 None
findall(pattern, string[, flags])	返回字符串中模式的所有匹配项组成的列表
split(pattern, string[, maxsplit=0])	根据模式匹配项分隔字符串
sub(pattern, repl, string[, count=0])	将字符串中 pattern 的所有匹配项用 repl 替换
escape(string)	将字符串中所有正则表达式特殊字符进行转义

其中,函数参数 flags 的值可以是 re.A(只匹配 ASCII 字符)、re.I(忽略大小写)、re.M(多行匹配模式)、re.S(使元字符"."匹配任意字符,包括换行符)、re.U(匹配 Unicode 字符)、re.X(忽略模式中的空格,并可以使用♯注释)的不同组合(使用＋或｜进行组合)。

(1) re.A:使得正则表达式中\w、\W、\b、\B、\d、\D、\s 和\S 等元字符只匹配 ASCII 字符,不匹配 Unicode 字符。

```
>>> import re
>>> re.findall('\d+', '1231234')
['1231234']
>>> re.findall('\d+', '1231234', re.A)
['123']
>>> re.findall('\w+', '1a2b3c1d2e3g4', re.A)
['1a2b3c', 'd', 'e', 'g']
>>> re.findall('\w+', '1a2b3c1d2e3g4')
['1a2b3c1d2e3g4']
```

(2) re.I:忽略大小写。

```
>>> re.findall('[a-z0-9]+', '1a2b3c1D2e3G4')
['1a2b3c', 'e']
>>> re.findall('[a-z0-9]+', '1a2b3c1D2e3G4', re.I)
['1a2b3c', 'D', 'e', 'G']
>>> re.findall('[a-z0-90-9]+', '1a2b3c1D2e3G4', re.I)
['1a2b3c1D2e3G4']
>>> re.findall('[a-z0-90-9]+', '1a2b3c1D2e3G4')
['1a2b3c1', '2e3', '4']
```

(3) re.M:多行匹配模式,^可以匹配每行开始,$可以匹配每行结束。默认情况下,分别匹配字符串的开始和结束。

```
>>> text = '''
abc1234
1234
abc
Python
董付国
'''
>>> re.findall(r'^\w+$', text)
```

```
[]
>>> re.findall(r'^.+$', text)
[]
>>> re.findall(r'^\w+$', text, re.M)
['abc1234', '1234', 'abc', 'Python', '董付国']
>>> re.findall(r'^.+$', text, re.M)
['abc1234', '1234', 'abc', 'Python', '董付国']
```

(4) re.S:单行模式,圆点可以匹配换行符。

```
>>> text = '''<p>Beautiful is better than ugly.
Explicit is better than implicit.
Simple is better than complex.</p>'''
>>> re.findall(r'<p>(.+?)</p>', text)
[]
>>> re.findall(r'<p>(.+?)</p>', text, re.S)
['Beautiful is better than ugly.\nExplicit is better than implicit.\nSimple is better than complex.']
>>> text = '''
good
bad
345a
abc456
'''
>>> re.findall(r'\w+', text)
['good', 'bad', '345a', 'abc456']
>>> re.findall(r'^\w+$', text)               #\w不能匹配换行符,^匹配整个字符串的开始
[]
>>> re.findall(r'^.+$', text)                #圆点也不能匹配换行符,$匹配整个字符串的结束
[]
>>> re.findall(r'^.+$', text, re.S)          #单行模式,此时圆点可以匹配换行符
['\ngood\nbad\n345a\nabc456\n']
>>> re.findall(r'^.+$', text, re.M)          #多行模式,^和$可以匹配每一行的开始和结束
['good', 'bad', '345a', 'abc456']
```

(5) re.X:允许正则表达式换行,并忽略其中的空白字符和#注释。

```
>>> text = 'abc123.4dfg8.88888hij9999.9'
>>> pattern = r'''\d+                        #数字
\.                                           #圆点
\d+'''
>>> re.findall(pattern, text)
[]
>>> re.findall(pattern, text, re.X)
['123.4', '8.88888', '9999.9']
```

(6) 多个flag可以使用+或|组合使用。

```
>>> text = '''abc123.4d
fg8.88888hi
j9999.9
000.00
asdf'''
>>> pattern = r'''^\d+                       #数字
```

```
\.                                              #圆点
\d+$'''
>>> re.findall(pattern, text)
[]
>>> re.findall(pattern, text, re.M)
[]
>>> re.findall(pattern, text, re.X)
[]
>>> re.findall(pattern, text, re.X+re.M)
['000.00']
```

4.2.3 直接使用 re 模块函数

可以直接使用 re 模块的函数来实现正则表达式操作。

```
>>> import re
>>> text = 'alpha. beta...gamma delta'
>>> re.split('[. ]+', text)                     #方括号中圆点前有一个空格
['alpha', 'beta', 'gamma', 'delta']
>>> re.split('[. ]+', text, maxsplit=2)         #分隔两次
['alpha', 'beta', 'gamma delta']
>>> pat = '[a-zA-Z]+'
>>> re.findall(pat, text)                       #查找所有单词
['alpha', 'beta', 'gamma', 'delta']
>>> pat = '{name}'
>>> text = 'Dear {name}...'
>>> re.sub(pat, 'Mr.Dong', text)                #字符串替换
'Dear Mr.Dong...'
>>> s = 'a s d'
>>> re.sub('a|s|d', 'good', s)                  #字符串替换
'good good good'
>>> re.escape('https://www.python.org')         #字符串转义
'https://www\\.python\\.org'
>>> print(re.match('done|quit', 'done'))        #匹配成功,返回 Match 对象
<_sre.SRE_Match object at 0x00B121A8>
>>> print(re.match('done|quit', 'done!'))       #匹配成功
<_sre.SRE_Match object at 0x00B121A8>
>>> print(re.match('done|quit', 'doe!'))        #匹配不成功,返回空值
None
>>> print(re.search('done|quit', 'd!one!done')) #匹配成功
<_sre.SRE_Match object at 0x0000000002D03D98>
>>> text = '''
good
bad
345a
abc456
'''
>>> re.findall(r'\w+', text)
['good', 'bad', '345a', 'abc456']
>>> re.findall(r'^\w+$', text)                  #\w 不能匹配换行符,^匹配整个字符串的开始
[]
>>> re.findall(r'^.+$', text)                   #圆点也不能匹配换行符,$匹配整个字符串的结束
```

```
[]
>>> re.findall(r'^.+$', text, re.S)            #单行模式,此时圆点可以匹配换行符
['\ngood\nbad\n345a\nabc456\n']
>>> re.findall(r'^.+$', text, re.M)            #多行模式,^和$可以匹配每一行的开始和结束
['good', 'bad', '345a', 'abc456']
>>> s = "It's a very good good idea"
>>> re.sub(r'(\b\w+) \1', r'\1', s)            #处理连续的重复单词
"It's a very good idea"
>>> re.sub(r'((\w+))\1', r'\2', s)             #注意,这个结果不理想
"It's a very goodidea"
```

下面的代码使用不同的方法删除字符串中多余的空格,连续多个空格只保留一个。

```
>>> import re
>>> s = 'aaa     bb    c d e   fff    '
>>> re.sub('\s+', ' ', s)                      #直接使用 re 模块的字符串替换函数
'aaa bb c d e fff '
>>> re.split('\s+', s)                         #注意结果最后有个空字符串
['aaa', 'bb', 'c', 'd', 'e', 'fff', '']
>>> re.split('\s+', s.strip())                 #同时删除了字符串尾部的空格
['aaa', 'bb', 'c', 'd', 'e', 'fff']
>>> ' '.join(re.split('\s+', s.strip()))
'aaa bb c d e fff'
>>> re.sub('\s+', ' ', s).strip()
'aaa bb c d e fff'
>>> re.sub('\s+', ' ', s.strip())
'aaa bb c d e fff'
>>> s.split()                                  #也可以不使用正则表达式
['aaa', 'bb', 'c', 'd', 'e', 'fff']
>>> ' '.join(s.split())
'aaa bb c d e fff'
```

下面的代码实现了特定字符串的搜索。

```
>>> import re
>>> example = 'ShanDong Institute of Business and Technology is a very beautiful school.'
>>> re.findall('\\ba.+?\\b', example)          #以 a 开头的完整单词
['and', 'a ']
>>> re.findall('\\ba\w*\\b', example)          #\\b 是为了避免和转义字符\b 冲突
['and', 'a']
>>> re.findall('\Bo.+?\b', example)            #含字母 o 的单词中第一个非首字母 o 开始的剩余部分
['ong', 'ology', 'ool']
>>> re.findall('\\b\w+?\\b', example)          #所有单词
['ShanDong', 'Institute', 'of', 'Business', 'and', 'Technology', 'is', 'a ',
'very', 'beautiful', 'school']
>>> re.findall(r'\b\w+?\b', example)           #使用原始字符串,减少需要输入的符号数量
['ShanDong', 'Institute', 'of', 'Business', 'and', 'Technology', 'is', 'a ',
'very', 'beautiful', 'school']
>>> re.split('\s', example)                    #使用任何空白字符分隔字符串
['ShanDong', 'Institute', 'of', 'Business', 'and', 'Technology', 'is', 'a',
'very', 'beautiful', 'school.']
>>> re.findall('\d\.\d\.\d+', 'Python 2.7.18,Python 3.12.0')
```

```
['2.7.18', '3.12.0']
>>>re.findall('ab(.+?)c', 'ab12cdeab345cdab67')    #findall()只返回子模式匹配的内容
['12', '345']
```

下面的两个函数分别使用 re 模块的 findall() 和 split() 函数实现从字符串中查找最长数字子串的功能。

```
import re
def longest1(s):
    '''查找所有连续数字子串'''
    t = re.findall('\d+', s)
    if t:
        return max(t, key=len)
    return 'No'

def longest2(s):
    '''使用非数字作为分隔符'''
    t = re.split('[^\d]+', s)
    if t:
        return max(t, key=len)
    return 'No'
```

4.2.4 使用正则表达式对象

首先使用 re 模块的 compile() 函数将正则表达式编译生成正则表达式对象,然后再使用正则表达式对象提供的方法进行字符串处理,可以提高字符串处理速度,适合多次使用同一个正则表达式的场合。

正则表达式对象的 match(string[,pos[,endpos]]) 方法在字符串开头或指定位置进行匹配,模式必须出现在字符串开头或指定位置;search(string[,pos[,endpos]]) 方法在整个字符串或指定范围中进行搜索;findall(string[,pos[,endpos]]) 方法在字符串指定范围中查找所有符合正则表达式的字符串并以列表形式返回。

```
>>>import re
>>>pattern = re.compile(r'\bB\w+\b', re.I)    #以 B 或 b 开头的单词
>>>pattern.findall('Beautiful is better than ugly.')
['Beautiful', 'better']
>>>pattern = re.compile(r'\b\w+y\b')           #以 y 结尾的单词
>>>pattern.findall('Although practicality beats purity.')
['practicality', 'purity']
>>>pattern.findall('Beautiful is better than ugly.')
['ugly']
>>>pattern = re.compile(r'\b\w*a\w*\b')        #包含 a 的单词
>>>pattern.match('Readability counts.')
<re.Match object; span=(0, 11), match='Readability'>
>>>pattern.match('Complex is better than complicated.')
                                               #第一个单词不包含 a,匹配不成功,没有返回值
>>>pattern.search('Complex is better than complicated.')
                                               #在字符串中查找第一个包含 a 的单词,返回 Match 对象
```

```
<re.Match object; span=(18, 22), match='than'>
>>> pattern.findall('Complex is better than complicated.')        #返回所有包含a的单词
['than', 'complicated']
```

正则表达式对象的 sub(repl,string[,count=0]) 和 subn(repl,string[,count=0]) 方法用于实现字符串替换功能。

```
>>> example = '''Beautiful is better than ugly.
Explicit is better than implicit.
Simple is better than complex.
Complex is better than complicated.
Flat is better than nested.
Sparse is better than dense.
Readability counts.'''
>>> pattern = re.compile(r'\bb\w*\b', re.I)
>>> pattern.sub('*', example)                      #将以字母b和B开头的单词替换为*
*is *than ugly.
Explicit is *than implicit.
Simple is *than complex.
Complex is *than complicated.
Flat is *than nested.
Sparse is *than dense.
Readability counts.
>>> pattern.sub('*', example, 1)                   #只替换一次
*is better than ugly.
Explicit is better than implicit.
Simple is better than complex.
Complex is better than complicated.
Flat is better than nested.
Sparse is better than dense.
Readability counts.
>>> pattern = re.compile(r'\bb\w*\b')
>>> pattern.sub('*', example, 1)                   #将第一个以字母b开头的单词替换为*
Beautiful is *than ugly.
Explicit is better than implicit.
Simple is better than complex.
Complex is better than complicated.
Flat is better than nested.
Sparse is better than dense.
Readability counts.
```

正则表达式对象的 split(string[,maxsplit=0]) 方法用于实现字符串分隔。

```
>>> example = r'one,two,three.four/five\six? seven[eight]nine|ten'
>>> pattern = re.compile(r'[,./\\?[\]\|]')         #指定多个可能的分隔符
>>> pattern.split(example)
['one', 'two', 'three', 'four', 'five', 'six', 'seven', 'eight', 'nine', 'ten']
>>> example = r'one1two2three3four4five5six6seven7eight8nine9ten'
>>> pattern = re.compile(r'\d+')                   #使用数字作为分隔符
>>> pattern.split(example)
['one', 'two', 'three', 'four', 'five', 'six', 'seven', 'eight', 'nine', 'ten']
>>> example = r'one two    three   four,five.six.seven,eight,nine9ten'
```

```
>>> pattern = re.compile(r'[\s,.\d]+')          #允许分隔符重复
>>> pattern.split(example)
['one', 'two', 'three', 'four', 'five', 'six', 'seven', 'eight', 'nine', 'ten']
```

4.2.5 子模式与 Match 对象

使用圆括号表示一个子模式,圆括号内的内容作为一个整体处理,例如"(red)+"可以匹配 redred、redredred 等多个重复 red 的情况。

```
>>> telNumber = '''Suppose my Phone No. is 0535-1234567,
yours is 010-12345678, his is 025-87654321.'''
>>> pattern = re.compile(r'(\d{3,4})-(\d{7,8})')
>>> pattern.findall(telNumber)                  #findall()返回子模式匹配的内容
[('0535', '1234567'), ('010', '12345678'), ('025', '87654321')]
```

正则表达式对象的 match()方法和 search()方法以及 re 模块的同名函数匹配成功后都会返回 Match 对象。Match 对象的主要方法有 group()(返回匹配的一个或多个子模式内容)、groups()(返回一个包含匹配的所有子模式内容的元组)、groupdict()(返回包含匹配的所有命名子模式内容的字典)、start()(返回指定子模式内容的起始位置)、end()(返回指定子模式内容结束位置的下一个位置)、span()(返回一个包含指定子模式内容起始位置和结束位置的下一个位置的元组)等。下面的代码使用 re 模块的 search()方法返回的 Match 对象来删除字符串中指定内容,可以自行尝试使用 re.sub()函数改写。

```
>>> email = 'dongfuguo2005@1remove_this26.com'
>>> m = re.search('remove_this', email)
>>> email[:m.start()] + email[m.end():]
'dongfuguo2005@126.com'
```

下面的代码演示了 Match 对象的 group()方法的用法。

```
>>> m = re.match(r'(\w+) (\w+)', 'Isaac Newton, physicist')
>>> m.group(0)          #返回整个模式内容
'Isaac Newton'
>>> m.group(1)          #返回第 1 个子模式内容(肉眼可见的第一对圆括号)
'Isaac'
>>> m.group(2)          #返回第 2 个子模式内容
'Newton'
>>> m.group(1, 2)       #返回指定的多个子模式匹配的内容
('Isaac', 'Newton')
>>> m = re.match(r'(?P<first_name>\w+) (?P<last_name>\w+)', 'Malcolm Reynolds')
>>> m.group('first_name')
'Malcolm'
>>> m.group('last_name')
'Reynolds'
```

下面的代码演示了 Match 对象的 groups()方法的用法。

```
>>> m = re.match(r'(\d+)\.(\d+)', '24.1632')
>>> m.groups()
('24', '1632')
```

下面的代码演示了 Match 对象的 groupdict()方法。

```
>>> m = re.match(r'(?P<first_name>\w+) (?P<last_name>\w+)', 'Malcolm Reynolds')
>>> m.groupdict()
{'first_name': 'Malcolm', 'last_name': 'Reynolds'}
```

下面的代码使用正则表达式提取字符串中的电话号码,主要演示正则表达式对象和 Match 对象方法的用法,可以参考前面的内容使用 findall()方法改写。

```
import re

telNumber = 'Suppose my Phone No. is 0535-1234567, yours is 010-12345678, his is 025-87654321.'
pattern = re.compile(r'(\d{3,4})-(\d{7,8})')
index = 0
while True:
    matchResult = pattern.search(telNumber, index)
    if not matchResult:
        break
    print('-'*30)
    print('Success:')
    for i in range(3):
        print('Searched content:', matchResult.group(i),
              'Start from:', matchResult.start(i), 'End at:', matchResult.end(i),
              'Its span is:', matchResult.span(i))
    index = matchResult.end(2)
```

上面程序的运行结果(部分)为

```
------------------------------
Success:
Searched content: 0535-1234567   Start from: 24 End at: 36   Its span is: (24, 36)
Searched content: 0535   Start from: 24 End at: 28   Its span is: (24, 28)
Searched content: 1234567   Start from: 29 End at: 36   Its span is: (29, 36)
```

使用子模式扩展语法可以实现更加复杂的字符串处理,语法如表 4-4 所示。

表 4-4 子模式扩展语法

语 法	功 能 说 明
(?P<groupname>)	为子模式命名
(?iLmsux)	设置匹配标志,可以是几个字母的组合,每个字母的含义与编译标志相同
(?:…)	匹配但不记录匹配的内容
(?P=groupname)	表示在此之前的命名为 groupname 的子模式在当前位置又出现一次
(?#…)	表示注释
(?=…)	用于正则表达式之后,表示如果"="后的内容在字符串中出现则匹配,但不返回"="之后的内容
(?!…)	用于正则表达式之后,表示如果"!"后的内容在字符串中不出现则匹配,但不返回"!"之后的内容
(?<=…)	用于正则表达式之前,其他与(?=…)含义相同
(?<!…)	用于正则表达式之前,其他与(?!…)含义相同

下面通过几个示例来演示子模式扩展语法的应用。

```
>>> import re
>>> exampleString = '''There should be one -- and preferably only one -- obvious way to do it.
Although that way may not be obvious at first unless you're Dutch.
Now is better than never.
Although never is often better than right now.'''
>>> pattern = re.compile(r'(?<=\w\s)never(?=\s\w)')  #查找不在句子开头和结尾的单词 never
>>> matchResult = pattern.search(exampleString)
>>> matchResult.span()
(172, 177)
>>> pattern = re.compile(r'(?<=\w\s)never')          #查找前面有内容的单词 never
>>> matchResult = pattern.search(exampleString)
>>> matchResult.span()
(156, 161)
>>> pattern = re.compile(r'(?<=is\s)better\sthan')   #查找前面是 is 的 better than 组合
>>> matchResult = pattern.search(exampleString)
>>> matchResult.span()
(141, 155)
>>> matchResult.group(0)                             #组 0 表示整个模式
'is better than'
>>> matchResult.group(1)
' than'
>>> pattern = re.compile(r'(?i)\bn\w+\b')            #查找以 n 或 N 开头的所有单词
>>> index = 0
>>> while True:
    matchResult = pattern.search(exampleString, index)
    if not matchResult:
        break
    print(matchResult.group(0), ':', matchResult.span(0))
    index = matchResult.end(0)

not : (92, 95)
Now : (137, 140)
never : (156, 161)
never : (172, 177)
now : (205, 208)
>>> pattern = re.compile(r'(?<!not\s)be\b')          #查找前面没有单词 not 的单词 be
>>> index = 0
>>> while True:
    matchResult = pattern.search(exampleString, index)
    if not matchResult:
        break
    print(matchResult.group(0), ':', matchResult.span(0))
    index = matchResult.end(0)

be : (13, 15)
>>> exampleString[13:20]                             #验证结果是否正确
'be one-'
>>> pattern = re.compile(r'(\b\w*(?P<f>\w)(?P=f)\w*\b)')
                                                     #查找具有连续相同字母的单词
>>> index = 0
>>> while True:
    matchResult = pattern.search(exampleString, index)
    if not matchResult:
        break
```

```
        print(matchResult.group(0), ':', matchResult.group(2))
        index = matchResult.end(0) + 1
unless : s
better : t
better : t
>>> s
'aabc abcd abbcd abccd abcdd'
>>> p = re.compile(r'(\b\w*(?P<f>\w)(?P=f)\w*\b)')
>>> p.findall(s)                              #有连续相同字母的单词
[('aabc', 'a'), ('abbcd', 'b'), ('abccd', 'c'), ('abcdd', 'd')]
```

4.2.6 正则表达式应用案例精选

例 4-4 编写程序,提取 Python 程序中的类名、函数名及变量名等标识符。

将下面的代码保存为 FindIdentifiersFromPyFile.py,在命令提示符环境中使用如下命令格式"Python FindIdentifiersFromPyFile.py 目标文件名",查找并输出目标文件中的标识符。

```
import re
import os
import sys

classes = {}
functions = []
variables = {'normal':{}, 'parameter':{}, 'infor':{}}
'''This is a test string:
atest, btest = 3, 5
to verify that variables in comments will be ignored by this algorithm
'''

def _identifyClassNames(index, line):
    '''parameter index is the line number of line,
     parameter line is a line of code of the file to check'''
    pattern = re.compile(r'(?<=class\s)\w+(?=.*?:)')
    matchResult = pattern.search(line)
    if not matchResult:
        return
    className = matchResult.group(0)
    classes[className] = classes.get(className, [])
    classes[className].append(index)

def _identifyFunctionNames(index, line):
    pattern = re.compile(r'(?<=def\s)(\w+)\((.*?)\)(?=:)')
    matchResult = pattern.search(line)
    if not matchResult:
        return
    functionName = matchResult.group(1)
    functions.append((functionName, index))
    parameters = matchResult.group(2).split(r', ')
    if parameters[0] == '':
```

```python
            return
        for v in parameters:
            variables['parameter'][v] = variables['parameter'].get(v, [])
            variables['parameter'][v].append(index)

    def _identifyVariableNames(index, line):
        #find normal variables, including the case: a, b = 3, 5
        pattern = re.compile(r'\b(.*?)(?=\s=)')
        matchResult = pattern.search(line)
        if matchResult:
            vs = matchResult.group(1).split(r', ')
            for v in vs:
                #consider the case 'if variable == value'
                if 'if ' in v:
                    v = v.split()[1]
                #consider the case: 'a[3] = 3'
                if '[' in v:
                    v = v[0:v.index('[')]
                variables['normal'][v] = variables['normal'].get(v, [])
                variables['normal'][v].append(index)
        #find the variables in for statements
        pattern = re.compile(r'(?<=for\s)(.*?)(?=\sin)')
        matchResult = pattern.search(line)
        if matchResult:
            vs = matchResult.group(1).split(r', ')
            for v in vs:
                variables['infor'][v] = variables['infor'].get(v, [])
                variables['infor'][v].append(index)

def output():
    print('='*30)
    print('The class names and their line numbers are:')
    for key, value in classes.items():
        print(key, ':', value)
    print('='*30)
    print('The function names and their line numbers are:')
    for i in functions:
        print(i[0], ':', i[1])
    print('='*30)
    print('The normal variable names and their line numbers are:')
    for key, value in variables['normal'].items():
        print(key, ':', value)
    print('-'*20)
    print('The parameter names and their line numbers in functions are:')
    for key, value in variables['parameter'].items():
        print(key, ':', value)
    print('-'*20)
    print('The variable names and their line numbers in for statements are:')
    for key, value in variables['infor'].items():
        print(key, ':', value)

#suppose the lines of comments less than 50
def comments(index):
```

```python
        for i in range(50):
            line = allLines[index+i].strip()
            if line.endswith('"""') or line.endswith("'''"):
                return i+1

if __name__ == '__main__':
    fileName = sys.argv[1]
    if not os.path.isfile(fileName):
        print('Your input is not a file.')
        sys.exit(0)
    if not fileName.endswith(('.py','.pyw')):
        print('Sorry. I can only check Python source file.')
        sys.exit(0)
    allLines = []
    with open(fileName, 'r') as fp:
        allLines = fp.readlines()
    index = 0
    totalLen = len(allLines)
    while index < totalLen:
        line = allLines[index]
        #strip the blank characters at both end of line
        line = line.strip()
        #ignore the comments starting with '#'
        if line.startswith('#'):
            index += 1
            continue
        #ignore the comments between ''' or """
        if line.startswith('"""') or line.startswith("'''"):
            index += comments(index)
            continue
        #identify identifiers
        _identifyClassNames(index+1, line)
        _identifyFunctionNames(index+1, line)
        _identifyVariableNames(index+1, line)
        index += 1
    output()
```

例 4-5 编写程序,并用该程序检查另一个 Python 程序的代码风格是否符合规范。

本例代码主要检查 Python 程序的一些基本规范,例如,运算符两侧是否有空格,是否每次只导入一个模块,在不同的功能模块之间是否有空行,等等。

```python
import sys
import re

def checkFormats(lines, desFileName):
    fp = open(desFileName, 'w')
    for i, line in enumerate(lines):
        print('='*30)
        print('Line:', i+1)
        if line.strip().startswith('#'):
            print(' '*10+'Comments.Pass.')
```

```python
                fp.write(line)
                continue
            flag = True
            #check operator symbols
            symbols = [',', '+', '-', '*', '/', '//', '**', '>>', '<<', '+=', '-=','*=', '/=']
            temp_line = line
            for symbol in symbols:
                pattern = re.compile(r'\s*'+re.escape(symbol)+r'\s*')
                temp_line = pattern.split(temp_line)
                sep = ' '+symbol+' '
                temp_line = sep.join(temp_line)
            if line != temp_line:
                flag = False
                line = temp_line
                print(' '*10+'You may miss some blank spaces in this line.')
            #check import statement
            if line.strip().startswith('import'):
                if ',' in line:
                    flag = False
                    print(' '*10+"You'd better import one module at a time.")
                    temp_line = line.strip()
                    modules = temp_line[temp_line.index(' ')+1:]
                    pattern = re.compile(r'\s*,\s*')
                    modules = pattern.split(modules)
                    temp_line = ''
                    for module in modules:
                        temp_line += line[:line.index('import')] +'import' +module +'\n'
                    line = temp_line
                pri_line = lines[i-1].strip()
                if pri_line and (not pri_line.startswith('import')) and \
                   (not pri_line.startswith('#')):
                    flag = False
                    print(' '*10+'You should add a blank line before this line.')
                    line = '\n' +line
                after_line = lines[i+1].strip()
                if after_line and (not after_line.startswith('import')):
                    flag = False
                    print(' '*10+'You should add a blank line after this line.')
                    line = line +'\n'
            #check if there is a blank line before new functional code block
            #including the class/function definition
            if line.strip() and not line.startswith(' ') and i>0:
                pri_line = lines[i-1]
                if pri_line.strip() and pri_line.startswith(' '):
                    flag = False
                    print(' '*10+"You'd better add a blank line before this line.")
                    line = '\n' +line
            if flag:
                print(' '*10 +'Pass.')
            fp.write(line)
    fp.close()

if __name__ == '__main__':
```

```
fileName = sys.argv[1]
fileLines = []
with open(fileName, 'r') as fp:
    fileLines = fp.readlines()
desFileName = fileName[:-3] + '_new.py'
checkFormats(fileLines, desFileName)
#check the ratio of comment lines to all lines
comments = [line for line in fileLines if line.strip().startswith('#')]
ratio= len(comments)/len(fileLines)
if ratio <= 0.3:
    print('='*30)
    print('Comments in the file is less than 30%.')
    print('Perhaps you should add some comments at appropriate position.')
```

例 4-6 查找文本中 ABAC 和 AABB 形式的词语。

例 4-6

```
from re import findall

text = '''行尸走肉、金蝉脱壳、百里挑一、金玉满堂、
背水一战、霸王别姬、天上人间、不吐不快、海阔天空、
情非得已、满腹经纶、兵临城下、春暖花开、插翅难逃、
黄道吉日、天下无双、偷天换日、两小无猜、卧虎藏龙、
珠光宝气、簪缨世族、花花公子、绘声绘影、国色天香、
相亲相爱、八仙过海、金玉良缘、掌上明珠、皆大欢喜\
浩浩荡荡、平平安安、秀秀气气、斯斯文文、高高兴兴'''

pattern = r'(((.).\3.)|((.)\5(.)\6))'
for item in findall(pattern, text):
    print(item[0])
```

本 章 小 结

（1）在 Python 中，字符串使用单引号、双引号、三单引号或三双引号作为界定符，并且不同的界定符之间可以互相嵌套。

（2）字符串属于有序不可变序列，不支持任何方法直接修改字符串的内容。

（3）在格式化字符串时，优先考虑使用 format() 方法和 f-字符串。

（4）Python 3.x 全面支持中文，可以使用汉字作变量名。

（5）字符串支持使用 replace()、maketrans() 和 translate() 等方法以及正则表达式进行内容替换操作，这些方法都返回新字符串，不对原字符串做任何修改。

（6）字符串的 split()、rsplit() 方法分别用于以指定字符为分隔符从字符串左端和右端开始将其分隔为多个字符串，返回包含分隔结果的列表；join() 方法用于连接可迭代对象中的字符串，并在相邻两个字符串之间插入指定连接符。

（7）对用户输入的字符串进行 eval() 操作时可能会有安全漏洞，应首先对用户输入的内容进行必要的检查和过滤，或者使用标准库 ast 中的函数 literal_eval()。

（8）在 string 模块中定义了多个字符串常量，包括数字字符、标点符号、全部英文字母、大写英文字母、小写英文字母等。

（9）正则表达式是字符串处理的有力工具和技术，可以快速实现字符串的复杂处理

任务。

(10) 可以直接使用 re 模块的函数进行字符串处理,也可以将模式编译为正则表达式对象,然后使用正则表达式对象的方法来处理字符串。

(11) 正则表达式中的子模式是作为一个整体来对待的,使用子模式扩展语法可以实现更加复杂的字符串处理任务。

(12) 正则表达式对象的 match(string[,pos[,endpos]])方法用于在字符串开头或指定位置进行搜索,模式必须出现在字符串开头或指定位置;search(string[,pos[,endpos]])方法用于在整个字符串或指定范围中进行搜索。若匹配成功,这两个方法都返回 Match 对象,Match 对象的主要方法有 group()、groups()、groupdict()、start()、end()、span()等。

(13) 正则表达式对象的 findall(string[,pos[,endpos]])方法用于在字符串中查找所有符合正则表达式的字符串并以列表形式返回。如果正则表达式中有子模式,则只返回子模式匹配的内容,re 模块的同名函数也具有这个特点。

习　　题

1. 假设有一段英文,其中有单独的字母 I 误写为 i,编写程序进行纠正。
2. 假设有一段英文,其中有单词中间的字母 i 误写为 I,编写程序进行纠正。
3. 有一段英文文本,其中有单词连续重复了两次,编写程序检查重复的单词并只保留一个。例如,文本内容为"This is is a desk.",程序输出为"This is a desk."。
4. 编写程序,用户输入一段英文,然后输出这段英文中所有长度为 3 个字母的单词。

第 5 章　函数设计与使用

在实际开发中,有很多操作是完全相同或者是非常相似的,仅仅是要处理的数据不同而已,因此,经常会在不同的位置多次执行相似甚至完全相同的代码块。从软件设计和代码复用的角度来讲,直接将该代码块复制到多个相应的位置然后进行简单修改绝对不是一个好主意。虽然这样使得多份复制的代码可以彼此独立地进行修改,但这样不仅增加了代码量,使得程序文件变大,也增加了代码理解和代码维护的难度,更重要的是为代码测试和纠错带来很大的困难。一旦被复制的代码块在将来某天被发现存在问题需要修改,必须对所有的复制都做同样正确的修改,这在实际中是很难完成的一项任务。由于代码量的大幅增加,导致代码之间的关系更加复杂,很可能在修补旧漏洞的同时又引入新错误。因此,应尽量避免使用直接复制代码块的方式来实现复用。

解决上述问题一种常用的方式是设计和编写函数;另一种是设计和编写面向对象程序设计中的类。本章介绍函数设计与使用,第 6 章介绍面向对象程序设计。将可能需要反复执行的代码封装为函数,并在需要执行该段代码功能的地方进行调用,不仅可以实现代码的复用,更重要的是可以保证代码的一致性,只需要修改该函数代码则所有调用位置均得到体现。当然,在实际开发中,需要对函数进行良好的设计和优化才能充分发挥其优势。在编写函数时,有很多原则需要参考和遵守,例如,不要在同一个函数中执行太多的功能,尽量只让其完成一个高度相关且大小合适的功能,以提高模块的内聚性。另外,尽量减少不同函数之间的隐式耦合,例如,减少全局变量的使用,使得函数之间仅通过调用和参数传递来显式体现其相互关系。

在编写函数时,函数体中代码的编写与前面章节介绍的内容基本一样,只是对代码进行了封装并增加了函数调用、传递参数、返回计算结果等外围接口,这也正是本章讲解的重点。由于 Python 程序是解释执行的,如果编写的函数或代码有问题,只有在被调用和执行时才可能被发现,甚至包括某些语法错误。另外,还有可能传递某些类型的参数时执行正确,而传递另一些类型的参数时却出现错误。出现这样的情况有多种可能的原因,例如,不同的参数值可能会使得函数执行不同的路径,或者不同的参数类型所支持的操作和运算符不同,等等。所以,在进行代码测试时一定要注意,一次或几次运行正常并不表示编写的代码没有问题,必须进行尽可能完全的测试,尽量满足各种覆盖性要求,对所有的执行路径都要测试,在代码发布之前发现和解决更多的潜在问题。

5.1　函数定义与调用

在 Python 中,定义函数的语法如下:

```
def 函数名([形参列表]):
    '''注释'''
    函数体
```

在 Python 中首先使用 def 关键字定义函数,然后是一个空格和函数名,接下来是一对圆括号,在圆括号内是形参列表,如果有多个参数则使用逗号分隔,圆括号之后是一个冒号和换行,最后是必要的注释和函数体代码。定义函数时需要注意以下事项。

(1) 函数形参不需要声明其类型,也不需要指定函数返回值类型,解释器会自动推断。
(2) 即使该函数不需要接收任何参数,定义和调用时也必须保留一对空的圆括号。
(3) 圆括号后面的冒号必不可少。
(4) 函数体相对于 def 关键字必须保持一定的空格缩进。
(5) Python 允许嵌套定义函数,可参见 5.8 节的讨论。

例如,下面的函数用来计算斐波那契数列中小于参数 n 的所有值:

```
def fib(n):                                    #n 为形参
    a, b = 1, 1
    while a < n:
        print(a, end=' ')
        a, b = b, a+b
    print()
```

该函数的调用方式为

```
fib(1000)                                      #1000 为实参
```

在定义函数时,开头部分的注释并不是必需的,但是如果为函数的定义加上一段注释,可以为用户提供友好的提示和使用帮助。例如,把上面生成斐波那契数列的函数定义修改为下面的形式,在函数开头加上一段注释。在调用该函数时,输入左侧圆括号,立刻会得到该函数的使用说明,如图 5-1 所示。

```
>>> def fib(n):
        '''accept an integer n.
        return the numbers less than n in Fibonacci sequence.'''
        a, b = 1, 1
        while a < n:
            print(a, end=' ')
            a, b = b, a+b
        print()

>>> fib(
        (n)
        accept an integer n.
        return the numbers less than n in Fibonacci sequence.
```

图 5-1　使用注释为用户提示函数使用说明

在定义函数时,可以声明形参类型和函数返回值类型,但并不真正约束和检查,可以接收任意类型的实参,也可以返回任意类型的值。例如:

```
def func(i:int, j:int) -> int:
    return i + j
print(func(3, 5))
```

如果在函数体中有调用该函数自身的代码,这称作递归函数。在编写递归函数时,应保证每次递归时问题性质不变但规模越来越小,并且当问题规模小到一定程度时可以直接解决而不需要继续递归。例 5-17 和 7.5 节遍历目录树的代码演示了递归函数的用法。

5.2 形参与实参

函数定义时圆括号内是使用逗号分隔的形参(parameters)列表,一个函数可以没有形参,但是定义和调用时一对圆括号必须有,表示这是一个函数并且不接收参数。函数调用时向其传递实参(arguments),将实参的引用传递给形参。

在定义函数时,对参数个数并没有限制,如果有多个形参,则需要使用逗号进行分隔。例如,下面的函数用来接收两个参数,输出其中的最大值,模拟内置函数 max() 的功能。

```
def printMax(a, b):
    if a >= b:
        print(a)
    else:
        print(b)
```

当然,这里只是为了演示,而忽略了一些细节,如果输入的参数不支持比较运算,则会出错,可以参考第 8 章中介绍的异常处理结构来解决这个问题。

在函数内直接修改形参的值不会影响实参。例如:

```
>>> def addOne(a):
    print(a)
    a += 1                          #这里实际是修改了形参 a 的引用,a= a + 1
    print(a)
>>> a = 3
>>> addOne(a)
3
4
>>> a                               #实参的值和引用不变
3
```

从运行结果可以看出,在函数内修改了形参 a 的值,但是当函数运行结束以后,实参 a 的值并没有被修改,可以参考 5.5 节中关于变量作用域的讨论。在有些情况下,可以通过一定的方式在函数内修改实参的值,例如下面的代码:

```
>>> def modify(v):                  #使用下标修改列表元素值
    v[0] = v[0] + 1                 #没有修改形参的引用
>>> a = [2]
>>> modify(a)
>>> a
[3]
>>> def modify(v, item):            #使用原地操作的 append()方法为列表增加元素
    v.append(item)                  #没有修改形参的引用
```

```
>>> a = [2]
>>> modify(a, 3)
>>> a
[2, 3]
>>> def modify(d):                    #使用下标修改字典元素值或为字典增加元素
    d['age'] = 38                     #没有修改形参的引用
>>> a = {'name':'Dong', 'age':37, 'sex':'Male'}
>>> modify(a)
>>> a
{'age': 38, 'name': 'Dong', 'sex': 'Male'}
```

也就是说，如果传递给函数的是 Python 可变对象，并且在函数内使用下标或对象自身原地操作的方法为可变对象增加、删除元素或修改元素值时，修改后的结果是可以反映到函数之外的，实参也得到相应的修改。

5.3 参数类型

在 Python 中，函数参数的形式有很多种，主要可以分为普通位置参数、默认值参数、关键参数、可变长度参数等。Python 函数的定义也非常灵活，在定义函数时不需要指定参数的类型，形参的类型完全由调用者传递的实参类型及 Python 解释器的理解和推断来决定；同样，也不需要指定函数的返回值类型。函数的返回值类型由 return 语句返回值的类型来决定，如果函数中没有 return 语句或者没有执行到 return 语句而返回或者执行了不带任何值的 return 语句，函数都默认为返回空值 None。

没有任何特殊说明的参数为位置参数，实参按顺序依次传递给形参，要求实参和形参的数量和顺序都一致。

5.3.1 默认值参数

在定义函数时，Python 支持默认值参数，即在定义函数时为形参设置默认值。在调用带有默认值参数的函数时，可以不用为设置了默认值的形参传递实参，此时函数将会直接使用函数定义时设置的默认值。默认值参数与 5.3.3 节介绍的可变长度参数可以实现类似于函数重载的目的。带有默认值参数的函数定义语法如下：

```
def 函数名(…, 形参名=默认值):
    函数体
```

调用带有默认值参数的函数时，可以不对默认值参数进行赋值，也可以通过显式赋值来替换其默认值，具有较大的灵活性。如果需要，可以使用"函数名.__defaults__"随时查看函数所有默认值参数的当前值，其返回值为一个元组，其中的元素依次表示每个默认值参数的当前值。例如下面的函数定义：

```
>>> def say(message, times=1):
    print((message+' ') * times)
>>> say.__defaults__
(1,)
```

调用该函数时，如果只为第一个参数传递实参，则第二个参数使用默认值 1；如果为第二个参数传递实参,则不再使用默认值1,而是使用调用者显式传递的值。

```
>>> say('hello')
hello
>>> say('hello', 3)
hello hello hello
>>> say('hi', 7)
hi hi hi hi hi hi hi
```

在定义带默认值参数的函数时,默认值参数必须全部出现在位置参数右侧,任何一个默认值参数右侧都不能再出现没有默认值的普通位置参数。例如下面的示例,前两个函数不符合这一要求,从而导致函数定义失败,如图5-2所示。

```
>>> def f(a=3,b,c=5):
        print a,b,c

SyntaxError: non-default argument follows default argument
>>> def f(a=3,b):
        print a,b

SyntaxError: non-default argument follows default argument
>>> def f(a,b,c=5):
        print a,b,c

>>>
```

图5-2 带默认值参数的函数定义

默认值参数的值是在函数定义时确定的,然后默认值参数的引用不再变化,调用函数且不给默认值参数传递实参时将一直使用这个引用。对于列表、字典这样可变类型的默认值参数,这一点可能会导致很严重的逻辑错误,而这种错误或许会耗费较多的精力来定位和纠正。例如:

```
def demo(newitem, old_list=[]):
    old_list.append(newitem)
    return old_list
print(demo('5', [1, 2, 3, 4]))
print(demo('aaa', ['a', 'b']))
print(demo('a'))
print(demo('b'))
```

运行上面的代码,仔细看看结果,是否能发现问题呢?然后把代码修改为下面的样子,再运行,看看区别在哪里。仔细阅读本节前面的内容,应该会发现答案。

```
def demo(newitem, old_list=None):
    if old_list is None:
        old_list = []
    new_list = old_list[:]
    new_list.append(newitem)
    return new_list

print(demo('5', [1, 2, 3, 4]))
print(demo('aaa', ['a', 'b']))
print(demo('a'))
print(demo('b'))
```

下面代码再一次演示了函数参数默认值是在函数定义时确定的。

```
>>> i = 3
>>> def f(n=i):                    #参数 n 的值仅取决于 i 的当前值
        print(n)

>>> f()
3
>>> i = 5                          #函数定义后修改 i 的值不影响参数 n 的默认值
>>> f()
3
```

5.3.2 关键参数

通过关键参数可以按参数名传递实参，实参顺序可以和形参顺序不一致，但不影响参数的传递结果，避免了用户需要牢记参数位置和顺序的麻烦，使得函数的调用和参数传递更加灵活方便。

```
>>> def demo(a, b, c=5):
        print(a, b, c)
>>> demo(3, 7)                     #位置参数 a 和 b,参数 c 使用默认值
3 7 5
>>> demo(c=8, a=9, b=0)            #关键参数
9 0 8
```

对于 Python 3.8 及更高版本，在定义函数时，可以使用单个斜线或单个星号作为参数，这两个符号不是真正的参数，只用于对其他参数进行约束。其中，单个斜线表示前面的所有参数都必须以位置参数的形式进行传递，单个星号表示后面的所有参数都必须以关键参数的形式进行传递。

```
>>> def func(a, /, *, b, c):       #参数 a 必须以位置参数的形式传递
                                   #参数 b 和 c 必须以关键参数的形式传递
        return a +b +c
>>> func(a=3, b=4, c=5)            #参数 a 不能使用关键参数的形式传递,出错
TypeError: func() got some positional- only arguments passed as keyword arguments: 'a'
>>> func(3, 4, 5)                  #参数 b 和 c 不能使用位置参数的形式传递,出错
TypeError: func() takes 1 positional argument but 3 were given
>>> func(3, b=4, c=5)              #符合要求,成功调用函数
12
```

5.3.3 可变长度参数

可变长度参数在定义函数时主要有两种形式：*parameter 和**parameter,前者用于接收任意多个位置实参并将其放在一个元组中，后者接收多个关键参数并将其放入字典中。

下面的代码演示了第一种形式可变长度参数的用法，无论调用该函数时传递了多少位置实参，一律将其放入元组中，元组长度由实参个数确定。

```
>>> def demo(*p):
        print(p)
>>> demo(1, 2, 3)
(1, 2, 3)
```

```
>>> demo(1, 2, 3, 4, 5, 6, 7)
(1, 2, 3, 4, 5, 6, 7)
```

下面的代码演示了第二种形式可变长度参数的用法,在调用该函数时自动将接收的关键参数转换为字典,字典长度由实参个数确定。

```
>>> def demo(**p):
        for item in p.items():
            print(item)
>>> demo(x=1, y=2, z=3)                            #关键参数
('x', 1)
('y', 2)
('z', 3)
```

下面的代码演示了定义函数时几种不同形式的参数混合使用的用法。虽然 Python 完全支持这样做,但是除非真的很必要,否则不要这样用,因为这会使得代码非常混乱而严重降低可读性,并导致程序查错非常困难。另外,一般而言,一个函数如果可以接收很多参数,很可能是函数设计得不好,例如,函数功能过多。需要进行必要的拆分和重新设计,以满足高内聚的要求,同时也利于代码阅读和维护。

```
>>> def func_4(a, b, c=4, *aa, **bb):
        print((a, b, c))
        print(aa)
        print(bb)
>>> func_4(1, 2, 3, 4, 5, 6, 7, 8, 9, xx='1', yy='2', zz=3)
(1, 2, 3)
(4, 5, 6, 7, 8, 9)
{'xx': '1', 'yy': '2', 'zz': 3}
```

5.3.4

5.3.4 参数传递时的序列解包

为含多个形参的函数传递参数时,可以使用 Python 列表、元组、集合、字典以及其他可迭代对象作为实参,并在实参名前加一个星号,Python 解释器将自动进行解包,然后传递给多个位置形参。如果使用字典对象作为实参,则默认使用字典的"键";如果需要将字典中"键:值"对作为参数,则需要使用 items()方法;如果需要将字典的"值"作为参数,则需要调用字典的 values()方法。最后,务必保证实参中元素个数与形参个数相等,否则将出现错误。

```
>>> def demo(a, b, c):               >>> dic = {1:'a', 2:'b', 3:'c'}
        print(a+b+c)                 >>> demo(*dic)
>>> seq = [1, 2, 3]                  6
>>> demo(*seq)                       >>> demo(*dic.values())
6                                    abc
>>> tup = (1, 2, 3)                  >>> Set = {1, 2, 3}
>>> demo(*tup)                       >>> demo(*Set)
6                                    6
```

如果使用字典作为函数实参,在前面使用两个星号进行解包时,会把字典解包成为关键参数进行传递,字典的"键"作为参数名,字典的"值"作为参数的值。

```
>>> def demo(a, b, c):
    print(a+b+c)
>>> demo(**{'a':97, 'b':98, 'c':99})            #每个"键"都必须在形参列表中
294
```

5.4　return 语句

return 语句用于结束函数的执行，同时还可以通过 return 语句从函数中返回一个任意类型的值。不论 return 语句出现在函数的什么位置，一旦得到执行将直接结束函数。如果函数没有 return 语句、有 return 语句但没有执行或者执行了不返回任何值的 return 语句，Python 将认为该函数以 return None 结束，即返回空值。

在调用函数和方法（见第 6 章）时，一定要注意有没有返回值，以及是否会对参数的值进行修改。例如第 2 章介绍过的列表对象方法 sort() 属于原地操作，没有返回值；而内置函数 sorted() 返回排序后的列表，并不对原列表做任何修改。

```
>>> a_list = [1, 2, 3, 4, 9, 5, 7]
>>> print(sorted(a_list))                       #返回排序后的新列表
[1, 2, 3, 4, 5, 7, 9]
>>> print(a_list)                               #不影响原列表的内容
[1, 2, 3, 4, 9, 5, 7]
>>> print(a_list.sort())                        #原地排序，没有返回值
None
>>> print(a_list)
[1, 2, 3, 4, 5, 7, 9]
```

5.5　变量作用域

变量起作用的代码范围称为变量的作用域，不同作用域内同名变量之间互不影响。

在 Python 中，主要有局部变量、nonlocal 变量和全局变量这三类，范围依次从近到远。在访问一个变量时，首先会使用局部变量，如果没有同名的局部变量则尝试使用外层函数中的 nonlocal 变量，如果不存在外层函数或者同名的 nonlocal 变量则尝试使用全局变量，如果全局变量也不存在则再尝试使用内置命名空间中的标识符，如果仍不存在则提示错误。本书重点介绍局部变量和全局变量。

一个变量在函数外定义和在函数内定义，其作用域是不同的，函数内使用赋值语句直接创建的变量为局部变量，在函数外创建的变量为全局变量。一般而言，局部变量的引用速度比全局变量快，应优先考虑使用。除非真的有必要，否则应尽量避免使用全局变量，因为全局变量会增加不同函数之间的隐式耦合度，从而降低代码可读性，同时也使得代码测试和纠错变得很困难。

在函数内定义的普通变量只在该函数内起作用，称为局部变量。当函数运行结束后，在该函数内定义的局部变量被自动删除而不可访问。在函数内使用关键字 global 定义的全局变量当函数结束以后仍然存在并且可以访问。

如果想在函数内修改一个在函数外定义的变量的值，那么这个变量就不能是局部的，其

作用域必须为全局的,能够同时作用于函数内外,称为全局变量,可以通过 global 声明或定义。这分两种情况。

（1）一个变量已在函数外定义,如果在函数内需要修改这个变量的值,可以在函数内用关键字 global 声明使用这个全局变量,明确声明要使用已定义的同名全局变量。

（2）在函数内直接使用 global 关键字将一个变量声明为全局变量,如果在函数外没有定义该全局变量,在调用这个函数之后,将自动增加新的全局变量,应避免这样做。

或者说,也可以这么理解：在函数内如果只引用某个变量的值而没有为其赋新值,该变量为(隐式的)全局变量;如果在函数内任意位置有为变量赋值的操作,该变量即被认为是(隐式的)局部变量,除非在函数内显式地用关键字 global 进行声明。

下面的示例代码演示了局部变量和全局变量的用法。

```
>>> def demo():
        global x                    #声明或创建全局变量
        x = 3                       #修改全局变量的值
        y = 4                       #局部变量
        print(x, y)
>>> x = 5                           #在函数外定义全局变量 x
>>> demo()                          #本次调用修改了全局变量 x 的值
3 4
>>> x
3
>>> y                               #局部变量在函数运行结束之后自动删除
NameError: name 'y' is not defined
>>> del x                           #删除了全局变量 x
>>> x
NameError: name 'x' is not defined
>>> demo()                          #本次调用创建了全局变量
3 4
>>> x
3
>>> y                               #局部变量在函数调用和执行结束后自动删除,在函数外不可访问
NameError: name 'y' is not defined
```

在函数内任意位置只要有为变量赋值的语句,那么在整个函数内该变量都是局部变量。在这条赋值语句之前不能有引用变量值的操作,否则会引发代码异常,除非在函数开始处使用关键字 global 声明该变量为全局变量。

```
>>>x = 3
>>>def f():
        print(x)                    #本意是先输出全局变量 x 的值,但是不允许这样做
        x = 5                       #有赋值操作,因此在整个作用域内 x 都是局部变量
        print(x)
>>>f()                              #略去异常的详细信息
UnboundLocalError: local variable 'x' referenced before assignment
```

如果局部变量与全局变量具有相同的名字,那么该局部变量会在自己的作用域内隐藏同名的全局变量,例如下面的代码：

```
>>> def demo():
        x = 3                          #创建了局部变量,并自动隐藏了同名的全局变量
>>> x = 5
>>> demo()                             #不会影响全局变量的值
>>> x
5
```

5.6　lambda 表达式

lambda 表达式常用于声明匿名函数,即没有函数名的临时使用的小函数。lambda 表达式只可以包含一个表达式,不允许使用选择结构、循环结构、函数定义等语法,但在表达式中可以调用其他函数,并支持默认值参数和关键参数,该表达式的计算结果就是函数的返回值。

lambda 表达式属于可调用对象之一,常用于内置函数 sorted()、max()、min()和列表方法 sort()的 key 参数,内置函数 map()、filter()和标准库函数 reduce()的第一个参数,以及其他可以使用函数的地方。

下面的代码演示了不同情况下 lambda 表达式的应用。

```
>>> f = lambda x, y, z: x+y+z                       #可以给 lambda 表达式起名字
>>> print(f(1, 2, 3))                               #可以像普通函数一样调用
6
>>> g = lambda x, y=2, z=3: x+y+z                   #含默认值参数
>>> print(g(1))
6
>>> print(g(2, z=4, y=5))                           #调用时使用关键参数
11
>>> L = [(lambda x: x**2), (lambda x: x**3), (lambda x: x**4)]
>>> print(L[0](2), L[1](2), L[2](2))                #使用没有名字的 lambda 表达式
4 8 16
>>> D = {'f1':(lambda: 2+3), 'f2':(lambda: 2*3), 'f3':(lambda: 2**3)}
>>> print(D['f1'](), D['f2'](), D['f3']())
5 6 8
>>> L = [1, 2, 3, 4, 5]
>>> print(map((lambda x: x+10), L))                 #使用没有名字的 lambda 表达式
[11, 12, 13, 14, 15]
>>> L
[1, 2, 3, 4, 5]
>>> def demo(n):
        return n*n
>>> demo(5)
25
>>> a_list = [1, 2, 3, 4, 5]
>>> list(map(lambda x: demo(x), a_list))            #包含函数调用并且没有名字的 lambda 表达式
[1, 4, 9, 16, 25]
>>> data = list(range(20))
>>> import random
>>> random.shuffle(data)
```

```
>>> data
[4, 3, 11, 13, 12, 15, 9, 2, 10, 6, 19, 18, 14, 8, 0, 7, 5, 17, 1, 16]
>>> data.sort(key=lambda x: x)            #用在列表的 sort()方法中
>>> data
[0, 1, 2, 3, 4, 5, 6, 7, 8, 9, 10, 11, 12, 13, 14, 15, 16, 17, 18, 19]
>>> data.sort(key=lambda x: len(str(x)))  #按转换为字符串后的长度排序
>>> data
[0, 1, 2, 3, 4, 5, 6, 7, 8, 9, 10, 11, 12, 13, 14, 15, 16, 17, 18, 19]
>>> data.sort(key=lambda x: len(str(x)), reverse=True)
>>> data
[10, 11, 12, 13, 14, 15, 16, 17, 18, 19, 0, 1, 2, 3, 4, 5, 6, 7, 8, 9]
>>> import random
>>> x = [[random.randint(1,10) for j in range(5)] for i in range(5)]
                                          #使用列表推导式创建列表
                                          #包含 5 个子列表的列表
                                          #每个子列表中包含 5 个 1~10 的随机数
>>> for item in x:                        #略去输出结果
    print(item)                           #这个 for 循环可以替换为 print(*x, sep='\n')
>>> y = sorted(x, key=lambda item: (item[1], item[4]))
                                          #按子列表中第 2 个元素升序、第 5 个元素升序排序
>>> for item in y:                        #略去输出结果
    print(item)
```

在使用 lambda 表达式时，要注意变量作用域带来的问题，在下面的代码中变量 x 是在外部作用域中定义的，对 lambda 表达式而言不是局部变量，从而导致出现错误。

5.6

```
>>> r = []
>>> for x in range(10):
    r.append(lambda: x**2)
>>> r[0]()                                #每次调用时 x 的值都是 9
81
```

若修改为下面的代码，则可以得到正确的结果。

```
>>> r = []
>>> for x in range(10):
    r.append(lambda n=x: n**2)            #默认值参数，见 5.3.1 节
>>> r[0]()
0
>>> r[1]()
1
>>> r[5]()
25
```

5.7 案例精选

例 5-1　编写函数计算圆的面积。

```
from math import pi as PI                 #圆周率常数

def area(r):
    if isinstance(r,(int,float)) and r>0: #确保接收的参数为大于 0 的数字
```

```
        return PI*r*r
    else:
        return ('You must give me a positive integer or float as radius.')
print(area(3))
```

例 5-2 编写函数,接收任意多个实数,返回一个元组,其中第一个元素为所有参数的平均值,其他元素为所有参数中大于平均值的实数。

```
def demo(*para):
    avg = sum(para) / len(para)
    g = [i for i in para if i>avg]
    return (avg,) + tuple(g)

print(demo(1, 2, 3, 4))
```

例 5-2

例 5-3 编写函数,接收字符串参数,返回一个元组,其中第一个元素为大写字母个数,第二个元素为小写字母个数。

```
def demo(s):
    result = [0, 0]
    for ch in s:
        if 'a'<= ch <= 'z':
            result[1] += 1
        elif 'A'<= ch <= 'Z':
            result[0] += 1
    return tuple(result)

print(demo('aaaabbbbC'))
```

例 5-4 编写函数,接收包含 20 个整数的列表 lst 和一个整数 k 作为参数,返回新列表。处理规则为:将列表 lst 中下标 k(不包括)之前的元素逆序,下标 k 及 k 之后的元素逆序,然后将整个列表 lst 中的所有元素逆序。

例 5-4

```
def demo(lst,k):
    x = lst[:]                                    #切片得到副本,不影响原列表
    x[:k] = reversed(x[:k])
    x[k:] = reversed(x[k:])
    x.reverse()
    return x

lst = list(range(1, 21))
print(demo(lst, 5))
```

例 5-5 使用秦九韶算法快速求解多项式的值。

```
from functools import reduce

def func(factors, x):
    return reduce(lambda a, b: a*x+b, factors)    #reduce()函数见 5.8 节

factors = (3, 8, 5, 9, 7, 1)
print(func(factors, 1))
```

例 5-5

例 5-6 编写函数,接收一个包含若干整数的列表参数 lst,返回一个元组,其中第一个元素为列表 lst 中的最小值,其余元素为最小值在列表 lst 中的下标。

```
import random

def demo(lst):
    m = min(lst)
    result = (m,) +tuple((index for index,value in enumerate(lst) if value==m))
    return result

x = [random.randint(1,20) for i in range(50)]
print(x, demo(x), sep= '\n')
```

例 5-7

例 5-7 编写函数,接收一个整数 t 为参数,打印杨辉三角的前 t 行。

```
def demo(t):
    print([1])
    print([1, 1])
    line = [1, 1]
    for i in range(2, t):
        r = []
        for j in range(0, len(line)-1):        #生成下一行首尾之间的数字
            r.append(line[j]+line[j+1])
        line = [1] + r + [1]                   #接上首尾数字,得到下一行所有数字
        print(line)

demo(10)
```

例 5-8

例 5-8 编写函数,接收一个正偶数为参数,输出两个素数,并且这两个素数之和等于原来的正偶数。如果存在多组符合条件的素数,则全部输出。

```
def isPrime(n):
    m = int(n**0.5) + 1
    for i in range(2, m):
        if n%i == 0:
            return False
    return True

def demo(n):
    if isinstance(n, int) and n>0 and n%2 == 0:
        for i in range(3, n//2+1):                  #可以优化,见微课视频
            if i%2 == 1 and isPrime(i) and isPrime(n-i):
                print(i, '+', n-i, '=', n)

demo(60)
```

例 5-9 编写函数,接收两个正整数作为参数,返回一个数组,其中第一个元素为最大公约数,第二个元素为最小公倍数。不允许使用 math 模块的 gcd()和 lcm()函数。

```
def demo(m,n):
    p = m * n
    while m != 0:
        r = n % m
```

```
        n, m = m, r
    return(n, p//n)

print(demo(20, 30))
```

例 5-10　编写函数，接收一个所有元素值互不相等的整数列表 x 和一个整数 n，要求将值为 n 的元素作为支点，将列表中所有值小于 n 的元素全部放到 n 的前面，所有值大于 n 的元素全部放到 n 的后面。

```
import random

def demo(x, n):
    i = x.index(n)                          #获取指定元素在列表中的索引
    x[0], x[i] = x[i], x[0]                 #将指定元素与第 0 个元素交换
    key = x[0]
    
    i = 0
    j = len(x) - 1
    while i < j:
        while i<j and x[j]>=key:            #从后向前寻找第一个比指定元素小的元素
            j -= 1
        x[i] = x[j]
        
        while i<j and x[i]<=key:            #从前向后寻找第一个比指定元素大的元素
            i += 1
        x[j] = x[i]
    
    x[i] = key

x = list(range(1, 10))
random.shuffle(x)
print(x)
demo(x, 4)
print(x)
```

例 5-11　编写函数，计算字符串匹配的准确率。以打字练习程序为例，假设 origin 为原始内容，userInput 为用户输入的内容，下面的代码用于测试用户输入的准确率。

例 5-11

```
def rate(origin, userInput):
    if not (isinstance(origin, str) and isinstance(userInput, str)):
        print('The two parameters must be strings.')
        return
    if len(origin) < len(userInput):
        print('Sorry. I suppose the second parameter string is shorter.')
        return
    right = 0                               #精确匹配的字符个数
    for origin_char, user_char in zip(origin, userInput):
        if origin_char == user_char:
            right += 1
```

```
        return right/len(origin)

origin = 'Readability counts.'
userInput = 'Readability counts.'
print(rate(origin, userInput))                    #输出测试结果
```

忽略参数的合法性检查,上面的函数 rate()还可以写成下面的形式:

```
def rate(origin, userInput):
    right = sum(map(lambda oc, uc: oc==uc, origin, userInput))
    return right/len(origin)
```

例 5-12 编写函数,模拟猜数游戏。系统随机产生一个数,玩家来猜,系统根据玩家的猜测进行提示,玩家可以根据系统的提示对下一次的猜测进行适当调整。

例 5-12

```
from random import randint

def guess(maxValue=10, maxTimes=3):
    value = randint(1, maxValue)                  #随机生成一个整数
    for i in range(maxTimes):
        prompt = 'Start to guess:' if i==0 else 'Guess again:'
        try:                                      #使用异常处理结构,防止输入不是数字
            x = int(input(prompt))
        except:
            print('Must input an integer between 1 and ', maxValue)
        else:
            if x == value:                        #猜对了
                print('Congratulations!')
                break
            elif x > value:
                print('Too big')
            else:
                print('Too little')
    else:                                         #次数用完还没猜对,游戏结束
        print('Game over. FAIL.')
        print('The value is ', value)
```

例 5-13 编写函数,计算形式如 a+aa+aaa+aaaa+…+aaa…aaa 的表达式前 n 项的值,其中 a 为小于 10 的自然数。

例 5-13

```
def demo1(a, n):
    assert type(a) == int and 0<=a<10, 'a must be integer between 1 and 9'
    a = str(a)
    return sum(eval(a*i) for i in range(1, n+1))

def demo2(a, n):
    a = str(a)
    return sum(map(lambda i:eval(a*i), range(1,n+1)))

print(demo1(1, 3))
print(demo2(5, 4))
```

例 5-14 定义函数,接收包含若干整数的列表,对列表中的全部整数进行拼接,返回能够生成的最小整数。

```python
from itertools import permutations

def min_number(lst):
    #枚举全排列得到的所有整数,返回最小的一个
    return (min(map(lambda item: int(''.join(map(str, item))),
                permutations(lst, len(lst)))))

data =[[321, 30, 32, 300], [72,725], [321, 34, 32, 344]]
for lst in data:
    print(lst, min_number(lst), sep=':')
```

运行结果:

```
[321, 30, 32, 300]: 3003032132
[72, 725]:72572
[321, 34, 32, 344]: 3213234344
```

例 5-15 编写程序模拟抓狐狸的小游戏。假设有一排 5 个洞口,小狐狸最开始的时候在其中一个洞口,然后人随机打开一个洞口,如果里面有小狐狸就抓到了。如果洞口里没有小狐狸就明天再来抓,但是小狐狸会在第二天有人来抓之前跳到隔壁洞口里。

例 5-15

```python
from random import choice, randint

def catchFox(n=5, maxTimes=10):
    #狐狸的初始位置,洞口编号为 1~n
    currentPosition = randint(1, n)
    for i in range(maxTimes):
        x = int(input('请输入要打开的洞口编号(1~{}):'.format(n)))
        if x == currentPosition:
            print('成功!')
            break
        #到头,往回跳
        if currentPosition == 1:
            currentPosition += 1
        elif currentPosition == n:
            currentPosition -= 1
        else:
            #中间位置,随机左右跳
            currentPosition += choice((-1,1))
        print('狐狸的当前位置:', currentPosition)
    else:
        print('失败。')

catchFox()
```

例 5-16 编写程序,模拟报数游戏。有 n 个人围成一圈,顺序编号,从第一个人开始由 1~k(例如 k = 3)报数,报到 k 的人退出圈子,然后圈子缩小,从下一个人继续游戏,问最后留下的是原来的第几号。

```python
from itertools import cycle

def demo(lst, k):
```

```
        #切片,以免影响原来的数据
        t_lst = lst[:]
        #游戏一直进行到只剩下最后一个人
        while len(t_lst) >1:
            #创建 cycle 对象
            c = cycle(t_lst)
            #由 1~k 报数
            for i in range(k):
                t = next(c)
            #一个人出局,圈子缩小
            index = t_lst.index(t)
            t_lst = t_lst[index+1:] +t_lst[:index]
        #游戏结束
        return t_lst[0]

lst = list(range(1, 11))
print(demo(lst, 3))
```

例 5-17 假设一段楼梯共 15 个台阶,小明一步最多能上 3 个台阶,那么小明上这段楼梯一共有多少种方法?

例 5-17

```
def climbStairs1(n):
    #递推法
    a, b, c = 1, 2, 4
    for i in range(n-3):
        c, b, a = a+b+c, c, b
    return c

def climbStairs2(n):
    #递归法
    first3 = {1:1, 2:2, 3:4}
    if n in first3.keys():
        return first3[n]
    else:
        return climbStairs2(n-1) +climbStairs2(n-2) +climbStairs2(n-3)
```

5.8 高级话题

在本章的最后,让我们来看几个高级话题,包括内置函数 map()、标准库函数 reduce()、内置函数 filter()、Python 字节码、生成器、函数嵌套定义与可调用对象等知识。

(1) 内置函数 map() 可以将一个可调用对象依次作用到一个或多个可迭代对象的元素上,返回一个 map 对象作为结果,其中每个元素是原可迭代对象中元素经过可调用对象处理后的结果,该函数不对原可迭代对象做任何修改。

```
>>> list(map(str, range(5)))             #接收一个参数的函数可以映射到一个可迭代对象
['0', '1', '2', '3', '4']
>>> def add5(v):
        return v +5
>>> list(map(add5, range(10)))           #map 对象属于迭代器对象,可转换为列表
[5, 6, 7, 8, 9, 10, 11, 12, 13, 14]
```

```
>>> def add(x, y):
        return x + y
>>> list(map(add, range(5), range(5, 10)))     #接收两个参数的函数可以映射到两个
                                                #可迭代对象
[5, 7, 9, 11, 13]
```

（2）标准库函数 reduce() 可以将一个双参数函数以迭代的方式从左到右依次作用到一个可迭代对象的所有元素上。

```
>>> from functools import reduce
>>> seq = [1, 2, 3, 4, 5, 6, 7, 8, 9]
>>> reduce(lambda x, y: x+y, seq)
45
>>> def add(x, y):
        return x + y
>>> reduce(add, range(10))
45
```

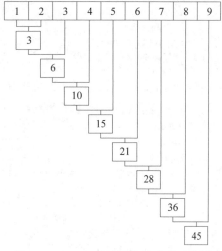

上面的代码运行过程如图 5-3 所示。

类似的运算并不局限于数值类型，例如下面的代码使用前面定义的函数 add() 实现了字符串连接。

```
>>> reduce(add, map(str, range(10)))
'0123456789'
```

图 5-3　reduce() 函数执行过程示意图

（3）内置函数 filter() 将一个单参数函数作用到一个可迭代对象上，返回该可迭代对象中使得单参数函数返回值等价于 True 的那些元素组成的 filter 对象（属于迭代器对象）。

```
>>> seq = ['foo', 'x41', '?!', '***']
>>> def func(x):
        return x.isalnum()
>>> list(filter(func, seq))                    #迭代器对象实际上并不包含元素，而是按需产生
['foo', 'x41']
>>> [x for x in seq if x.isalnum()]            #使用列表推导式实现同样功能
['foo', 'x41']
>>> list(filter(str.isalnum, seq))             #类的方法也属于可调用对象
['foo', 'x41']
>>> list(filter(None, range(-3,3)))            #None 表示只保留等价于 True 的元素
[-3, -2, -1, 1, 2]
```

（4）包含 yield 语句的函数用于创建生成器对象（属于迭代器对象），特点是惰性求值，按需产生数据，占用内存空间小，适用于大数据处理。下面的代码演示了如何使用生成器函数来生成无限长的斐波那契数列。

```
>>> def f():
        a, b = 1, 1
        while True:
            yield a                             #给出一个值，并暂停代码的执行
```

生成器函数

```
            a, b = b, a + b
>>> a = f()
>>> for i in range(10):
    print(next(a), end=' ')              #next()函数向生成器索要一个值,恢复代码执行
1 1 2 3 5 8 13 21 34 55
```

上面定义的生成器函数还可以这样使用:

```
>>> for i in f():
    if i > 100:
        break
    print(i, end=' ')
1 1 2 3 5 8 13 21 34 55 89
```

(5) 使用 dis 模块的功能可以查看函数的字节码指令。

```
>>> def add(n):
    n += 1
    return n
>>> import dis
>>> dis.dis(add)           #Python 3.10 及之前版本的字节码
    2          0 LOAD_FAST         0 (n)
               2 LOAD_CONST        1 (1)
               4 INPLACE_ADD
               6 STORE_FAST        0 (n)
    3          8 LOAD_FAST         0 (n)
              10 RETURN_VALUE
>>> dis.dis(add)           #Python 3.11 及之后版本的字节码
    1          0 RESUME            0
    2          2 LOAD_FAST         0 (n)
               4 LOAD_CONST        1 (1)
               6 BINARY_OP        13 (+=)
              10 STORE_FAST        0 (n)
    3         12 LOAD_FAST         0 (n)
              14 RETURN_VALUE
```

(6) 函数嵌套定义与可调用对象。

嵌套函数

在 Python 中,函数是可以嵌套定义的。除函数、lambda 表达式、类、方法之外,任何包含 __call__()方法的类的对象也是可调用的。下面的代码演示了函数嵌套定义的情况:

```
def linear(a, b):
    def result(x):
        return a*x + b
    return result                        #返回内层定义的函数
```

下面的代码演示了可调用对象类的定义,面向对象程序设计的内容见第 6 章。

```
class linear:
    def __init__(self, a, b):
        self.a, self.b = a, b
    def __call__(self, x):
        return self.a*x + self.b
```

使用上面的两种方式中任何一个,都可以通过以下的方式来定义一个可调用对象:

```
taxes = linear(0.3, 2)                              #参数分别传给 a 和 b
```
然后通过下面的方式来调用该对象：
```
taxes(5)                                            #参数传给 x
```
（7）修饰器。

修饰器（decorator）是函数嵌套定义的一个重要应用。修饰器本质上也是一个函数，只不过这个函数接收其他函数作为参数并对其进行一定的改造之后返回新函数。

Python 面向对象程序设计中的静态方法、类方法、属性等都是通过修饰器实现的，见第 6 章，Python 中还有很多这样的用法。

下面的代码演示了修饰器的定义与使用方法，定义其他函数调用之前或之后需要执行的通用代码，可作用于其他任何函数，提高代码复用度。

```
def before(func):                                   #定义修饰器
    def wrapper(*args, **kwargs):
        print('Before function called.')
        return func(*args, **kwargs)
    return wrapper                                  #返回内层定义的函数名

def after(func):                                    #定义修饰器
    def wrapper(*args, **kwargs):
        result = func(*args, **kwargs)
        print('After function called.')
        return result
    return wrapper

@before
@after
def test():                                         #同时使用两个修饰器改造函数,距离 test()近的先起作用
    print(3)
test()                                              #调用被修饰的函数
```

下面的代码复用了用户名检查功能的代码，关于面向对象程序设计的知识可参考第 6 章。

```
def check_permission(func):
    def wrapper(*args, **kwargs):
        if kwargs.get('username') != 'admin':
            raise Exception('Sorry. You are not allowed.')
        return func(*args, **kwargs)
    return wrapper

class ReadWriteFile(object):
    #把函数 check_permission 作为修饰器使用
    @check_permission
    def read(self, username, filename):
        with open(filename,'r') as fp:
            return fp.read()
```

```
        def write(self, username, filename, content):
            with open(filename,'w') as fp:
                fp.write(content)
        #把函数check_permission作为普通函数使用,创建新的成员方法
        write = check_permission(write)

t = ReadWriteFile()
print('Originally…')
print(t.read(username='admin', filename=r'd:\sample.txt'))
print('Now, try to write to a file…')
t.write(username='admin', filename=r'd:\sample.txt', content='\nhello world')
print('After calling to write…')
print(t.read(username='admin', filename=r'd:\sample.txt'))
```

(8) 偏函数(partial function)和函数柯里化(function currying)是函数式编程中常用的技术。有时在复用已有函数时可能需要固定其中的部分参数,这除了可以通过默认值参数来实现外,还可以使用偏函数。例如,有个函数用于实现3个数字相加:

```
def add3(a, b, c):
    return a+b+c
```

如果现在需要一个类似的函数,与上面的函数add3()的区别仅在于参数b固定为一个数字(例如666),这时就可以使用偏函数的技术来复用上面的函数。

```
def add2(a, c):
    return add3(a, 666, c)
print(add2(1, 1))
```

或者使用标准库functools提供的 **partial()** 方法创建指定函数的偏函数。

```
from functools import partial

add2 = partial(add3, b=666)
print(add2(a=1, c=1))
```

下面的代码根据标准库math中的最大公约数函数gcd()构造了新的最大公约数函数,用于计算某个固定值与其他任意多个整数的最大公约数。

```
from math import gcd
from functools import partial

print(partial(gcd, 3)(5))
print(partial(gcd, 3)(15, 21, 48))
```

函数柯里化除了可以实现偏函数类似的功能外,还可以利用单参数函数来实现多参数函数,这要归功于Python对函数嵌套定义和lambda表达式的支持。

```
def func(a):
    return lambda b: a+b
print(func(3)(5))
```

或者

```
def func(a):
```

```
    def funcNested(b):
        return a+b
    return funcNested
print(func(3)(5))
```

（9）反编译 Python 字节码。

可以使用 Python 扩展库 uncompyle6 或其他类似模块完成这个功能。使用 pip 工具安装 uncompyle6 之后，使用类似下面的代码对 .pyc 文件进行反编译得到源代码：

```
uncompyle6.uncompyle_file('__pycache__\\Stack.cpython-310.opt-1.pyc',
                          open('__pycache__\\Stack.py', 'w'))
```

另外，http://tool.lu/pyc/这个网站可以在线上传一个 .pyc 文件，然后得到 Python 源代码（个别地方可能不是非常准确，需要自己稍做调整），并且还提供了一定的代码美化功能，能够自动处理代码布局和排版规范。

本 章 小 结

（1）函数是用来实现代码复用的常用方式之一。

（2）定义函数时使用关键字 def。

（3）可以在函数定义的开头部分使用一对三引号增加一段注释来为用户提示函数使用说明。

（4）定义函数时不需要指定其形参类型，而是根据调用函数时传递的实参自动推断。

（5）测试函数时，一次或几次运行正确并不能说明函数的设计与实现没有问题，应进行尽可能全面的测试。

（6）在函数内直接修改形参的引用不会影响实参。

（7）如果传递给函数的是 Python 可变对象，并且在函数内使用下标或对象自身的原地操作方法为可变对象增加、删除元素或修改元素值时，修改后的结果是可以反映到函数之外的，即实参也得到了相应的修改。

（8）定义函数时可以为形参设置默认值，如果调用该函数时不为默认值参数传递实参，将自动使用默认值。

（9）如果使用默认值参数，必须保证默认值参数出现在函数形参列表的最后，默认值参数后面不能出现不带默认值的普通位置参数。

（10）默认值参数的引用是在定义函数时确定的，对于列表、字典这样可变类型的默认值参数，可能会导致很严重的逻辑错误。

（11）传递参数时可以使用关键参数，避免牢记参数顺序的麻烦。

（12）定义函数时，形参前面加一个星号表示可以接收多个位置实参并将其放置到一个元组中，形参前面加两个星号表示可以接收多个关键参数并将其放置到字典中。

（13）为含多个形参的函数传递参数时，可以使用 Python 列表、元组、集合、字典及其他可迭代对象作为实参，并在实参名前加一个星号，Python 解释器将自动解包为位置参数，然后传递给多个形参。

（14）lambda 表达式可以用来创建只包含一个表达式的匿名函数。

（15）在 lambda 表达式中可以调用其他函数，并支持默认值参数和关键参数。

（16）定义函数时不需要指定其返回值的类型，而是由 return 语句来决定，如果函数中没有执行 return 语句或执行了不返回任何值的 return 语句，则认为该函数返回空值 None。

（17）在函数内定义的普通变量只在该函数内起作用，称为局部变量。当函数运行结束后，在该函数内定义的局部变量被自动删除。在函数内定义的全局变量（不建议这样做）当函数结束以后仍然存在并且可以访问。

（18）在函数内可以通过 global 关键字来声明或者定义全局变量。

（19）从 Python 3.8 开始，自定义函数时可以使用单个斜线作为参数表示前面的参数必须为位置参数，单个星号作为参数表示后面的参数必须为关键参数。

习　　题

1. 在函数内可以通过关键字_____来定义全局变量。

2. 如果函数中没有 return 语句或者 return 语句不带任何返回值，那么该函数的返回值为_____。

3. 包含_____语句的函数可以用于创建生成器。

4. 调用带默认值参数的函数时，不能为默认值参数传递任何值，必须使用函数定义时设置的默认值（判断对错）。

5. lambda 表达式只能用于创建匿名函数，不能为这样的函数起名（判断对错）。

6. 运行 5.3.1 节倒数第 2、3 段的示例代码，查看结果并分析原因。

7. 编写函数，判断一个整数是否为素数，并编写主程序调用该函数。

8. 编写函数，接收一个字符串，分别统计大写字母、小写字母、数字、其他字符的个数，并以元组的形式返回结果。

9. 在 Python 程序中，局部变量会隐藏同名的全局变量吗？编写代码进行验证。

10. 编写函数，可以接收任意多个整数并输出其中的最大值和所有整数之和。

11. 编写函数，模拟内置函数 sum()。

12. 编写函数，模拟内置函数 sorted()。

13. 编写函数，模拟内置函数 map()。

14. 编写函数，模拟内置函数 filter()。

第 6 章 面向对象程序设计

面向对象程序设计(Object Oriented Programming,OOP)的思想主要针对大型软件设计而提出,使得软件设计更加灵活,能够更好地支持代码复用和设计复用,并且使得代码具有更好的可读性和可扩展性。将数据以及对数据的操作封装在一起,组成一个相互依存、不可分割的整体,即对象(或称实例)。对于相同类型的对象进行分类、抽象后,得出共同的特征和行为而形成类,面向对象程序设计的关键就是如何合理地定义和组织这些类以及类之间的关系。

Python 完全支持面向对象程序设计的思想,是真正面向对象的高级动态编程语言,完全支持面向对象的基本功能,如封装、继承、多态,以及对基类方法的覆盖或重写。Python 中对象的概念很广泛,一切内容都可以称为对象。创建类时用变量形式表示的对象属性称为数据成员或成员属性,用函数形式表示的对象行为称为成员方法,成员属性和成员方法统称为类的成员。

6.1 类的定义与使用

6.1.1 类定义语法

Python 使用 class 关键字来定义类,class 关键字之后是一个空格,然后是类的名字,再后是一个冒号,最后换行并定义类的内部实现。类名的首字母一般要大写,当然也可以按照自己的习惯定义类名,但是一般推荐参考惯例来命名,并在整个系统的设计和实现中保持风格一致,这一点对于团队合作尤其重要。例如:

```
class Car:                              #定义类
    def infor(self):                    #成员方法
        print("This is a car")
```

定义了类之后,可以用来实例化对象,并通过"对象名.成员"的方式来访问其中的数据成员或成员方法,例如下面的代码:

```
car = Car()
car.infor()
```

可以使用内置函数 isinstance()来测试一个对象是否为某个类的实例,返回 True 或 False,下面的代码演示了 isinstance()的用法。

```
isinstance(car, Car)
```

Python 关键字 pass 表示空语句,可以用在类和函数的定义中或者选择结构、循环结构中。当暂时没有确定如何实现功能,或者为以后的软件升级预留空间时,可以使用该关键字来"占位"。例如下面的代码都是合法的:

```
class A:
    pass
def demo():
    pass
if 5 > 3:
    pass
```

6.1.2 self 参数

类的所有实例方法都必须至少有一个名为 self 的参数,并且必须是方法的第一个形参(如果有多个形参),self 参数代表对象本身。在类的实例方法中访问实例属性时需要以 self 为前缀,但在外部通过对象名调用对象方法时并不需要传递这个参数,如果在外部通过类名调用对象方法则需要显式为 self 参数传递该类的一个对象,参考 6.2 节的讨论。

6.1.3 类成员与实例成员

这里主要指数据成员,或者广义上的属性。可以说属性有两种:一种是实例属性;另一种是类属性。实例属性一般在构造方法__init__()中定义,定义和使用时必须以 self 作为前缀;类属性是在类中所有方法之外定义的数据成员。在主程序中(或类的外部),实例属性属于实例(对象),只能通过对象名访问;类属性属于类,可以通过类名或对象名访问。

在 Python 中,可以动态地为类和对象增加成员,这一点是和很多面向对象程序设计语言不同的,也是 Python 动态类型特点的一种重要体现。

```
class Car:
    price = 100000                          #定义类属性
    def __init__(self, c):                  #构造方法
        self.color = c                      #定义实例属性

car1 = Car('Red')
car2 = Car('Blue')
print(car1.color, Car.price)
Car.price = 110000                          #修改类属性
Car.name = 'QQ'                             #增加类属性
car1.color = 'Yellow'                       #修改实例属性
print(car2.color, Car.price, Car.name)
print(car1.color, Car.price, Car.name)
def setSpeed(self, s):
    self.speed = s
import types
car1.setSpeed = types.MethodType(setSpeed, car1)   #动态为对象增加成员方法
car1.setSpeed(50)                                  #调用对象的成员方法
```

Python 类型的动态性使得我们可以动态地为自定义类及其对象增加新的属性和行为,俗称混入(mixin)机制,这在大型项目开发中会非常方便和实用。

例如,某系统中的所有用户分类非常复杂,不同用户组具有不同的行为和权限,并且可能会经常改变。这时可以独立地定义一些行为,然后根据需要来为不同的用户设置相应的行为能力。

```
>>> import types
>>> class Person(object):
    def __init__(self, name):
        assert isinstance(name, str), 'name must be string'
        self.name = name
>>> def sing(self):
    print(self.name+' can sing.')
>>> def walk(self):
    print(self.name+' can walk.')
>>> def eat(self):
    print(self.name+' can eat.')
>>> zhang = Person('zhang')
>>> zhang.sing()                                    #用户不具有该行为
AttributeError: 'Person' object has no attribute 'sing'
>>> zhang.sing = types.MethodType(sing, zhang)      #动态增加一个新行为
>>> zhang.sing()
zhang can sing.
>>> zhang.walk()
AttributeError: 'Person' object has no attribute 'walk'
>>> zhang.walk = types.MethodType(walk, zhang)
>>> zhang.walk()
zhang can walk.
>>> del zhang.walk                                  #删除用户行为
>>> zhang.walk()
AttributeError: 'Person' object has no attribute 'walk'
```

6.1.4 私有成员与公有成员

Python 没有对私有成员提供严格的访问保护机制。如果成员名以两个下画线开头但不以两个下画线结束，则表示是私有成员。私有成员在类的外部不能直接访问，需要通过调用对象的公有成员方法来访问，或者通过 Python 支持的特殊方式来访问。

6.1.4

私有成员是为了数据封装和保密而设的，一般只能在类的成员方法中使用，虽然 Python 支持一种特殊的方式从外部访问私有成员，但是并不推荐这样做。公有成员是可以公开使用的，既可以在类的内部进行访问，也可以在外部程序中使用。

```
>>> class A:
    def __init__(self, value1=0, value2=0):     #构造方法，见 6.4 节
        self.value1 = value1
        self.__value2 = value2                  #创建私有数据成员
    def setValue(self, value1, value2):
        self.value1 = value1
        self.__value2 = value2                  #在成员方法中访问私有数据成员
    def show(self):
        print(self.value1)
        print(self.__value2)
>>> a = A()
>>> a.value1
0
>>> a._A__value2                                #在外部访问对象的私有数据成员，不建议这样做
0
```

在 IDLE 或其他开发环境中,在对象或类名后面加上一个圆点".",稍等 1s 则会自动列出其所有公有成员,如图 6-1 所示,模块也具有同样的特点。如果在圆点"."后面再加一个下画线,则会列出该对象或类的所有成员,包括私有成员,如图 6-2 所示。

图 6-1　列出对象所有公有成员

图 6-2　列出对象所有成员

在 Python 中,带下画线的变量名和方法名有特殊的含义,尤其是在类的定义中。用下画线作为变量名和方法名前缀和后缀来表示类的特殊成员。

(1) _xxx:以单下画线开头的对象叫作保护成员,模块中这样的对象默认不能用 from module import *导入。

(2) __xxx__:前后各两个下画线表示系统预定义的特殊成员,不能随意定义和增加,只能修改或重新实现其功能。

(3) __xxx:以双下画线开头的类中的私有成员,只有该类对象自己能访问,子类对象也不能直接访问,但在对象外部可以通过"对象名._类名__xxx"的特殊形式来访问。Python 中不存在严格意义上的私有成员。

另外,交互模式中一个下画线"_"表示解释器中最后一次显示的内容或最后一个表达式正确执行的输出结果。例如:

```
>>> 3 + 5
8
>>> _ + 2
10
```

在程序中,下画线可以作为匿名变量使用,往往用于语法上需要一个变量但实际上并不需要关心该变量的值的场合。例如:

```
div, _ = divmod(2024, 20)                              #只关心整商,不关心余数
for _ in range(3):                                      #for 循环中不使用循环变量
    print('only a test')
data =[(1,2,3), (1,2,3,4,5,6), (1,2,3,4,5,6,7)]
for a, b, *_ in data:                                   #收集第 3 个以及后面的值,但不使用
    print(a, b)
```

6.2　方　　法

在类中定义的方法可以细分为 5 类:公有方法、私有方法、静态方法、类方法和抽象方法。公有方法通过对象名直接调用;私有方法不能通过对象名直接调用,只能在属于对象的方法中通过 self 调用或在外部通过 Python 支持的特殊方式来调用。如果通过类名来调用

属于对象的公有方法,需要显式为该方法的 self 参数传递一个对象名,用于明确指定访问哪个对象的数据成员。静态方法和类方法都可以通过类名和对象名调用,但不能直接访问属于对象的成员,只能访问属于类的成员。一般将 cls 作为类方法的第一个参数名,并且在调用类方法时不需要为该参数传递值。例如:

6.2

```
>>> class Root:
    __total = 0                          #属于类的私有数据成员
    def __init__(self, v):
        self.__value = v                 #属于对象的私有数据成员
        Root.__total += 1                #在对象方法中可以访问类的成员
    def show(self):
        print('self.__value:', self.__value)
        print('Root.__total:', Root.__total)
    @classmethod
    def classShowTotal(cls):             #类方法
        print(cls.__total)
    @staticmethod
    def staticShowTotal():               #静态方法
        print(Root.__total)
>>> r = Root(3)
>>> r.classShowTotal()                   #通过对象来调用类方法
1
>>> r.staticShowTotal()                  #通过对象来调用静态方法
1
>>> r.show()                             #通过对象来调用对象的公有成员方法
self.__value: 3
Root.__total: 1
>>> rr = Root(5)
>>> Root.classShowTotal()                #通过类调用类方法
2
>>> Root.staticShowTotal()               #通过类调用静态方法
2
>>> Root.show()                          #试图通过类直接调用实例方法,失败
TypeError: unbound method show() must be called with Root instance as first argument (got nothing instead)
>>> Root.show(r)                         #可以通过这种形式调用方法并访问实例成员
self.__value: 3
Root.__total: 2
>>> Root.show(rr)                        #通过类调用实例方法时为 self 参数显式传递对象
self.__value: 5
Root.__total: 2
```

抽象方法只能在抽象类中定义,抽象类不能实例化,这样做时会报错并提示"TypeError: Can't instantiate abstract class AbstractBase with abstract method...",必须创建抽象类的派生类并实现全部抽象方法之后才能对派生类进行实例化。

```
import abc

class AbstractBase(abc.ABC):             #创建抽象类
    def __init__(self, v):
        self.value = v
    def modify(self, v):
        self.value = v
    @abc.abstractmethod
```

```
    def show(self):                          #定义抽象方法,可以没有具体实现
        pass

class Child(AbstractBase):                   #继承抽象类,创建派生类
    def show(self):                          #实现抽象方法
        print(self.value)

c = Child(3)                                 #实例化,创建派生类的对象
c.modify(5)
c.show()
```

6.3 属　　性

属性综合了私有数据成员和公有成员方法的优点。在 Python 3.x 中,属性得到了完整的实现,支持全面的保护机制。例如下面的代码,如果设置属性为只读,则无法修改其值,也无法为对象增加与属性同名的新成员,同时,也无法删除对象属性。

```
>>> class Test:
    def __init__(self, value):
        self.__value = value
    @property                                #修饰器,定义属性
    def value(self):                         #只读属性,无法修改和删除
        return self.__value                  #返回私有数据成员的值
>>> t = Test(3)
>>> t.value
3
>>> t.value = 5                              #只读属性不允许修改值
AttributeError: can't set attribute
>>> del t.value                              #试图删除对象的只读属性,失败
AttributeError: can't delete attribute
>>> t.value
3
```

下面的代码把属性设置为可读、可修改,而不允许删除。

```
>>> class Test:
    def __init__(self, value):
        self.__value = value
    def __get(self):                         #访问属性时调用的方法
        return self.__value
    def __set(self, v):                      #修改属性时调用的方法
        self.__value = v
    value = property(__get, __set)           #修饰器,定义可读可写属性
    def show(self):                          #普通成员方法
        print(self.__value)
>>> t = Test(3)
>>> t.value                                  #允许读取属性值
3
>>> t.value = 5                              #允许修改属性值
>>> t.value
```

```
5
>>> t.show()                                    #属性对应的私有变量也得到了相应的修改
5
>>> del t.value                                 #试图删除属性,失败
AttributeError: can't delete attribute
```

也可以将属性设置为可读、可修改、可删除。

6.3

```
>>> class Test:
    def __init__(self, value):
        self.__value = value
    def __get(self):
        return self.__value
    def __set(self, v):
        self.__value = v
    def __del(self):                             #删除属性时调用的方法
        del self.__value                         #删除对应的私有数据成员
    value = property(__get, __set, __del)
    def show(self):
        print(self.__value)
>>> t = Test(3)
>>> t.value
3
>>> t.value = 5
>>> t.value
5
>>> del t.value                                  #删除对象属性及其对应的私有数据成员
>>> t.value
AttributeError: 'Test' object has no attribute '_Test__value'
>>> t.show()
AttributeError: 'Test' object has no attribute '_Test__value'
>>> t.value = 1                                  #为对象动态增加属性和对应的私有数据成员
>>> t.show()
1
>>> t.value
1
```

下面的代码定义了一个矩形类,支持设置矩形的宽度和高度,以及获取矩形的宽度、高度和面积,除了前面已经介绍过的定义属性的方式,还演示了另一种定义属性的方式。

```
class Rectangle:
    def __init__(self, w, h):                    #构造方法,名字是固定的
        self.width = w                           #调用属性的setter方法进行赋值
        self.height = h
    @property
    def width(self):
        return self.__width                      #读取属性的值时,自动调用这个方法
    @width.setter
    def width(self, w):                          #修改属性的值时,自动调用这个方法
```

```
            assert isinstance(w, (int,float)) and w>0, '矩形宽度必须大于0'
            self.__width = w
        @width.delete
        def width(self):                                        #删除属性时,自动调用这个方法
            del self.__width

        def __get_height(self):
            return self.__height
        def __set_height(self, h):
            assert isinstance(h, (int,float)) and h>0, '矩形宽度必须大于0'
            self.__height = h
        def __del_height(self):
            del self.__height
        #使用property()函数定义属性,分别设置读取、修改、删除时调用的方法
        height = property(__get_height, __set_height, __del_height)

        @property
        def area(self):                                         #定义只读属性
            return self.__width * self.__height

r1 = Rectangle(3, 5)
print(r1.area)
r2 = Rectangle(4, 6)
r2.width = 5
r2.height = 7
print(r2.area)
```

*6.4 特殊方法与运算符重载

6.4.1 常用特殊方法

Python 类有大量的特殊方法,其中比较常用的是构造方法和析构方法。不管类名是什么,Python 中类的构造方法都是 __init__(),一般用于为数据成员设置初值或进行其他必要的初始化工作,在创建对象时被自动调用和执行,可以通过为构造方法定义默认值参数来实现类似于其他语言中构造方法重载的目的。如果用户没有设计构造方法,Python 将提供一个默认的构造方法用来进行必要的初始化工作。Python 中类的析构方法是 __del__(),一般用于释放对象占用的资源,在 Python 删除对象时被自动调用和执行。如果用户没有编写析构方法,Python 将提供一个默认的析构方法进行必要的清理工作。

在 Python 中,除了构造方法和析构方法之外,还预定义了大量的特殊方法支持更多的功能,实现对运算符或内置函数的支持,例如,运算符重载就是通过在类中重写特殊方法来实现的。表 6-1 列出了其中一部分 Python 类特殊成员。

表 6-1 Python 类特殊成员(部分)

特 殊 成 员	功 能 说 明
__new__()	类的静态方法,可用于确定是否要创建对象
__init__()	构造方法,创建对象时自动调用
__del__()	析构方法,删除对象时自动调用

续表

特 殊 成 员	功 能 说 明
__add__()	+
__sub__()	-
__mul__()	*
__truediv__()	/
__floordiv__()	//
__mod__()	%
__pow__()	**
__eq__()、__ne__()、__lt__()、__le__()、__gt__()、__ge__()	==、!=、<、<=、>、>=
__lshift__()、__rshift__()	<<、>>
__and__()、__or__()、__invert__()、__xor__()	&、|、~、^
__iadd__()、__isub__()	+=、-=，很多其他运算符也有与之对应的复合赋值分隔符
__pos__()	一元运算符+，正号
__neg__()	一元运算符-，负号
__contains__()	与成员测试运算符 in 对应
__radd__()、__rsub__()	反射加法、反射减法，一般与普通加法和减法具有相同的功能，但操作数的位置或顺序相反，很多其他运算符也有与之对应的反射运算符
__abs__()	与内置函数 abs() 对应
__bool__()	与内置函数 bool() 对应，要求该方法必须返回 True 或 False
__bytes__()	与内置函数 bytes() 对应，要求该方法必须返回字节串
__complex__()	与内置函数 complex() 对应，要求该方法必须返回复数
__dir__()	与内置函数 dir() 对应
__divmod__()	与内置函数 divmod() 对应
__float__()	与内置函数 float() 对应，要求该方法必须返回实数
__hash__()	与内置函数 hash() 对应，要求该方法必须返回整数，同时实现了 __hash__() 和 __eq__() 的类的对象为可哈希对象，这两个方法设置为 None 时为不可哈希对象
__int__()	与内置函数 int() 对应，要求该方法必须返回整数
__iter__()	与内置函数 iter() 对应，实现了该方法的类的对象为可迭代对象
__len__()	与内置函数 len() 对应，要求该方法必须返回整数
__next__()	与内置函数 next() 对应，同时实现了该方法和 __iter__() 的类为迭代器
__reduce__()	提供对 reduce() 函数的支持
__reversed__()	与内置函数 reversed() 对应
__round__()	与内置函数 round() 对应

续表

特殊成员	功能说明
__str__()	与内置函数 str() 对应，要求该方法必须返回 str 类型的数据
__repr__()	与内置函数 repr() 对应，要求该方法必须返回 str 类型的数据
__getitem__()	按照索引获取值
__setitem__()	按照索引赋值
__delattr__()	删除对象的指定属性
__getattr__()	获取对象指定属性的值，对应成员访问运算符"."
__getattribute__()	获取对象指定属性的值，如果同时定义了该方法与__getattr__()，那么__getattr__()将不会被调用，除非在__getattribute__()中显式调用__getattr__()或者抛出 AttributeError 异常
__setattr__()	设置对象指定属性的值
__base__	该类的基类
__class__	返回对象所属的类
__dict__	对象所包含的属性与值的字典
__subclasses__()	返回该类的所有子类
__call__()	包含该特殊方法的类的实例为可调用对象
__get__() __set__() __delete__()	定义了这 3 个特殊方法中任何一个的类称作描述符，描述符对象一般作为其他类的属性来使用，这 3 个方法分别在获取属性、修改属性值或删除属性时被调用

6.4.2 案例精选

例 6-1　自定义一个数组类，支持数组与数字之间的四则运算，数组之间的加法运算、内积运算和大小比较，数组元素访问和修改，以及成员测试等功能。

例 6-1

```
class MyArray:
    def __isNumber(self, n):
        return isinstance(n, (int, float, complex))

    def __init__(self, *args):
        if not args:
            self.__value = []
        else:
            for arg in args:
                if not self.__isNumber(arg):
                    raise Exception('All elements must be numbers')
            self.__value = list(args)

    #重载运算符+，数组中每个元素都与数字 n 相加或两个数组相加，返回新数组
    def __add__(self, n):
```

```python
        if self.__isNumber(n):
            #数组中每个元素都与数字n相加
            b = MyArray()
            b.__value = [item+n for item in self.__value]
            return b
        elif isinstance(n, MyArray):
            #两个等长的数组对应元素相加
            if len(n.__value) == len(self.__value):
                c = MyArray()
                c.__value = list(map(lambda x,y: x+y, self.__value, n.__value))
                return c
            else:
                print('Length not equal')
        else:
            print('Not supported')

    #重载运算符-,数组中每个元素都与数字n相减,返回新数组
    def __sub__(self, n):
        if not self.__isNumber(n):
            raise Exception('not supported.')
        b = MyArray()
        b.__value = [item-n for item in self.__value]
        return b

    #重载运算符*,数组中每个元素都与数字n相乘,返回新数组
    def __mul__(self, n):
        if not self.__isNumber(n):
            raise Exception('not supported.')
        b = MyArray()
        b.__value = [item*n for item in self.__value]
        return b

    #重载运算符/,数组中每个元素都与数字n相除,返回新数组
    def __truediv__(self, n):
        if not self.__isNumber(n):
            raise Exception('not supported.')
        b = MyArray()
        b.__value = [item/n for item in self.__value]
        return b

    #重载运算符//,数组中每个元素都与数字n整除,返回新数组
    def __floordiv__(self, n):
        if not isinstance(n, int):
            raise Exception(n, ' is not an integer')
        b = MyArray()
        b.__value = [item//n for item in self.__value]
        return b

    #重载运算符%,数组中每个元素都与数字n求余数,返回新数组
    def __mod__(self, n):
        if not self.__isNumber(n):
            raise Exception('not supported.')
        b = MyArray()
```

```python
        b.__value = [item%n for item in self.__value]
        return b

    #重载运算符**,数组中每个元素都与数字 n 进行幂计算,返回新数组
    def __pow__(self, n):
        if not self.__isNumber(n):
            raise Exception('not supported.')
        b = MyArray()
        b.__value = [item**n for item in self.__value]
        return b

    def __len__(self):
        return len(self.__value)

    #直接使用该类对象作为表达式来查看对象的值
    def __repr__(self):
        return repr(self.__value)

    #支持使用 print()函数查看对象的值
    def __str__(self):
        return str(self.__value)

    #追加元素
    def append(self, v):
        assert self.__isNumber(v), 'Only number can be appended.'
        self.__value.append(v)

    #获取指定下标的元素值,支持使用列表或元组指定多个下标
    def __getitem__(self, index):
        length = len(self.__value)
        #如果指定单个整数作为下标,则直接返回元素值
        if isinstance(index, int) and 0<=index<length:
            return self.__value[index]
        #使用列表或元组指定多个整数下标
        elif isinstance(index, (list,tuple)):
            for i in index:
                if not (isinstance(i,int) and 0<=i<length):
                    return 'index error'
            result = []
            for item in index:
                result.append(self.__value[item])
            return result
        else:
            return 'index error'

    #修改元素值,支持使用列表或元组指定多个下标,同时修改多个元素值
    def __setitem__(self, index, value):
        length = len(self.__value)
        #如果下标合法,则直接修改元素值
        if isinstance(index, int) and 0<=index<length and self.__isNumber(value):
            self.__value[index] = value
        #支持使用列表或元组指定多个下标
        elif isinstance(index, (list,tuple)):
```

```python
            for i in index:
                if not (isinstance(i,int) and 0<=i<length):
                    raise Exception('index error')
            #如果下标和给的值都是列表或元组,并且个数一样
            #则分别为多个下标的元素修改值
            if isinstance(value, (list,tuple)):
                if len(index) == len(value):
                    for i, v in zip(index, value):
                        self.__value[i] = v
                else:
                    raise Exception('values and index must be the same length')
            #如果指定多个下标和一个普通数值,则把多个元素修改为相同的值
            elif isinstance(value, (int,float,complex)):
                for i in index:
                    self.__value[i] = value
            else:
                raise Exception('value error')
        else:
            raise Exception('index error')

    #支持成员测试运算符 in,测试数组中是否包含某个元素
    def __contains__(self, v):
        return v in self.__value

    #模拟向量内积
    def dot(self, v):
        if not isinstance(v, MyArray):
            raise Exception(v, ' must be an instance of MyArray.')
        if len(v) != len(self.__value):
            raise Exception('The size must be equal.')
        return sum([i*j for i,j in zip(self.__value, v.__value)])

    #重载运算符==,测试两个数组是否相等
    def __eq__(self, v):
        assert isinstance(v, MyArray), 'wrong type'
        return self.__value == v.__value

    #重载运算符<,比较两个数组大小
    def __lt__(self, v):
        assert isinstance(v, MyArray), 'wrong type'
        return self.__value < v.__value

if __name__ == '__main__':
    print('Please use me as a module.')
```

将上面的程序保存为 MyArray.py 文件放在当前目录中,可以作为 Python 模块导入并使用其中的数组类。

```
>>> from MyArray import MyArray
```

```
>>> x = MyArray(1, 2, 3, 4, 5, 6)
>>> y = MyArray(6, 5, 4, 3, 2, 1)
>>> len(x)
6
>>> x + 5
[6, 7, 8, 9, 10, 11]
>>> x * 3
[3, 6, 9, 12, 15, 18]
>>> x.dot(y)
56
>>> x.append(7)
>>> x
[1, 2, 3, 4, 5, 6, 7]
>>> x.dot(y)
Exception: The size must be equal.
>>> x[9] = 8
Exception: Index error
>>> x / 2
[0.5, 1.0, 1.5, 2.0, 2.5, 3.0, 3.5]
>>> x // 2
[0, 1, 1, 2, 2, 3, 3]
>>> x % 3
[1, 2, 0, 1, 2, 0, 1]
>>> x[2]
3
>>> 'a' in x
False
>>> 3 in x
True
>>> x < y
True
>>> x = MyArray(1, 2, 3, 4, 5, 6)
>>> x + y
[7, 7, 7, 7, 7, 7]
```

例 6-2 自定义支持关键字 with 的类。如果自定义类中实现了特殊方法 __enter__()和 __exit__(),那么该类的对象就可以像内置函数 open()返回的文件对象一样支持关键字 with 来实现资源的自动管理。

```python
class myOpen:
    def __init__(self, fileName, mode='r'):
        self.fp = open(fileName, mode)
    def __enter__(self):
        return self.fp
    def __exit__(self, exceptionType, exceptionVal, trace):
        self.fp.close()

with myOpen('test.txt') as fp:
    print(fp.read())
```

例 6-3 为自定义类实现关系运算符支持默认排序规则。

```python
from random import randrange, shuffle

class Country:
```

```python
    #构造方法,初始化对象
    def __init__(self, name, area):
        self.__setName(name)
        self.__setArea(area)
    #检查并设置国家名称
    def __setName(self, name):
        assert isinstance(name, str), '国家名称必须是字符串'
        self.__name = name
    def __setArea(self, area):
        assert isinstance(area, int), '面积必须是整数'
        self.__area = area
    #返回国家名称
    def getName(self):
        return self.__name
    def getArea(self):
        return self.__area
    #支持<运算符
    def __lt__(self, otherCountry):
        return self.__area < otherCountry.__area
    def __str__(self):
        return str((self.__name, self.__area))

#创建国家对象并添加至列表
countries = []
countryNames = list('abcdefghij')
shuffle(countryNames)
for name in countryNames:
    country = Country(name, randrange(10**2, 10**5))
    countries.append(country)

#输出原始数据,不做任何排序
print('原始数据'.center(20, '='))
for country in countries:
    print(country)
#使用内置函数 sorted()排序,默认调用对象的__lt__()方法,按面积排序
print('默认排序'.center(20, '='))
for country in sorted(countries):
    print(country)
#按国家名字进行排序,自定义排序规则
print('按国家名字排序'.center(20, '='))
for country in sorted(countries, key=lambda c:c.getName()):
    print(country)
#按国家面积进行排序,自定义排序规则
print('按国家面积排序'.center(20, '='))
for country in sorted(countries, key=lambda c:c.getArea()):
    print(country)
```

6.5 继 承

继承是为代码复用和设计复用而设计的,是面向对象程序设计的重要特性之一。当设计一个新类时,如果可以继承一个已有的设计良好的类然后进行二次开发,无疑会大幅减少

开发工作量。在继承关系中,已有的、设计好的类称为父类或基类,新设计的类称为子类或派生类。派生类可以继承父类的公有成员,但是不能继承其私有成员。如果需要在派生类中调用基类的方法,可以使用内置函数 super() 或者通过"基类名.方法名()"的方式来实现。

例 6-4　在派生类中调用基类方法。首先设计 Person 类,然后以 Person 为基类派生 Teacher 类,分别创建 Person 类和 Teacher 类的对象,并在派生类对象中调用基类方法。

例 6-4

```python
class Person(object):
    def __init__(self, name='', age=20, sex='man'):
        self.setName(name)
        self.setAge(age)
        self.setSex(sex)

    def setName(self, name):
        if not isinstance(name, str):
            print('name must be string.')
            return
        self.__name = name

    def setAge(self, age):
        if not isinstance(age, int):
            print('age must be integer.')
            return
        self.__age = age

    def setSex(self, sex):
        if sex not in ('man','woman'):
            print('sex must be "man" or "woman"')
            return
        self.__sex = sex

    def show(self):
        print('Name:', self.__name)
        print('Age:', self.__age)
        print('Sex:', self.__sex)

class Teacher(Person):                                  #派生类
    def __init__(self, name='', age=30, sex='man', department='Computer'):
        super(Teacher, self).__init__(name, age, sex)
        ##or, use another method like below:
        #Person.__init__(self, name, age, sex)
        self.setDepartment(department)

    def setDepartment(self, department):
        if not isinstance(department, str):
            print('department must be a string.')
            return
        self.__department = department

    def show(self):
        super(Teacher, self).show()
```

```
            print('Department:', self.__department)
if __name__ == '__main__':
    zhangsan = Person('Zhang San', 19, 'man')
    zhangsan.show()
    lisi = Teacher('Li Si',32, 'man', 'Math')
    lisi.show()
    lisi.setAge(40)
    lisi.show()
```

为了更好地理解 Python 类的继承机制,再来看下面的代码,并认真体会构造方法、私有方法和普通公开方法的继承原理。

继承

```
>>> class A(object):
    def __init__(self):
        self.__private()
        self.public()
    def __private(self):
        print('__private() method in A')
    def public(self):
        print('public() method in A')
>>> class B(A):                                 #注意,类 B 没有定义构造方法
    def __private(self):
        print('__private() method in B')
    def public(self):
        print('public() method in B')
>>> b = B()
__private() method in A
public() method in B
>>> dir(b)
['_A__private', '_B__private', '__class__',…]
>>> class C(A):
    def __init__(self):                         #显式定义构造方法,覆盖了 A 类中的构造方法
        self.__private()
        self.public()
    def __private(self):
        print('__private() method in C')
    def public(self):
        print('public() method in C')
>>> c = C()
__private() method in C
public() method in C
>>> dir(c)
['_A__private', '_C__private', '__class__',…]
```

在 Python 3.x 的多继承树中,如果在中间层某类有向上一层解析的迹象,则会先把本层右侧的其他类方法解析完,然后从本层最后一个解析的类方法中直接进入上一层并继续解析,也就是在从子类到超类的反向树中按广度优先解析。

如果在解析过程中,不再有向基类方向上一层解析的迹象,则同一层中右侧其他类方法不再解析。例如下面的代码(解析过程示意图见配套 PPT):

```python
class BaseClass(object):
    def show(self):
        print('BaseClass')

class SubClassA(BaseClass):
    def show(self):
        print('Enter SubClassA')
        super().show()
        print('Exit SubClassA')

class SubClassB(BaseClass):
    def show(self):
        print('Enter SubClassB')
        super().show()
        print('Exit SubClassB')

class SubClassC(BaseClass):
    def show(self):
        print('Enter SubClassC')
        super().show()
        print('Exit SubClassC')

class SubClassD(SubClassA, SubClassB, SubClassC):
    def show(self):
        print('Enter SubClassD')
        super().show()
        print('Exit SubClassD')

d = SubClassD()
d.show()
print(SubClassD.mro())
```

运行结果如下，本书配套 PPT 和微信公众号"Python 小屋"提供了更多 MRO(Method Resolution Order,方法解析顺序)示例。

```
Enter SubClassD
Enter SubClassA
Enter SubClassB
Enter SubClassC
BaseClass
Exit SubClassC
Exit SubClassB
Exit SubClassA
Exit SubClassD
[<class '__main__.SubClassD'>, <class '__main__.SubClassA'>, <class '__main__.SubClassB'>, <class '__main__.SubClassC'>, <class '__main__.BaseClass'>, <class 'object'>]
```

创建派生类时,会自动调用基类的特殊方法__init_subclass__(),基类也是通过这个特殊方法知道自己被继承的。

```python
class BaseClass:
    __total = 0
    def __init_subclass__(self):
```

```
            BaseClass.__total += 1
            print(f'这是我被第{BaseClass.__total}次继承!')

class SubClassA(BaseClass):
    pass

class SubClassB(BaseClass):
    pass
```

运行结果：

```
这是我被第 1 次继承!
这是我被第 2 次继承!
```

6.6 多　　态

多态（polymorphism）是指基类的同一个方法在不同派生类对象中具有不同的表现和行为。派生类继承了基类行为和属性之后，还会增加某些特定的行为和属性，同时还可能会对继承来的某些行为进行一定的改变，这都是多态的表现形式。

Python 大多数运算符可以作用于多种不同类型的操作数，并且对于不同类型的操作数往往有不同的表现，这本身就是多态，是通过特殊方法与运算符重载实现的。

```python
class Animal(object):                              #定义基类
    def show(self):
        print('I am an animal.')
class Cat(Animal):                                 #派生类,覆盖了基类的show()方法
    def show(self):
        print('I am a cat.')
class Dog(Animal):                                 #派生类
    def show(self):
        print('I am a dog.')
class Tiger(Animal):                               #派生类
    def show(self):
        print('I am a tiger.')
class Test(Animal):                                #派生类,没有覆盖基类的show()方法
    pass
x =[item() for item in (Animal, Cat, Dog, Tiger, Test)]
for item in x:                                     #遍历基类和派生类对象并调用show()方法
    item.show()
```

运行结果：

```
I am an animal.
I am a cat.
I am a dog.
I am a tiger.
I am an animal.
```

本 章 小 结

（1）面向对象程序设计的思想主要针对大型软件设计而提出，使得软件设计更加灵活，能够很好地支持代码复用和设计复用，并且使代码具有更好的可读性和可扩展性。

（2）定义类时使用关键字 class。

（3）可以动态地为类和对象增加成员。

（4）类中所有实例方法都至少包含一个 self 参数，并且必须是第一个参数，用来表示对象本身，通过对象名调用实例方法时不需要为 self 参数传递任何值。

（5）实例属性一般是指在构造方法 __init__() 中定义的，定义和访问时以 self 作为前缀；类属性是在类中所有方法之外定义的数据成员。

（6）如果通过类来调用属于对象的公有方法，需要显式为该方法的 self 参数传递一个对象，用来明确指定访问哪个对象的数据成员。

（7）在 Python 中，运算符重载是通过重新实现一些特殊方法来实现的。

（8）Python 支持多继承，如果多个父类中有相同名字的成员且在子类中使用该成员时没有指定其所属父类名，则 Python 解释器将从左向右按顺序进行搜索。

（9）在 Python 中，以下画线开头的变量名有特殊的含义，尤其是在类的定义中。

（10）在 IDLE 中，单个下画线表示上次语句正常执行的输出结果，在程序中表示匿名变量。

（11）Python 支持封装、继承、多态等面向对象程序设计核心特征。

习 题

1. 面向对象程序设计的三要素分别为_____、_____和_____。

2. 与运算符"**"对应的特殊方法名为_____，与运算符"//"对应的特殊方法名为_____。

3. 假设 a 为类 A 的对象且包含一个私有数据成员 __value，那么在类的外部通过对象 a 直接将其私有数据成员 __value 的值设置为 3 的语句可以写作_____。

4. 继承例 6-4 中的 Person 类生成 Student 类，编写新的方法用来设置学生专业，然后生成该类对象并显示信息。

5. 设计一个三维向量类，并实现向量的加法、减法以及向量与标量的乘法和除法运算。

6. 简单解释 Python 中以下画线开头的成员名的特点和含义。

7. 编写自定义类，模拟内置集合类。

8. 编写自定义类，模拟双端队列。

9. 编写自定义类，模拟稀疏矩阵。

10. 编写自定义类，模拟单链接表。

11. 编写自定义类，模拟双链表。

第 7 章 文 件 操 作

为了长期保存数据以便重复使用、修改和共享,必须将数据以文件的形式存储到外部存储介质(如磁盘、U 盘、光盘等)或云盘中。管理信息系统大多使用数据库来存储数据,而数据库最终还是要以文件的形式存储,应用程序的配置信息往往也是使用文件来存储的,图形、图像、音频、视频、可执行文件等也都是以文件的形式存储的。因此,文件操作在各类应用软件的开发中均占有重要的地位。

按数据的组织形式和解释方式可以把文件分为文本文件和二进制文件两大类。

1. 文本文件

文本文件内容是常规字符串(实际上在计算机中也是存储二进制数),由若干文本行组成,每行以换行符'\n'结尾。常规字符串是指记事本或其他文本编辑器能正常显示、编辑并且人类能够直接阅读和理解的字符串,如英文字母、汉字、数字字符串。文本文件可以使用gedit、记事本直接进行编辑,并自动完成字符串和字节串之间的编码与解码。

2. 二进制文件

二进制文件把对象内容以字节串(bytes)进行存储,无法用记事本或其他普通文本处理软件直接进行编辑,不能解码为字符串,通常也无法被人类直接阅读和理解,需要使用专门的软件进行读取、显示、修改或执行。常见的如图形图像文件、音视频文件、可执行文件、资源文件、各种数据库文件、各类 Office 文档等都属于二进制文件。

7.1 文 件 对 象

无论是文本文件还是二进制文件,其操作流程基本都是一致的,即首先打开文件并创建文件对象,然后通过该文件对象对文件内容进行读取、写入、删除、修改等操作,最后关闭并保存文件内容。Python 内置了文件对象,通过 open() 函数即可以指定模式打开指定文件并创建文件对象,语法为

7.1

```
open(file, mode='r', buffering=-1, encoding=None, errors=None, newline=None,
     closefd=True, opener=None)
```

其中,file 指定了要打开的文件路径,如果要打开的文件不在当前目录中,需要指定完整路径,为了减少完整路径中"\"符号的输入,可以使用原始字符串;mode(见表 7-1)指定了打开文件后的处理方式;buffering 指定了读写文件的缓存模式,数值 0 表示不缓存,数值 1 表示缓存,如大于 1 则表示缓冲区的大小,默认值-1 表示系统管理缓存;encoding 指定文本文件的编码格式,Windows 系统中该参数默认值为'GBk',以二进制模式打开文件时不能使用该参数。如果执行正常,open() 函数返回一个文件对象,通过该文件对象可以对文件内容进行各种操作,如果指定文件不存在、访问权限不够、磁盘空间不够或其他原因导致创建文件对象失败则抛出异常。例如,下面的代码分别以读、写方式打开了两个文件并创建了与之对应的文件对象。

```
f1 = open('file1.txt', 'r')
f2 = open('file2.txt', 'w')
```

当对文件内容操作完以后,一定要关闭文件,以保证所做的任何修改都得到保存。

```
f1.close()
```

一般更推荐使用 with 关键字管理文件对象,(见 8.4.2 节和本章例题)。文件对象常用属性如表 7-2 所示,文件对象常用方法如表 7-3 所示。

表 7-1 文件打开模式

模式	说明
r	读模式
w	写模式,文件存在时会清空原内容
a	追加模式
b	二进制模式(可与其他模式组合使用)
+	读写模式(可与其他模式组合使用)

表 7-2 文件对象常用属性

属性	说明
closed	判断文件是否关闭,若文件已被关闭,则返回 True
mode	返回文件的打开模式
name	返回文件的名称

表 7-3 文件对象常用方法

方法	功能说明
flush()	把缓冲区的内容写入文件,但不关闭文件
close()	把缓冲区的内容写入文件,同时关闭文件
read([size])	从文件中读取 size 字节(二进制模式)或字符(文本模式)的内容作为结果返回,如果省略 size,则表示一次性读取所有内容
readline()	从文本模式打开的文本文件中读取一行内容作为结果字符串返回
readlines()	把文本文件中的每行文本作为一个字符串存入列表中,返回该列表
seek(offset[,whence])	把文件指针移动到新的位置,offset 表示相对于 whence 的偏移量,单位是字节。whence 为 0 表示从文件头开始计算,1 表示从当前位置开始计算,2 表示从文件尾开始计算,默认为 0
tell()	返回文件指针的当前位置,单位是字节
truncate([size])	删除从当前指针位置到文件末尾的内容。如果指定了 size,则不论指针在什么位置都只留下前 size 字节,其余的删除
write(s)	把 s 的内容写入文件
writelines(s)	把列表 s 中的字符串逐个写入文本文件,不添加换行符

7.2　文本文件内容操作案例精选

在本节中,主要通过几个例题来演示文本文件内容的读写操作。对于 read()、write() 以及其他读写方法,每次都是从当前位置开始读写,并且当读写操作完成之后都会自动移动文件指针。如果需要对文件指针进行定位,可以使用 seek()方法;如果需要获知文件指针当前位置,可以使用 tell()方法。

例 7-1 向文本文件中写入内容。

```
s = '文本文件的读取方法\n 文本文件的写入方法\n'
with open('sample.txt', 'a+') as fp:              #可根据需要修改 mode 参数
    fp.write(s)
```

使用上下文管理关键字 with 可以自动管理资源,不论何种原因跳出 with 块,总能保证文件被正确关闭,并且可以在代码块执行完毕后自动还原进入该代码块时的现场。

例 7-2 读取并显示文本文件的前 5 个字符。

```
with open('sample.txt', encoding='utf8') as fp:   #需要根据实际情况修改编码格式
    print(fp.read(5))                             #从当前位置开始读
```

例 7-3 读取并显示文本文件所有行。

```
with open('sample.txt', encoding='utf8') as fp:
    for line in fp:
        print(line)
```

例 7-3

例 7-4 移动文件指针,然后读取并显示文本文件中的内容。

seek()方法将文件指针定位到文件中指定字节的位置。读取时遇到无法解码的字符会抛出异常。

例 7-4

```
>>>s = '微信公众号: Python 小屋'
>>>fp = open('sample.txt', 'w')          #默认使用 GBK 编码格式
>>>fp.write(s)                            #写入字符串,返回写入字符的数量
14
>>>fp.close()                             #保存并关闭文件
>>>fp = open('sample.txt')                #默认以读模式打开文件
>>>fp.read(3)                             #读取 3 个字符
'微信公'
>>>fp.read(1)                             #继续读取 1 个字符
'众'
>>>fp.seek(14)                            #把文件指针定位到第 14 字节
                                          #GBK 编码格式中每个汉字占 2 字节
14
>>>fp.read(1)
't'
>>>fp.seek(3)                             #这次定位不是一个汉字编码开始的位置
3
>>>fp.read()                              #无法解码,抛出异常
UnicodeDecodeError: ' gbk ' codec can ' t decode byte 0xab in position 2: illegal multibyte sequence
>>>fp.close()
>>>with open('sample.txt', 'w', encoding='utf8') as fp:
    fp.write(s)
14
>>>fp = open('sample.txt', encoding= 'utf8')  #使用 UTF-8 编码格式
>>>fp.read(3)                                  #读取前 3 个字符
```

```
'微信公'
>>> fp.read(1)                                          #继续读取 1 个字符
'众'
>>> fp.seek(6)                                          #定位到第 6 字节
                                                        #UTF-8 编码格式中每个汉字占 3 字节
6
>>> fp.read(1)                                          #从当前位置开始读取 1 个字符
'公'
>>> fp.seek(2)                                          #这次定位不是一个汉字编码开始的位置
2
>>> fp.read()                                           #无法解码,抛出异常
UnicodeDecodeError: 'utf-8' codec can't decode byte 0xae in position 0:invalid start byte
```

例 7-5 读取文本文件 data.txt(每行一个整数)中所有整数,将其按升序排序后再写入文本文件 data_asc.txt 中。

例 7-5

```
with open('data.txt', 'r') as fp:
    data = fp.readlines()
data = [int(line.strip()) for line in data]
data.sort()
data = [str(i)+'\n' for i in data]
with open('data_asc.txt', 'w') as fp:
    fp.writelines(data)
```

例 7-6 编写程序,保存为 demo6.py,运行后生成文件 demo6_new.py,其中的内容与 demo6.py 一致,但是在每行的行尾加上了行号。

例 7-6

```
filename = 'demo6.py'
with open(filename, 'r') as fp:
    lines = fp.readlines()
maxLength = len(max(lines, key=len))
lines = [line.rstrip().ljust(maxLength)+'#' +str(index)+'\n'
        for index, line in enumerate(lines)]
with open(filename[:-3]+'_new.py', 'w') as fp:
    fp.writelines(lines)
```

例 7-7 批量修改文本文件编码格式为 UTF-8。

例 7-7

```
from os import listdir
from chardet import detect                              #扩展库,需要先安装

fns = (fn for fn in listdir() if fn.endswith('.txt'))   #当前目录中所有 TXT 文件
for fn in fns:
    with open(fn, 'rb+') as fp:                         #可读写的二进制模式
        content = fp.read()
        encoding = detect(content)['encoding']          #判断编码格式
        content = content.decode(encoding).encode('utf8')  #格式转换
        fp.seek(0)                                      #写回文件
        fp.write(content)
```

7.3 二进制文件操作案例精选

数据库文件、图像文件、可执行文件、音视频文件、Office文档等均属于二进制文件。二进制文件不能使用记事本或其他文本编辑软件正常读写,也无法通过Python的文件对象直接读取和理解二进制文件的内容。必须正确理解二进制文件结构和序列化规则,才能准确地理解其中内容并且设计正确的反序列化规则。所谓序列化,简单地说就是把Python对象在不丢失其类型信息的情况下转成二进制形式的过程,对象序列化后的字节串经过正确的反序列化过程应该能够准确无误地得到原来的对象。

Python中常用的序列化模块有pickle、struct、json、marshal和shelve。本节主要介绍pickle和struct模块在对象序列化和二进制文件操作方面的应用,其他模块请参考有关文档或配套PPT。

7.3.1 使用pickle模块

7.3.1

pickle是较为常用并且速度非常快的二进制文件序列化模块,下面通过两个示例来了解一下如何使用pickle模块进行对象序列化和二进制文件读写。

例7-8 使用pickle模块序列化不同类型的Python对象并写入二进制文件。

```
import pickle

i = 9999**99
a = 3.1415926
s = 'Python 小屋'
lst = [[1, 2, 3], [4, 5, 6], [7, 8, 9]]
tu = (-5, 10, 8)
coll = {4, 5, 6}
dic = {'a':'apple', 'b':'banana', 'g':'grape', 'o':'orange'}
data = [i, a, s, lst, tu, coll, dic]

with open('sample_pickle.dat', 'wb') as fp:
    try:
        pickle.dump(len(data), fp)              #表示后面将要写入的数据个数
        for item in data:
            pickle.dump(item, fp)
    except:
        print('写文件异常!')                     #如果写文件异常则跳到此处执行
```

例7-9 读取例7-8中写入二进制文件的内容并还原为Python对象。

```
import pickle

with open('sample_pickle.dat', 'rb') as fp:
    n = pickle.load(fp)                         #读出文件中的数据个数
    for i in range(n):
        x = pickle.load(fp)                     #读取并反序列化一个对象
        print(x)
```

7.3.2 使用 struct 模块

struct 也是比较常用的对象序列化和二进制文件读写模块，下面通过两个示例来简单介绍使用 struct 模块对二进制文件进行读写的用法，在网络编程领域的应用见《Python 网络程序设计》（微课版）（董付国，清华大学出版社）。

例 7-10 使用 struct 模块写入二进制文件。

```
import struct

n, x, b, s = 1300000000, 3.14, True, 'abc测试'
sn = struct.pack('if?', n, x, b)            #序列化，得到字节串
with open('sample_struct.dat', 'wb') as fp:
    fp.write(sn)                            #写入字节串
    fp.write(s.encode())                    #字符串直接编码为字节串写入
```

例 7-11 使用 struct 模块读取例 7-10 写入二进制文件的内容。

```
import struct

with open('sample_struct.dat', 'rb') as fp:
    sn = fp.read(9)                         #整数、实数各占4字节，逻辑值占1字节
    tu = struct.unpack('if?', sn)           #struct 模块不能处理太大的数字
    n, x, b1 = tu
    print('n=', n, 'x=', x, 'b1=', b1)
    s = fp.read(9).decode()                 #UTF-8编码，英文字符占1字节，汉字占3字节
    print('s=', s)
```

7.4 文件级操作

7.4.1 os 与 os.path 模块

os 模块除了提供使用操作系统功能之外，还提供了大量文件级操作的函数，如表 7-4 所示。os.path 模块提供了大量用于路径判断、切分、连接及文件属性查看的函数，如表 7-5 所示。

表 7-4 os 模块常用文件级操作的函数

函 数	功 能 说 明
chmod(path,mode, * ,dir_fd=None, follow_symlinks=True)	改变文件的访问权限
remove(path)	删除指定的文件，要求文件不能有只读、隐藏等属性
rename(src,dst)	重命名文件或目录
stat(path)	返回文件的所有属性
listdir(path)	返回 path 目录下的文件和目录列表，只包含直接子节点
startfile(filepath [,operation])	使用关联的应用程序打开指定文件或 URL

表 7-5 os.path 模块常用文件级操作的函数

函　　数	功　能　说　明
basename(p)	返回路径中最后一个分隔符后面的部分
dirname(p)	返回路径中最后一个分隔符前面的部分
exists(path)	判断路径是否存在
getatime(filename)	返回文件的最后访问时间对应的纪元秒数
getctime(filename)	返回文件的创建时间，纪元秒数
getmtime(filename)	返回文件的最后修改时间，纪元秒数
getsize(filename)	返回文件的大小，单位是字节
isabs(path)	判断 path 是否为绝对路径
isdir(path)	判断 path 是否为目录
isfile(path)	判断 path 是否为文件
join(path, *paths)	连接两个或多个 path
split(path)	对路径进行分割，返回元组
splitext(path)	从路径中分割文件的扩展名，返回元组
splitdrive(path)	从路径中分割驱动器的名称，返回元组

下面的代码可以列出当前目录下所有扩展名为 pyc 的文件，其中用到了列表推导式，可以查阅前面的 2.1.9 节了解相关知识。

```
>>> import os
>>> print([fname for fname in os.listdir()
          if os.path.isfile(fname) and fname.endswith('.pyc')])
```

下面的代码将当前目录中所有扩展名为 html 的文件重命名为扩展名为 htm 的文件。

```
import os

file_list =[filename for filename in os.listdir() if filename.endswith('.html')]
for filename in file_list:
    newname = filename[:-4] + 'htm'
    os.rename(filename, newname)
    print(filename+"更名为: "+newname)
```

7.4.2 shutil 模块

shutil 模块也提供了大量的函数支持文件和文件夹操作，详细的函数列表可以使用 dir(shutil)查看。

```
>>> import shutil
>>> dir(shutil)
```

例如，下面的代码使用该模块的 copyfile()函数复制文件。

```
>>> import shutil
>>> shutil.copyfile('C:\\dir.txt', 'C:\\dir1.txt')
```

下面的代码将 C:\Python312\Dlls 文件夹以及该文件夹中所有文件压缩至 D:\a.zip 文件。

```
>>> shutil.make_archive('D:\\a', 'zip', 'C:\\Python312', 'Dlls')
'D:\\a.zip'
```

下面的代码将刚压缩得到的文件 D:\a.zip 解压缩至 D:\a_unpack 文件夹。

```
>>> shutil.unpack_archive('D:\\a.zip', 'D:\\a_unpack')
```

下面的代码删除刚刚解压缩得到的文件夹。

```
>>> shutil.rmtree('D:\\a_unpack')
```

7.5 目录操作

除了支持文件操作，os 和 os.path 模块还提供了大量的目录操作函数，os 模块常用目录操作函数与成员如表 7-6 所示，可以通过 dir(os.path) 查看 os.path 模块更多关于目录操作的函数。

表 7-6 os 模块常用目录操作函数与成员

成 员	功 能 说 明
mkdir(path[, mode=0o777])	创建目录
makedirs(path1/path2…, mode=511)	创建多级目录，十进制 511 与八进制 0o777 相等
rmdir(path)	删除目录
removedirs(path1/path2…)	删除多级目录
listdir(path)	返回指定目录下的文件和目录信息
getcwd()	返回当前工作目录
get_exec_path()	返回可执行文件的搜索路径
chdir(path)	把 path 设为当前工作目录
walk(top, topdown=True, onerror=None)	遍历目录树，该函数返回一个元组，包括 3 个元素：所有路径名、所有目录列表与文件列表
sep	当前操作系统所使用的路径分隔符
extsep	当前操作系统所使用的文件扩展名分隔符

下面的代码演示了如何使用 os 模块的函数来查看、改变当前工作目录，以及创建与删除目录，请自行运行代码并查看结果。

```
>>> import os
```

```
>>> os.getcwd()                                 #返回当前工作目录
>>> os.mkdir(os.getcwd()+'\\temp')              #创建目录
>>> os.chdir(os.getcwd()+'\\temp')              #改变当前工作目录
>>> os.getcwd()
>>> os.mkdir(os.getcwd()+'\\test')
>>> os.listdir('.')
>>> os.rmdir('test')                            #删除目录
```

遍历指定目录下所有子目录和文件时,可以使用递归的方法进行深度优先遍历,例如:

```
import os

def visitDir(path):
    if not os.path.isdir(path):
        print('Error:', path,' is not a directory or does not exist.')
        return
    for lists in os.listdir(path):
        sub_path = os.path.join(path, lists)
        print(sub_path)
        if os.path.isdir(sub_path):
            visitDir(sub_path)

visitDir('E:\\test')
```

或者使用广度优先的目录树遍历,例如:

```
from os import listdir
from os.path import join, isfile, isdir

def listDirWidthFirst(directory):
    dirs = [directory]
    #如果还有没遍历过的文件夹,继续循环
    while dirs:
        #遍历还没遍历过的第一项
        current = dirs.pop(0)
        #遍历该文件夹,如果是文件就直接输出显示
        #如果是文件夹,输出显示后,标记为待遍历项,放入列表 dirs 尾部
        for subPath in listdir(current):
            path = join(current, subPath)
            if isfile(path):
                print(path)
            elif isdir(path):
                print(path)
                dirs.append(path)
```

下面的代码使用 os 模块的 walk() 方法指定目录的遍历。

```
import os

def visitDir2(path):
    if not os.path.isdir(path):
        print('Error:', path,' is not a directory or does not exist.')
        return
    list_dirs = os.walk(path)
```

```
        for root, dirs, files in list_dirs:     #遍历该元组的目录和文件信息
            for d in dirs:
                print(os.path.join(root, d))     #获取完整路径
            for f in files:
                print(os.path.join(root, f))     #获取文件绝对路径

visitDir2('h:\\music')
```

7.6 案 例 精 选

例 7-12 计算 CRC32 值。下面的代码分别使用 zlib 和 binascii 模块的方法计算任意字符串的 CRC32 值,该代码经过简单修改,即可用于计算文件的 CRC32 值。

```
>>> import zlib
>>> print(zlib.crc32(b'1234'))
2165402659
>>> print(zlib.crc32(b'111'))
1298878781
>>> print(zlib.crc32('Python 小屋'.encode()))
3279126710
>>> import binascii
>>> binascii.crc32('Python 小屋'.encode())
3279126710
```

例 7-13 计算文本文件中最长行的长度。

```
with open('sample.txt', encoding='gbk') as fp:     #需要根据实际情况修改编码格式
    print(max((len(line.strip()) for line in fp)))     #可以使用readlines()方法改写
```

例 7-14 计算字符串 MD5 值。MD5 值可以用于判断文件发布之后是否被篡改,对于文件完整性保护具有重要意义,也经常用于数字签名。

```
>>>import hashlib
>>>hashlib.md5('12345'.encode()).hexdigest()
'827ccb0eea8a706c4c34a16891f84e7b'
>>>hashlib.md5('123456'.encode()).hexdigest()
'e10adc3949ba59abbe56e057f20f883e'
```

对上面的代码稍加完善,即可实现自己的 MD5 计算器,例如:

```
import os
import sys
import hashlib

fileName = sys.argv[1]
if os.path.isfile(fileName):
    with open(fileName, 'rb') as fp:
        data = fp.read()
        print(hashlib.md5(data).hexdigest())
```

将上面的代码保存为文件 CheckMD5OfFile.py,然后计算指定文件的 MD5 值,对该文件做微小修改后再次计算其 MD5 值。可以发现,即使只修改了一点点内容,MD5 值的变化

也是非常大的，如图 7-1 所示。

图 7-1 计算文件的 MD5 值

另外，也可以使用 ssdeep 工具来计算文件的模糊哈希值或分段哈希值，或者编写 Python 程序调用 ssdeep 提供的 API 函数来计算文件的模糊哈希值，模糊哈希值可以用来比较两个文件的相似百分比。

例 7-14

```
>>> from ssdeep import ssdeep
>>> s = ssdeep()
>>> print(s.hash_file(filename))
```

对于某些恶意软件来说，可能会对自身进行加壳或加密，真正运行时再脱壳或解密，这样会使磁盘文件的哈希值与内存中脱壳或解密后进程的哈希值相差很大。因此，根据磁盘文件和其相应的进程之间模糊哈希值的相似度可以判断该文件是否包含自修改代码，并以此来判断其为恶意软件的可能性。

例 7-15 判断一个文件是否为 GIF 图像文件。任何一种文件都具有专门的文件头结构，在文件头中存放了大量的信息，其中就包括该文件的类型。通过文件头信息判断文件类型的方法更加准确，不依赖于文件扩展名。

例 7-15

```
>>> def is_gif(fname):
        with open(fname, 'rb') as fp:
            first4 = fp.read(4)
        return first4 == b'GIF8'
>>> is_gif('a.gif')
True
>>> is_gif('a.png')
False
```

例 7-16 把指定文件夹中的所有文件名批量随机化，保持文件类型不变。

例 7-16

```
from string import ascii_letters
from os import listdir, rename
from os.path import splitext, join
from random import choices, randint

def randomFilename(directory):
    for fn in listdir(directory):
        #切分，得到文件名和扩展名
        name, ext = splitext(fn)
```

```
            n = randint(5, 20)
            #生成随机字符串作为新文件名
            newName = ''.join(choices(ascii_letters,k=n))
            #修改文件名
            rename(join(directory, fn), join(directory, newName+ext))
randomFilename('C:\\test')
```

例 7-17 使用 xlwt 模块写入 xls 格式的 Excel 文件，该模块可以使用 pip 来安装。

```
from xlwt import *

book = Workbook()
sheet1 = book.add_sheet("First")
al = Alignment()
al.horz = Alignment.HORZ_CENTER        #对齐方式
al.vert = Alignment.VERT_CENTER
borders = Borders()
borders.bottom = Borders.THICK         #边框样式
style = XFStyle()
style.alignment = al
style.borders = borders
row0 = sheet1.row(0)
row0.write(0, 'test', style=style)
book.save(r'D:\test.xls')
```

例 7-18 使用 xlrd 模块读取 xls 格式的 Excel 文件，xlrd 模块需要单独安装。

```
import xlrd

book = xlrd.open_workbook(r'D:\test.xls')
sheet1 = book.sheet_by_name('First')
row0 = sheet1.row(0)
print(row0[0].value)
```

例 7-19 使用 Pywin32 操作 Excel 文件。Pywin32 模块需要单独安装，这是一个功能非常强大的模块，提供了 Windows 底层 API 函数的封装，使得可以在 Python 中直接调用 Windows API 函数，支持大量的 Windows 底层操作。下面代码需要先导入 win32com。

```
xlApp = win32com.client.Dispatch('Excel.Application')   #打开 Excel 应用程序
xlBook = xlApp.Workbooks.Open('D:\\1.xls')              #打开 Excel 文件
xlSht = xlBook.Worksheets('sheet1')                     #获取工作表
aaa = xlSht.Cells(1, 2).Value
xlSht.Cells(2, 3).Value = aaa
xlBook.Close(SaveChanges=1)
del xlApp
```

例 7-20

例 7-20 检查 Word 文档的连续重复字。在 Word 文档中，经常会由于键盘操作不小心而使得文档中出现连续的重复字，如"用户的的资料"或"需要需要用户输入"之类的情况。下面的代码使用 Pywin32 模块中的 win32com 对 Word 文档进行检查并提示类似的重复汉字或标点符号，要求已安装 Office 或 WPS。

```
import sys
from win32com import client
```

```
filename = r'c:\Python可以这样学.doc'
word = client.Dispatch('Word.Application')      #启动 Word
doc = word.Documents.Open(filename)             #打开文件
content = str(doc.Content)                      #获取文件文本
doc.Close()                                     #关闭文件
word.Quit()                                     #退出 Word 应用程序

repeatedWords = []
lens = len(content)
for i in range(lens-2):
    ch, ch1, ch2 = content[i:i+3]
    if ('\u4e00'<= ch<='\u9fa5') or ch in (',', '。', '、'):
        if ch == ch1 and ch+ch1 not in repeatedWords:
            print(ch+ch1)
            repeatedWords.append(ch+ch1)
        elif ch == ch2 and ch+ch1+ch2 not in repeatedWords:
            print(ch+ch1+ch2)
            repeatedWords.append(ch+ch1+ch2)
```

扩展库 python-docx 封装了对 docx 格式 Word 文档的操作，使用 pip 安装之后，可以使用下面的代码实现类似的功能。更多关于 docx 文件操作的介绍，参考作者公众号"Python 小屋"。

```
from docx import Document

doc = Document('《Python 程序设计开发宝典》.docx')
contents = ''.join((p.text for p in doc.paragraphs))
words = set()
for index, ch in enumerate(contents[:-2]):
    if ch==contents[index+1] or ch==contents[index+2]:
        word = contents[index:index+3]
        if word not in words:
            words.add(word)
            print(word)
```

例 7-21 编写程序，进行文件夹增量备份。

程序功能与用法：指定源文件夹与目标文件夹，自动检测自上次备份以来源文件夹中内容的改变，包括修改的文件、新建的文件、新建的文件夹等，自动复制新增或修改过的文件到目标文件夹中，自上次备份以来没有修改过的文件将被忽略而不复制，从而实现增量备份。本例内容属于系统自动运维的范畴，更多相关的知识可参考 12.6 节。

```
import os
import sys
import shutil
import filecmp

def autoBackup(scrDir, dstDir):
    if ((not os.path.isdir(scrDir)) or (not os.path.isdir(dstDir)) or
        (os.path.abspath(scrDir)!=scrDir) or (os.path.abspath(dstDir)!=dstDir)):
        usage()
    for item in os.listdir(scrDir):
```

```python
            scrItem = os.path.join(scrDir, item)
            dstItem = scrItem.replace(scrDir,dstDir)
            if os.path.isdir(scrItem):
                #创建新增的文件夹,保证目标文件夹的结构与原始文件夹一致
                if not os.path.exists(dstItem):
                    os.makedirs(dstItem)
                    print('make directory'+dstItem)
                autoBackup(scrItem, dstItem)
            elif os.path.isfile(scrItem):
                #只复制新增或修改过的文件
                if ((not os.path.exists(dstItem)) or
                    (not filecmp.cmp(scrItem, dstItem, shallow = False))):
                    shutil.copyfile(scrItem, dstItem)
                    print('file:' +scrItem + '==>' +dstItem)

def usage():
    print('scrDir and dstDir must be existing absolute path of certain directory')
    print('For example:{0} c:\\olddir c:\\newdir'.format(sys.argv[0]))
    sys.exit(0)

if __name__ == '__main__':
    if len(sys.argv) != 3:
        usage()
    scrDir, dstDir = sys.argv[1], sys.argv[2]
    autoBackup(scrDir, dstDir)
```

例7-22 编写程序,统计指定文件夹大小以及文件和子文件夹数量。本例代码也属于系统运维范畴,可用于磁盘配额的计算。

```python
import os

totalSize, fileNum, dirNum = 0, 0, 0
def visitDir(path):
    global totalSize, fileNum, dirNum
    for lists in os.listdir(path):
        sub_path = os.path.join(path, lists)
        if os.path.isfile(sub_path):
            fileNum = fileNum +1                                    #统计文件数量
            totalSize = totalSize +os.path.getsize(sub_path)        #统计文件总大小
        elif os.path.isdir(sub_path):
            dirNum = dirNum +1                                      #统计文件夹数量
            visitDir(sub_path)                                      #递归遍历子文件夹

def main(path):
    if not os.path.isdir(path):
        print(f'Error:"{path}" is not a directory or does not exist.')
        return
    visitDir(path)

def sizeConvert(size):                                              #单位换算
    K, M, G = 1024, 1024**2, 1024**3
    if size >= G:
```

```python
        return str(size/G) + 'G Bytes'
    elif size >= M:
        return str(size/M) + 'M Bytes'
    elif size >= K:
        return str(size/K) + 'K Bytes'
    else:
        return str(size) + 'Bytes'

def output(path):
    print(f'The total size of {path} is:{sizeConvert(totalSize)}({totalSize}Bytes)')
    print(f'The total number of files in {path} is:{fileNum}')
    print(f'The total number of directories in {path} is:{dirNum}')

if __name__ == '__main__':
    path = r'd:\idapro6.5plus'
    main(path)
    output(path)
```

例 7-23 编写程序，统计指定目录所有 C++ 源程序文件中不重复代码的行数。

例 7-23

```python
from os import listdir
from os.path import isdir, join

notRepeatedLines = set()                               #保存非重复的代码行
file_num, code_num = 0, 0                              #文件数量与代码总行数
def linesCount(directory):
    global file_num, code_num
    for filename in listdir(directory):                #listdir()返回的列表中是相对路径
        temp = join(directory, filename)
        if isdir(temp):                                #递归遍历子文件夹
            linesCount(temp)
        elif temp.endswith('.cpp'):                    #只考虑.cpp文件
            file_num += 1
            with open(temp, 'r', encoding='utf8') as fp:
                for line in fp:
                    notRepeatedLines.add(line.strip()) #忽略每行两侧的空白字符
                    code_num += 1                      #记录所有代码行
linesCount('F:\教学课件\计算机图形学')
print('总行数：{0},非重复行数：{1}'.format(code_num, len(notRepeatedLines)))
print('文件数量：{0}'.format(file_num))
```

例 7-24 编写程序，递归删除指定文件夹中指定类型的文件。本例代码也属于系统运维范畴，可用于清理系统中的临时垃圾文件或其他指定类型的文件，稍加扩展还可以删除大小为 0 字节的文件。

```python
from os import remove, listdir
from os.path import isdir, join, splitext

filetypes = ['.tmp', '.log', '.obj', '.txt']           #指定要删除的文件类型

def delCertainFiles(directory):
    if not isdir(directory):
        return
```

```
        for filename in listdir(directory):
            temp = join(directory, filename)
            if isdir(temp):
                delCertainFiles(temp)
            elif splitext(temp)[1] in filetypes:        #检查文件类型
                remove(temp)
                print(temp, ' deleted.…')

directory = r'E:\new'
delCertainFiles(directory)
```

如果文件夹中有带特殊属性的文件或子文件夹，上面的代码可能会无法删除带特殊属性的文件，利用 os.chmod() 函数或 Python 扩展库 Pywin32 可以解决这一问题。

```
import os
import win32con
import win32api
From win32con import FILE_ATTRIBUTE_NORMAL

def del_dir(path):
    for file in os.listdir(path):
        file_or_dir = os.path.join(path,file)
        if os.path.isdir(file_or_dir) and not os.path.islink(file_or_dir):
            del_dir(file_or_dir)                 #递归删除子文件夹及其文件
        else:
            try:
                os.remove(file_or_dir)           #尝试删除该文件
            except:                              #无法删除,很可能是文件拥有特殊属性
                win32api.SetFileAttributes(file_or_dir, FILE_ATTRIBUTE_NORMAL)
                os.remove(file_or_dir)           #修改文件属性,设置为普通文件,再次删除
    os.rmdir(path)                               #删除目录
del_dir("E:\\old")
```

例 7-25　使用扩展库 openpyxl 读写 Excel 2007 及更高版本的 Excel 文件。

```
import openpyxl
from openpyxl import Workbook

fn = r'f:\test.xlsx'                             #文件名
wb = Workbook()                                  #创建工作簿
ws = wb.create_sheet(title='你好,世界')           #创建工作表
ws['A1'] = '这是第一个单元格'                      #单元格赋值
ws['B1'] = 3.1415926
wb.save(fn)                                      #保存 Excel 文件
wb = openpyxl.load_workbook(fn)                  #打开已有的 Excel 文件
ws = wb.worksheets[1]                            #打开指定索引的工作表
print(ws['A1'].value)                            #读取并输出指定单元格的值
ws.append([1, 2, 3, 4, 5])                       #添加一行数据
ws.merge_cells('F2:F3')                          #合并单元格
ws['F2'] = '=sum(A2:E2)'                         #写入公式
```

```
    for r in range(10, 15):
        for c in range(3, 8):
            ws.cell(row=r, column=c, value=r*c)            #写入单元格数据
wb.save(fn)
```

假设某学校所有课程每学期允许多次考试，学生可随时参加考试，系统自动将每次成绩添加到 Excel 文件（包含 3 列：姓名、课程、成绩）中，期末要求统计所有学生每门课程的最高成绩。下面的代码首先模拟生成随机成绩数据，然后进行统计分析。更多关于 xlsx 文件操作的案例见微信公众号"Python 小屋"。

例 7-25

```
import random
import openpyxl
from openpyxl import Workbook

#生成随机数据
def generateRandomInformation(filename):
    workbook = Workbook()
    worksheet = workbook.worksheets[0]
    worksheet.append(['姓名','课程','成绩'])
    #中文名字中的第一、第二、第三个字
    first, middle, last = '赵钱孙李', '伟昀琛东', '坤艳志'
    subjects = ('语文','数学','英语')              #课程名称
    for i in range(200):                          #随机生成 200 个数据
        name = random.choice(first)
        #按一定概率生成只有两个字的中文名字
        if random.randint(1, 100) > 50:
            name = name + random.choice(middle)
        name = name + random.choice(last)
        #依次把姓名、课程和成绩放入列表，然后追加到工作表尾部
        line = [name,random.choice(subjects),random.randint(0, 100)]
        worksheet.append(line)
    #保存数据,生成 Excel 2007 格式的文件
    workbook.save(filename)

def getResult(oldfile, newfile):
    #用于存放结果数据的字典
    result = dict()
    #打开原始数据
    workbook = openpyxl.load_workbook(oldfile)
    worksheet = workbook.worksheets[0]
    #遍历原始数据
    for row in list(worksheet.rows)[1:]:
        #姓名,课程,成绩
        name, subject, grade = row[0].value, row[1].value, row[2].value
        #获取当前姓名对应的课程和成绩
        #如果 result 字典中不包含,则返回空字典
        t = result.get(name, {})
        #获取当前学生当前课程的成绩,若不存在,则返回 0
        f = t.get(subject, 0)
```

```
            #只保留该学生该课程的最高成绩
            if grade > f:
                t[subject] = grade
                result[name] = t
    #创建 Excel 文件
    workbook1 = Workbook()
    worksheet1 = workbook1.worksheets[0]
    worksheet1.append(['姓名','课程','成绩'])
    #将 result 字典中的结果数据写入 Excel 文件
    for name, t in result.items():
        for subject, grade in t.items():
            worksheet1.append([name, subject, grade])
    workbook1.save(newfile)

if __name__ == '__main__':
    oldfile = r'd:\test.xlsx'
    newfile = r'd:\result.xlsx'
    generateRandomInformation(oldfile)
    getResult(oldfile, newfile)
```

例 7-26 编写代码，查看指定 zip 和 rar 压缩文件中的文件列表。

Python 标准库 zipfile 提供了对 zip 和 apk 文件的访问。

```
import zipfile

with zipfile.ZipFile(r'D:\Jakstab-0.8.3.zip') as fp:
    for f in fp.namelist():
        print(f)
```

Python 扩展库 rarfile（可通过 pip 工具安装）提供了对 rar 文件的访问。

```
import rarfile

with rarfile.RarFile(r'D:\asp网站.rar') as r:
    for f in r.namelist():
        print(f)
```

例 7-27

例 7-27 小学口算题库生成器。本例主要演示使用扩展库 python-docx 创建 Word 文档，使用 GUI 标准库 tkinter 设计界面的内容可参见第 9 章。

```
import random
import os
import tkinter
import tkinter.ttk
from docx import Document

columnsNumber = 4
def main(rowsNumber=20, grade=4):
    if grade < 3:
        operators = '+-'
        biggest = 20
    elif grade <= 4:
```

```python
        operators = '+-×÷'
        biggest = 100
    elif grade == 5:
        operators = '+-×÷('
        biggest = 100

    document = Document()
    #创建表格
    table = document.add_table(rows=rowsNumber, cols=columnsNumber)
    #遍历每个单元格,生成并写入题目
    for row in range(rowsNumber):
        for col in range(columnsNumber):
            first = random.randint(1, biggest)
            second = random.randint(1, biggest)
            operator = random.choice(operators)
            if operator != '(':
                if operator == '-':
                    #如果是减法口算题,确保结果为正数
                    if first < second:
                        first, second = second, first
                r = f'{first:2d}{operator}{second:2d}='
            else:
                #生成带括号的口算题,需要3个数字和2个运算符
                third = random.randint(1, 100)
                while True:
                    o1 = random.choice(operators)
                    o2 = random.choice(operators)
                    if '(' not in (o1,o2):
                        break
                rr = random.randint(1, 100)
                if rr > 50:
                    if o2 == '-':
                        if second < third:
                            second, third = third, second
                    r = f'{first:<2d}{o1}({second:<2d}{o2}{third:<2d})= '
                else:
                    if o1 == '-':
                        if first < second:
                            first, second = second, first
                    r = f'({first:<2d}{o1}{second:<2d}){o2}{third:<2d}= '
            #获取指定单元格并写入口算题
            cell = table.cell(row, col)
            cell.text = r
    document.save('kousuan.docx')
    os.startfile('kousuan.docx')

if __name__ == '__main__':
    app = tkinter.Tk()
```

```
            app.title('KouSuan------by Dong Fuguo')
            app['width'] = 300
            app['height'] = 150
            labelNumber = tkinter.Label(app, text='Number:', justify=tkinter.RIGHT, width=50)
            labelNumber.place(x=10, y=40, width=50, height=20)
            comboNumber = tkinter.ttk.Combobox(app, values=(100, 200, 300, 400, 500), width=50)
            comboNumber.place(x=70, y=40, width=50, height=20)

            labelGrade = tkinter.Label(app, text='Grade:', justify=tkinter.RIGHT, width=50)
            labelGrade.place(x=130, y=40, width=50, height=20)
            comboGrade = tkinter.ttk.Combobox(app, values=(1,2,3,4,5), width=50)
            comboGrade.place(x=200, y=40, width=50, height=20)

            def generate():
                number = int(comboNumber.get())
                grade = int(comboGrade.get())
                main(number, grade)
            buttonGenerate = tkinter.Button(app, text='GO', width=40, command=generate)
            buttonGenerate.place(x=130, y=90, width=40, height=30)

            app.mainloop()
```

例 7-28 编写程序,统计指定文件夹中所有 pptx 格式的 PowerPoint 文件中幻灯片数量。本例需要安装扩展库 Python-pptx,更多相关案例见微信公众号"Python 小屋"。

例 7-28

```
import os
import os.path
import pptx

total = 0
def pptCount(path):
    global total
    for subPath in os.listdir(path):
        subPath = os.path.join(path, subPath)
        if os.path.isdir(subPath):
            pptCount(subPath)
        elif subPath.endswith('.pptx'):
            print(subPath)
            presentation = pptx.Presentation(subPath)
            total += len(presentation.slides)

pptCount(r'F:\教学课件\Python 程序设计(第 4 版)')
print(total)
```

例 7-29 编写程序,检测 U 盘插入并自动复制全部文件。代码中使用到了扩展库 psutil,运行代码前需要先使用 pip 安装这个扩展库。

```
from time import sleep
from shutil import copytree
from psutil import disk_partitions

while True:
```

```
        sleep(3)
        for item in disk_partitions():          #检查所有驱动器
            if 'removable' in item.opts:        #发现可移动驱动器
                driver = item.device
                print('Found USB disk:', driver)
                break
            else:
                continue
        break

# 复制根目录
copytree(driver, r'D:\usbdriver')
print('all files copied.')
```

例 7-30 安装扩展库,编写程序,把文字型 PDF 文件转换为图片式 PDF 文件。

```
from os import remove
import fitz

fn = '测试文件.pdf'
with fitz.open(fn) as doc, fitz.open() as fpMerge:
    for page in doc:
        #水平和垂直方向分辨率变为 2 倍
        mat = fitz.Matrix(2, 2)
        #把当前页转换为图片
        pic = page.get_pixmap(matrix=mat)
        pic.save('t.png')
        #把图片转换为 PDF 文件
        t = fitz.open('pdf', fitz.open('t.png').convert_to_pdf())
        fpMerge.insert_pdf(t)
        remove('t.png')
    fpMerge.save('结果文件.pdf')
```

本 章 小 结

(1) 文件操作在各类软件开发中均占有重要的地位。

(2) 对二进制文件无法直接读取和理解其内容,必须了解其文件结构和所使用的序列化规则并使用正确的反序列化方法。

(3) Python 内置了文件对象,通过 open()函数即可以指定模式打开指定文件并创建文件对象,然后通过对象读写文件内容。

(4) Python 中常用的序列化模块有 struct、pickle、json、marshal 和 shelve。

(5) 文件对象的读写方法都会自动改变文件指针位置。

(6) os、os.path 和 shutil 模块提供了大量用于文件和文件夹操作的函数,包括文件和文件夹的移动、复制、删除、重命名,以及压缩与解压缩等。

习 题

1. 文件对象的_____方法用于把缓冲区的内容写入文件,但不关闭文件。

2. os.path 模块中的_____函数用于测试指定的路径是否为文件。

3. os 模块的_____函数用于返回包含指定文件夹中所有文件和子文件夹的列表。

4. 假设有一个英文文本文件,编写程序读取并输出其内容,并将其中的大写字母变为小写字母,小写字母变为大写字母。

5. 编写程序,使用 pickle 模块将包含学生成绩的字典序列化并保存为二进制文件,然后再读取内容并显示。

6. 编写程序,使用 shutil 模块中的 move() 函数进行文件移动。

7. 简单解释文本文件与二进制文件的区别。

8. 编写代码,将当前工作目录修改为"C:\",并验证,最后将当前工作目录恢复为原来的目录。

9. 编写程序,用户输入一个目录和一个文件名,搜索该目录及其子目录中是否存在该文件,如果存在,就输出这些文件的路径。

10. 查阅资料,编写程序,把给定的 docx 文件中所有简体中文转换为繁体字,并保存到新文件中。

11. 查阅资料,编写程序,输出指定文件夹中包含特定字符串的所有 docx、xlsx、pptx 文件。

12. 查阅资料,编写程序,输出 docx 文件中红色字体的文字和超链接地址与文本。

第 8 章 异常处理结构与程序调试、测试

简单地说,异常是指程序运行时引发的错误,引发错误的原因有很多,例如除以 0、下标越界、文件不存在、网络异常、类型错误、名字错误、字典键错误、磁盘空间不足等。如果这些错误得不到正确的处理会导致程序终止运行,而合理地使用异常处理结构可以使得程序更加健壮,具有更强的容错性,不会因为用户不小心的错误输入或其他运行时原因而造成程序终止。程序出现异常或错误之后是否能够调试程序并快速定位和解决存在的问题也是程序员综合水平和能力的重要体现之一。

8.1 基本概念

什么是异常呢?程序发生异常有什么表现呢?我们先来看几个示例。

```
>>> x, y = 10, 5
>>> a = x / y
>>> print(A)                    #拼写错误,Python 区分变量名等标识符字母的大小写
Traceback(most recent call last):
File "<pyshell#2>", line 1, in<module>
print(A)
NameError: name 'A' is not defined
>>> 10 * (1/0)                  #除以 0 错误
ZeroDivisionError: division by zero
>>> 4 + spam * 3                #使用了未定义的变量,与拼写错误的情形相似
NameError: name 'spam' is not defined
>>> '2' + 2                     #对象类型不支持特定的操作
TypeError: can only concatenate str (not "int") to str
```

在前面的章节中,出现过多次类似的信息,没错,这就是 Python 异常的标准表现形式。熟练运用异常处理结构对于提高程序的健壮性和容错性具有重要的作用,同时也可以把 Python 晦涩难懂的错误提示转换为友好的提示显示给最终用户。

异常处理是指因为程序执行过程中出错而在正常控制流之外采取的行为。严格来说,语法错误和逻辑错误不属于异常,但有些语法或逻辑错误往往会导致异常,例如,由于大小写拼写错误而试图访问不存在的对象,或者试图访问不存在的文件,等等。当 Python 检测到一个错误时,解释器就会指出当前程序流已无法继续执行,这时就出现了异常。当程序执行过程中出现错误时 Python 会自动引发异常,程序员也可以通过 raise 语句显式地引发异常。可以关注微信公众号"Python 小屋"发送消息"异常"学习更多。

尽管异常处理结构非常重要也非常有效,但不建议用来代替常规的检查,例如必要的 if…else…判断。在编程时应避免过多依赖异常处理结构来提高程序健壮性。

8.2 Python 内置异常类与自定义异常

下面较为完整地展示了 Python 内置异常类的继承层次。
```
BaseException
 +--SystemExit
 +--KeyboardInterrupt
 +--GeneratorExit
 +--Exception
      +--StopIteration
      +--ArithmeticError
      |    +--FloatingPointError
      |    +--OverflowError
      |    +--ZeroDivisionError
      +--AssertionError
      +--AttributeError
      +--BufferError
      +--EOFError
      +--ImportError
      +--LookupError
      |    +--IndexError
      |    +--KeyError
      +--MemoryError
      +--NameError
      |    +--UnboundLocalError
      +--OSError
      |    +--BlockingIOError
      |    +--ChildProcessError
      |    +--ConnectionError
      |    |    +--BrokenPipeError
      |    |    +--ConnectionAbortedError
      |    |    +--ConnectionRefusedError
      |    |    +--ConnectionResetError
      |    +--FileExistsError
      |    +--FileNotFoundError
      |    +--InterruptedError
      |    +--IsADirectoryError
      |    +--NotADirectoryError
      |    +--PermissionError
```

```
     |      +--ProcessLookupError
     |      +--TimeoutError
     +--ReferenceError
     +--RuntimeError
     |      +--NotImplementedError
     +--SyntaxError
     |      +--IndentationError
     |             +--TabError
     +--SystemError
     +--TypeError
     +--ValueError
     |      +--UnicodeError
     |             +--UnicodeDecodeError
     |             +--UnicodeEncodeError
     |             +--UnicodeTranslateError
     +--Warning
            +--DeprecationWarning
            +--PendingDeprecationWarning
            +--RuntimeWarning
            +--SyntaxWarning
            +--UserWarning
            +--FutureWarning
            +--ImportWarning
            +--UnicodeWarning
            +--BytesWarning
            +--ResourceWarning
```

如果需要，可以继承 Python 内置异常类来实现自定义的异常类，例如：

```python
class ShortInputException(Exception):
    def __init__(self, length, atleast):
        Exception.__init__(self)
        self.length = length
        self.atleast = atleast
try:
    s = input('请输入 -->')
    if len(s) < 3:
        raise ShortInputException(len(s), 3)
except EOFError:
    print('您输入了一个结束标记 EOF')
except ShortInputException as x:
    print(f'ShortInputException: 输入长度是{x.length}，长度至少应是{x.atleast}')
else:
    print('没有异常发生。')
```

再如下面的示例：

```
>>> class MyError(Exception):
    def __init__(self, value):
        self.value = value
    def __str__(self):
        return repr(self.value)
>>> try:
    raise MyError(2*2)
except MyError as e:
    print('My exception occurred, value:', e)    #按两次 Enter 键执行代码
My exception occurred, value: 4                  #此行为执行结果
>>> raise MyError('oops!')
Traceback (most recent call last):
  File "<stdin>", line 1, in ?
__main__.MyError: 'oops!'
```

如果自己编写的某个模块需要抛出多个不同但相关的异常，可以先创建一个基类，然后创建多个派生类分别表示不同的异常，再参考上述代码添加__str__()方法。

```
class Error(Exception):                          #创建基类
    pass

class InputError(Error):                         #派生类 InputError
    """Exception raised for errors in the input.
    Attributes:
        expression——input expression in which the error occurred
        message ——explanation of the error
    """
    def __init__(self, expression, message):
        self.expression = expression
        self.message = message

class TransitionError(Error):                    #派生类 TransitionError
    """Raised when an operation attempts a state transition that's not allowed.
    Attributes:
        previous——state at beginning of transition
        next——attempted new state
        message——explanation of why the specific transition is not allowed
    """
    def __init__(self, previous, next, message):
        self.previous = previous
        self.next = next
        self.message = message
```

8.3

8.3 异常处理结构语法应用

8.3.1 try…except…

异常处理结构中最常见且最基本的是 try…except…结构。其中，try 子句中的代码块包含可能出现异常的语句，except 子句用来捕获相应的异常，except 子句的代码块用来处理异常。如果 try 中的代码块没有出现异常，则继续往下执行异常处理结构后面的代码；如果

出现异常并且被 except 子句捕获，则执行 except 子句的异常处理代码；如果出现异常但没有被 except 捕获，则继续往外层抛出；如果所有层都没有捕获并处理该异常，则程序终止并将该异常抛给最终用户（这是最糟糕的情况，应尽量避免）。该结构语法如下：

```
try:
    try 块                                      #被监控的语句,可能会引发异常
except Exception[ as reason]:
    except 块                                   #处理异常的代码
```

如果需要捕获所有类型的异常，可以使用 Python 异常类的基类 BaseException（很少这样做），代码格式如下：

```
try:
    ...
except BaseException as e:
    except 块                                   #处理所有错误
```

上面的结构可以捕获所有异常，尽管这样做很安全，但是一般并不建议这样做。对于异常处理结构，一般的建议是尽量精准捕获可能会出现的异常，并且有针对性地编写代码进行处理，因为在实际应用开发中，很难使用同一段代码处理所有类型的异常。当然，为了避免遗漏导致没有得到处理的异常干扰程序的正常执行，在捕获了所有可能想到的异常之后，也可以使用异常处理结构的最后一个 except 来捕获 BaseException，见 8.3.3 节。

下面的代码演示了 try…except…结构的用法，代码运行后提示用户输入内容，如果输入的是数字，则循环结束，否则一直提示用户输入正确格式的内容。

```
while True:
    try:
        x = int(input("Please enter a number: "))
        break
    except ValueError:
        print("That was not a valid number.Try again...")
```

在使用时，except 子句可以在异常类名字后面指定一个变量，用于捕获异常的参数或更详细的信息。

```
try:
    raise Exception('spam', 'eggs')
except Exception as inst:
    print(type(inst))           #the exception instance
    print(inst.args)            #arguments stored in .args
    print(inst)                 #__str__ allows args to be printed directly
                                #but may be overridden in exception subclasses
    x, y = inst.args            #unpack args
    print('x=', x)
    print('y=', y)
```

8.3.2　try…except…else…

另一种常用的异常处理结构是 try…except…else…结构。如果 try 中的代码抛出了异常，并且被 except 捕获，则执行相应的异常处理代码，这种情况下不会执行 else 中的代码；

如果 try 中的代码没有抛出任何异常,则执行 else 中的代码。

```
a_list = ['China', 'America', 'England', 'France']
while True:
    n = input("请输入字符串的序号:")
    try:
        n = int(n)
        print(a_list[n])
    except (IndexError,ValueError):              #捕获两种异常,进行同样的处理
        print('列表元素的下标越界或格式不正确,请重新输入字符串的序号')
    else:
        break
```

8.3.3 try…except…except…except…

在实际开发中,同一段代码可能会抛出多个异常,需要针对不同的异常类型进行相应的处理。为了支持多个异常的捕获和处理,Python 提供了带有多个 except 的异常处理结构,类似于多分支选择结构。一旦某个 except 捕获了异常,后面剩余的 except 子句将不会再执行。该结构的语法为

```
try:
    try 块                                    #被监控的语句
except Exception1:
    except 块 1                               #处理异常 1 的语句
except Exception2:
    except 块 2                               #处理异常 2 的语句
```

下面的代码演示了该结构的用法:

使用这种形式的异常处理结构时,每个 except 子句捕获的异常应该是正交的(互相不包含),否则应按照从精准到模糊或者从子类到父类的顺序捕获和处理。

```
try:
    x = input('请输入被除数: ')
    y = input('请输入除数: ')
    z = float(x) / float(y)                   #这一句可能会引发多种异常
except ZeroDivisionError:
    print('除数不能为 0')
except TypeError:
    print('被除数和除数应为数值类型')
except NameError:
    print('变量不存在')
else:                                          #如果没有引发异常,就执行这里的代码
    print(x, '/', y, '=', z)
```

将要捕获的异常写在一个元组中,可以使用一个 except 语句捕获多种异常,并且共用同一段异常处理代码,当然,除非确定要捕获的多个异常可以使用同一段代码来处理,否则并不建议这样做。

```
try:
    f = open('myfile.txt')
    s = f.readline()
```

```
        i = int(s.strip())
except (OSError, ValueError, RuntimeError, NameError):
    pass
```

8.3.4　try…except…else…finally…

最后一种常用也是最完整的异常处理结构是 try…except…else…finally…结构（其中 except 可以有多个）。在该结构中，finally 子句中的语句块无论是否发生异常都会执行，常用来做一些清理工作以释放 try 子句中申请的资源。语法如下：

```
try:
    …
except 异常 1:
    …
except 异常 2:
    …
else:
    …
finally:
    …                                    #无论如何都会执行
```

这个结构有很多变形用法，例如下面的代码，无论是否发生异常，语句 print(5) 都会被执行。

```
>>> try:
    3 / 0
except:
    print(3)
finally:
    print(5)
3
5
```

如果 try 子句中的异常没有被捕获和处理，或者 except 子句中的代码出现了异常，那么这些异常将会在 finally 子句执行完后再次抛出。例如下面的代码，在 try 中的语句出现了异常但是没有得到处理，因此，finally 中的语句执行完以后再次抛出该异常。

```
>>> try:
    3 / 0
finally:
    print(5)
5
ZeroDivisionError: division by zero
```

下面的代码较为完整地演示了这个语法结构的用法。

```
>>> def divide(x, y):
    try:
        result = x / y
    except ZeroDivisionError:
        print("division by zero!")
    else:
        print("result is", result)
```

```
        finally:
            print("executing finally clause")
>>> divide(2, 1)
result is 2.0
executing finally clause
>>> divide(2, 0)
division by zero!
executing finally clause
>>> divide("2", "1")
executing finally clause
TypeError: unsupported operand type(s) for /: 'str' and 'str'
```

另外,finally 中的代码也可能抛出异常,例如下面的代码,使用异常处理结构的本意是为了防止文件读取操作出现异常而导致文件不能正常关闭,但是如果因为文件不存在而导致文件对象创建失败,那么 finally 子句中关闭文件对象的代码将会抛出异常从而导致程序终止运行,在此之前还会抛出异常提示文件不存在,请自行测试。

```
>>> try:
    f = open('test.txt', 'r')
    line = f.readline()
    print(line)
finally:
    f.close()
NameError: name 'f' is not defined
```

最后,在函数中使用带有 finally 子句的异常处理结构时,应避免在 finally 子句中使用 return 语句,否则可能出现出乎意料的错误,例如下面的代码:

```
>>> def demo_div(a, b):
    try:
        return a/b
    except:
        pass
    finally:
        return -1
>>> demo_div(1, 0)
-1
>>> demo_div(1, 2)                              #即使没有出现异常,也得到错误结果
-1
```

简单总结一下,可以把异常处理结构理解为"请求原谅比请求允许要容易"。也就是说,有些代码执行可能会出现错误,也可能不会出现错误,这主要由运行时的各种客观因素决定,此时建议使用异常处理结构。如果使用大量的选择结构来提前判断,仅当满足相应条件时才执行该代码,这些条件判断可能严重干扰正常的业务逻辑,也会严重降低代码的可读性。

8.4 断言与上下文管理

8.4.1 断言

断言语句的语法如下:

```
assert expression[, reason]
```

当判断表达式 expression 为真时，什么都不做；如果表达式为假，则抛出异常。

assert 语句一般用于对程序某个时刻必须满足的条件进行验证，仅当程序的特殊变量 __debug__ 为 True 时有效。当 Python 脚本以优化模式编译为字节码文件时，assert 语句将被移除以提高运行速度。

断言和异常处理结构经常结合使用，例如：

```
>>> try:
    assert 1 == 2, '1 is not equal 2!'
except AssertionError as reason:
    print('%s:%s'%(reason.__class__.__name__, reason))     #这里按两次 Enter 键
AssertionError:1 is not equal 2!
```

8.4.2 上下文管理

使用上下文管理语句 with 可以自动管理资源，在代码块执行完毕后自动还原进入该代码块之前的现场或上下文。不论何种原因跳出 with 块，也不论是否发生异常，总能保证资源被正确释放，大大简化了程序员的工作，常用于文件操作、网络通信、数据库连接、多线程与多进程编程之类的场合，详见本书第 7、10、13、14 章。

with 语句的语法如下：

```
with context_expr [as var]:
    with 块
```

*8.5 使用 IDLE 调试代码

当程序运行发生错误或者得到了非预期的结果时，是否能够熟练地对程序进行调试并快速定位和解决问题是体现程序员综合能力的重要标准之一。

几乎任何一种集成开发环境都提供了代码调试功能，Python 标准开发环境 IDLE 也不例外。使用 IDLE 的调试功能时，首先单击 IDLE 的 Debug→Debugger 菜单命令打开调试器窗口，然后打开并运行要调试的程序，最后切换到调试器窗口使用其中的调试按钮进行调试。如图 8-1 所示为 IDLE 调试器窗口及其功能简要介绍，可以使用调试按钮对程序进行

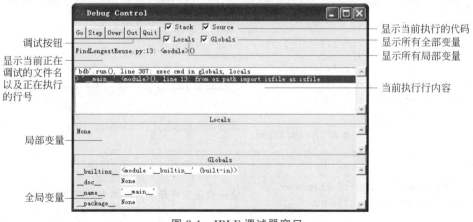

图 8-1 IDLE 调试器窗口

单步执行,实时查看变量的当前值并跟踪其变化过程,对于理解程序内部工作原理和发现程序中存在的问题非常有帮助。图 8-2 和图 8-3 是使用 IDLE 对例 5-2 的程序进行调试过程中的两个截图,单击调试窗口上的 Step 按钮进行单步执行,可以清晰地观察程序执行过程中数据的变化。关于可执行文件的分析和调试,可参考第 16 章。

图 8-2　程序调试截图(一)

图 8-3　程序调试截图(二)

*8.6　使用 pdb 模块调试程序

8.6.1　pdb 模块常用命令

pdb 是 Python 自带的交互式源代码调试模块,可以完成代码调试的绝大部分功能,包括设置/清除(条件)断点、启用/禁用断点、单步执行、查看栈帧、查看变量值、查看当前执行位置、列出源代码、执行任意 Python 代码或表达式等。pdb 还支持事后调试,可在程序控制下被调用,并且可以通过 pdb 和 cmd 接口对其进行扩展。pdb 模块常用调试命令如表 8-1 所示。

表 8-1　pdb 模块常用调试命令

简写/完整命令	用 法 示 例	解　　释
a(rgs)		显示当前函数中的参数
b(reak) [[filename:]lineno | function[,condition]]	b 173	在 173 行设置断点
	b function	在 function() 函数第一条可执行语句位置设置断点
	b	不带参数则列出所有断点,包括每个断点的触发次数、当前忽略计数以及与之关联的条件
	b 175, condition	设置条件断点,仅当 condition 的值为 True 时该断点有效
cl(ear) [filename:lineno | bpnumber [bpnumber …]]	cl	清除所有断点
	cl filename:lineno	删除指定文件指定行的所有断点
	cl 3 5 9	删除第 3、5、9 个断点

续表

简写/完整命令	用法示例	解　释
condition bpnumber [condition]	condition 3 a<b	仅当 a<b 时 3 号断点有效
	condition 3	将 3 号断点设置为无条件断点
continue		继续运行至下一个断点或脚本结束
disable [bpnumber] [bpnumber …]	disable 3 5	禁用第 3、5 个断点，禁用后断点仍存在，可以再次被启用
d(own)		在栈跟踪器中向下移动一个栈帧
enable [bpnumber] [bpnumber …]	enable n	启用第 n 个断点
h(elp) [command]		查看 pdb 帮助
ignore bpnumber [count]		为断点设置忽略计数，count 默认值为 0。若某断点的忽略计数不为 0，则每次触发时自动减 1，当忽略计数为 0 时该断点处于活动状态
j(ump)	j 20	跳至第 20 行继续运行，不能跳至块的中间位置
l(ist) [first [,last]]	l	列出脚本清单，默认列出 11 行
	l m, n	列出从第 m～n 行的脚本代码
	l m	列出从第 m 行开始的 11 行代码
n(ext)		执行下一条语句，遇到函数时不进入其内部
p(rint)	p i	打印变量 i 的值
q(uit)		退出 pdb 调试环境
r(eturn)		一直运行至当前函数返回
tbreak		设置临时断点，该类型断点只被中断一次，触发后该断点自动删除
step		执行下一条语句，遇到函数时进入其内部
u(p)		在栈跟踪器中向上移动一个栈帧
w(here)		查看当前栈帧
[!]statement		在 pdb 中执行语句，!与要执行的语句之间不需要空格，任何非 pdb 命令都被解释为 Python 语句并执行，甚至可以调用函数或修改当前上下文中变量的值
直接按 Enter 键		默认执行上一条命令

8.6.2　使用 pdb 模块调试 Python 程序

可以通过 3 种不同的形式来使用 pdb 模块提供的调试功能，分别为在交互模式下调试特定的代码块、在程序中显式插入断点把 pdb 作为模块来调试程序。

（1）在交互模式下使用 pdb 模块提供的功能可以直接调试语句块、表达式、函数等多种脚本,常用的调试方法有 4 种。

① pdb.run(statement[,globals[,locals]])——调试指定语句,可选参数 globals 和 locals 用来指定代码执行的环境,默认是 __main__ 模块的字典。

② pdb.runeval(expression[,globals[,locals]])——返回表达式的值,可选参数 globals 和 locals 的含义与上面的 run() 函数一样。

③ pdb.runcall(function[,argument, …])——调试指定函数。

④ pdb.post_mortem([traceback])——进入 traceback 对象的事后调试模式,如果没有指定 traceback 对象,则使用当前正在处理的一个异常。

例如,下面的代码演示了如何调试一个函数,其中(Pdb)为提示符,在后面输入并执行表 8-1 中介绍的命令即可。

```
>>> import pdb
>>> def f():
    x = 5
    print(x)
>>> pdb.runcall(f)
><pyshell#5>(2)f()
(Pdb) n
><pyshell#5>(3)f()
(Pdb) l
[EOF]
(Pdb) p x
5
(Pdb) n
5
--Return--
><pyshell#5>(3)f()->None
(Pdb) n
>>>
```

(2) 在程序中嵌入断点来实现调试功能。

在程序中导入 pdb 模块,使用 pdb.set_trace() 在需要的位置设置断点。执行程序时将自动打开 pdb 调试环境,即使该程序当前不处于调试状态。例如,下面的程序 IsPrime.py:

```
import pdb

n = 37
pdb.set_trace()                              #设置断点,自动进入 pdb 调试模式
for i in range(2, n):
    if n%i == 0:
        print('No')
        break
else:
    print('Yes')
```

由于使用 pdb 设置了断点,运行后自动打开调试模式,如图 8-4 所示。

在命令提示符环境中运行该程序同样自动打开调试模式,如图 8-5 所示。

```
========================= RESTART: C:/Python39/IsPrime.py
> c:\python39\isprime.py(5)<module>()
-> for i in range(2, n):
(Pdb) n
> c:\python39\isprime.py(6)<module>()
-> if n%i == 0:
(Pdb) n
> c:\python39\isprime.py(5)<module>()
-> for i in range(2, n):
(Pdb) p i
2
(Pdb) n
> c:\python39\isprime.py(6)<module>()
-> if n%i == 0:
(Pdb) n
> c:\python39\isprime.py(5)<module>()
-> for i in range(2, n):
(Pdb) n
> c:\python39\isprime.py(6)<module>()
-> if n%i == 0:
(Pdb) p i
4
(Pdb) a
(Pdb) l
  1     import pdb
  2
  3     n = 37
  4     pdb.set_trace()
  5     for i in range(2, n):
  6  ->     if n%i == 0:
  7             print('No')
  8             break
  9     else:
 10         print('Yes')
[EOF]
```

图 8-4　自动打开调试模式(一)

```
命令提示符 - python isprime.py

C:\Python39>python isprime.py
> c:\python39\isprime.py(5)<module>()
-> for i in range(2, n):
(Pdb) p i
*** NameError: name 'i' is not defined
(Pdb) n
> c:\python39\isprime.py(6)<module>()
-> if n%i == 0:
(Pdb) p i
2
(Pdb) l
  1     import pdb
  2
  3     n = 37
  4     pdb.set_trace()
  5     for i in range(2, n):
  6  ->     if n%i == 0:
  7             print('No')
  8             break
  9     else:
 10         print('Yes')
[EOF]
(Pdb)
```

图 8-5　自动打开调试模式(二)

(3) 使用命令行调试程序。

在命令行提示符下执行“python-m pdb 脚本文件名”,可以直接进入调试环境;当调试结束或程序正常结束以后,pdb 将重启该程序。把上面程序中 pdb 模块的导入和断点插入函数都删除,然后在命令提示符环境中使用调试模式运行,如图 8-6 所示。

```
命令提示符 - python -m pdb isprime.py

C:\Python39>python -m pdb isprime.py
> c:\python39\isprime.py(1)<module>()
-> n = 37
(Pdb) n
> c:\python39\isprime.py(2)<module>()
-> for i in range(2, n):
(Pdb) w
  c:\python39\lib\bdb.py(580)run()
-> exec(cmd, globals, locals)
  <string>(1)<module>()
> c:\python39\isprime.py(2)<module>()
-> for i in range(2, n):
(Pdb) n
> c:\python39\isprime.py(3)<module>()
-> if n%i == 0:
(Pdb) p i
2
(Pdb) p n
37
(Pdb)
```

图 8-6　以调试模式运行程序

*8.7　Python 单元测试

软件测试对于保证软件质量非常重要,尤其是系统升级过程中对代码的改动不应该影响原有功能,此时软件测试是未来重构代码的信心保证。几乎所有软件公司都有专门的测试团队来保证软件质量,但作为程序员,首先应该保证自己编写的代码准确无误地实现了预定功能。单元测试是保证模块质量的重要手段之一。从软件工程角度来讲,软件测试分为白盒测试和黑盒测试。其中,白盒测试主要通过阅读程序源代码来判断是否符合功能要求,对于复杂的业务逻辑白盒测试难度非常大,一般以黑盒测试为主,白盒测试为辅。黑盒测试不关心模块的内部实现方式,只关心其功能是否正确,通过精心设计完备的测试用例检验模块的输入输出是否正确来判断其是否符合预定的功能要求。

通过单元测试的方式来管理和使用设计好的测试用例,不仅可以避免人工输入可能引入的错误,还可以重复利用设计好的测试用例,具有很好的可扩展性。Python 标准库 unittest 提供了大量用于单元测试的类和方法,其中 TestCase 类的常用方法如表 8-2 所示。

表 8-2 TestCase 类的常用方法

方 法 名 称	功 能 说 明	方 法 名 称	功 能 说 明
assertEqual(a,b)	a == b	assertNotEqual(a,b)	a != b
assertTrue(x)	bool(x) is True	assertFalse(x)	bool(x) is False
assertIs(a,b)	a is b	assertIsNot(a,b)	a is not b
assertIsNone(x)	x is None	assertIsNotNone(x)	x is not None
assertIn(a,b)	a in b	assertNotIn(a,b)	a not in b
assertIsInstance(a,b)	isinstance(a,b)	assertNotIsInstance(a,b)	not isinstance(a,b)
assertAlmostEqual(a, b)	round(a-b, 7) == 0	assertNotAlmostEqual(a, b)	round(a-b,7)!= 0
assertGreater(a,b)	a > b	assertGreaterEqual(a, b)	a >= b
assertLess(a,b)	a < b	assertLessEqual(a, b)	a <= b
assertRegex(s,r)	r.search(s)	assertNotRegex(s, r)	not r.search(s)
setUp()	每项测试开始之前自动调用该方法	tearDown()	每项测试完成之后自动调用该方法

TestCase 类中 setUp() 和 tearDown() 这两个方法比较特殊,分别在每个测试之前和之后自动调用,常用来执行数据库连接的创建与关闭、文件的打开与关闭等操作。

例 8-1 编写单元测试程序。以第 2 章自定义栈的代码为例,演示如何利用 unittest 库对 Stack 类中的入栈、出栈、改变大小以及满/空测试等方法进行测试,并将测试结果写入文件 test_Stack_result.txt。

```
import Stack
import unittest

class TestStack(unittest.TestCase):
    def setUp(self):
        #测试之前以追加模式打开指定文件
        self.fp = open('D:\\test_Stack_result.txt', 'a+')

    def tearDown(self):
```

```python
            #测试结束后关闭文件
            self.fp.close()

    def test_isEmpty(self):
        try:
            s = Stack.Stack()
            #确保 isEmpty()方法返回结果为 True
            self.assertTrue(s.isEmpty())
            self.fp.write('isEmpty passed\n')
        except Exception as e:
            self.fp.write('isEmpty failed\n')

    def test_empty(self):
        try:
            s = Stack.Stack(5)
            for i in ['a', 'b', 'c']:
                s.push(i)
            #测试清空栈操作是否正常工作
            s.empty()
            self.assertTrue(s.isEmpty())
            self.fp.write('empty passed\n')
        except Exception as e:
            self.fp.write('empty failed\n')

    def test_isFull(self):
        try:
            s = Stack.Stack(3)
            s.push(1)
            s.push(2)
            s.push(3)
            self.assertTrue(s.isFull())
            self.fp.write('isFull passed\n')
        except Exception as e:
            self.fp.write('isFull failed\n')

    def test_pushpop(self):
        try:
            s = Stack.Stack()
            s.push(3)
            #确保入栈后立刻出栈得到原来的元素
            self.assertEqual(s.pop(), 3)
            s.push('a')
            self.assertEqual(s.pop(), 'a')
            self.fp.write('push and pop passed\n')
        except Exception as e:
            self.fp.write('push or pop failed\n')

    def test_setSize(self):
        try:
            s = Stack.Stack(8)
            for i in range(8):
                s.push(i)
            self.assertTrue(s.isFull())
```

```python
            #测试扩大栈空间是否正常工作
            s.setSize(9)
            s.push(8)
            self.assertTrue(s.isFull())
            self.assertEqual(s.pop(), 8)
            #测试缩小栈空间是否正常工作
            s.setSize(4)
            self.assertTrue(s.isFull())
            self.assertEqual(s.pop(), 3)
            self.fp.write('setSize passed\n')
        except Exception as e:
            self.fp.write('setSize failed\n')

if __name__ == '__main__':
    #启动测试,自动执行所有以 test_开头的方法
    unittest.main()
```

最后需要说明的是：①测试用例的设计应该是完备的，应保证覆盖尽可能多的情况，尤其是要覆盖边界条件，对目标模块的功能进行充分测试，避免漏测；②测试用例以及测试代码的设计与编写也可能存在 bug，通过测试并不代表目标代码没有错误，但是一般而言，不能通过测试的模块代码是存在问题的；③再好的测试方法和测试用例也无法保证能够发现所有错误，只能通过改进和综合多种测试方法并且精心设计测试用例来发现尽可能多的潜在问题；④除了功能测试，还应对程序进行性能测试与安全性测试，甚至还需要进行规范性测试以保证代码的可读性和可维护性。

8.8　文档测试

Python 标准库 doctest 可以搜索程序中类似于交互式 Python 代码的文本片段，并运行这些交互式代码来验证是否符合预期结果和功能，常用于 Python 程序的模块测试。

```
def add(value1, value2):
    #下面一对三引号之间是测试代码,doctest 会搜索这些代码并执行
    #并且根据执行结果与预期结果的匹配程度来测试代码是否正确
    '''add two int/float/str/list/tuple
    >>>add(3, 5)
    8
    >>>add(3.0, 5.0)
    8.0
    >>>add([1,2], [3, 4])
    [1, 2, 3, 4]
    >>>add((1,), (2, 3, 4))
    (1, 2, 3, 4)
    >>>add(1, [3])
    Traceback (most recent call last):
    ...
    TypeError: value1 and value2 must be of the same type
    >>>add(1, '2')
    Traceback (most recent call last):
    ...
```

```
    TypeError: value1 and value2 must be of the same type
    >>> add([1], (2,))
    Traceback (most recent call last):
      ...
    TypeError: value1 and value2 must be of the same type
    >>> add('1234', [1,2,3,4])
    Traceback (most recent call last):
      ...
    TypeError: value1 and value2 must be of the same type
    >>> add({1,2,3}, {3,4,5})
    {1, 2, 3, 4, 5}
    >>> add({1:1}, {2:2})
    Traceback (most recent call last):
      ...
    TypeError: value1 and value2 must be the type of int,float,str,list,tuple or set
    '''
    #下面是正式的功能代码
    if type(value1) not in (int, float, str, list, tuple, set):
        raise TypeError('value1 and value2 must be the type of int,float,str,list,tuple or set')
    if type(value1) != type(value2):
        raise TypeError('value1 and value2 must be of the same type')
    if type(value1) == set:
        return value1 | value2
    else:
        return value1 + value2

if __name__ == '__main__':
    import doctest
    doctest.testmod()
    print(add(3,5))
```

8.9 性 能 测 试

一个高质量的程序,不仅仅要求功能是正确的,还应该是优雅的、高效的、健壮的、安全的。关于 Python 编码规范可以参考第 1 章的内容,代码优化和健壮编程的内容在本书每章的案例中都有所涉及,第 10、12、13、14、16、18 章中部分案例介绍了安全编码的内容。本节简单介绍 Python 程序的性能测试,包括代码运行时间测试和内存占用情况测试。

(1) 使用 time 模块测试代码运行时间。另外,timeit 模块用法见 4.1.2 节。

```
import time

def demo():
    start = time.time()
    for i in range(9999999):
        1+1
    end = time.time()
    print(end-start)

demo()
```

(2) 内存占用情况测试。使用 pip 安装 Python 扩展库 memory_profiler,然后编写并运行下面的代码:

```python
from memory_profiler import profile

@profile
def memory_test():
    test = []
    test.append([8] * 100000)
    test.append([8] * 200000)
    test.append([8] * 300000)
    test.append([8] * 400000)
    test.append([8] * 500000)

memory_test()
```

运行结果如图 8-7 所示。

```
Line #    Mem usage    Increment   Line Contents
================================================
    3     44.6 MiB     44.6 MiB    @profile
    4                              def memory_test():
    5     44.6 MiB     0.0 MiB         test = []
    6     45.4 MiB     0.8 MiB         test.append([8]*100000)
    7     46.9 MiB     1.5 MiB         test.append([8]*200000)
    8     49.2 MiB     2.3 MiB         test.append([8]*300000)
    9     52.3 MiB     3.1 MiB         test.append([8]*400000)
   10     56.1 MiB     3.8 MiB         test.append([8]*500000)
```

图 8-7　内存占用情况测试结果

本 章 小 结

(1) 程序出现异常或错误后是否能够调试程序并快速定位和解决存在的问题是程序员综合水平和能力的重要体现之一。

(2) 异常处理结构可以提高程序的容错性和健壮性,但不建议过多依赖异常处理结构。

(3) 可以继承 Python 内建异常类来实现自定义的异常类。

(4) 可以使用 BaseException 来捕获所有异常,但不建议这样做。

(5) 异常处理结构中主要的关键字有 try、except、else 和 finally。

(6) 异常处理结构中也可以使用 else 子句,当没有异常发生时执行 else 子句中的代码块。

(7) 断言语句 assert 一般用于对程序某个时刻必须满足的条件进行验证。

(8) 上下文管理语句 with 在代码块执行完毕后能够自动还原进入代码块之前的现场或上下文,无论是否发生异常总能保证资源被正确释放。

(9) 异常处理结构的 finally 子句中的代码也可能抛出异常。

(10) 可以通过 3 种方式使用 pdb 模块的调试功能:在交互模式下使用 pdb 模块的方法调试指定语句或函数等、在程序中显式插入断点、执行 Python 程序时指定 pdb 调试模块。

(11) 白盒测试主要通过阅读程序源代码来判断是否符合功能要求,对于复杂的业务逻辑白盒测试难度非常大，一般以黑盒测试为主，白盒测试为辅。

（12）黑盒测试不关心模块的内部实现方式，只关心其功能是否正确，通过精心设计完备的测试用例检验模块的输入输出是否正确来判断其是否符合预定的功能要求。

（13）Python 标准库 unittest 提供了大量用于单元测试的类和方法。

（14）单元测试标准库 unittest 中 TestCase 类的 `setUp()` 和 `tearDown()` 方法分别在每项测试开始之前和完成之后自动调用，常用来执行数据库连接的创建与关闭、文件的打开与关闭等操作，避免编写过多的重复代码。

（15）测试用例及测试代码的设计与编写也可能存在 bug，通过测试并不代表目标代码没有错误，但是一般而言，不能通过测试的模块代码是存在问题的。

（16）除了功能测试，还应对程序进行性能测试与安全性测试，甚至还需要进行规范性测试以保证代码可读性和可维护性。

习 题

1. Python 内建异常类的基类是_____。
2. 断言语句的语法为_____。
3. Python 上下文管理语句的关键字为_____。
4. Python 异常处理结构有哪几种形式？
5. 异常和错误有什么区别？
6. 使用 pdb 模块进行 Python 程序调试主要有哪几种方式？

第 9 章 tkinter 应用开发

常用 GUI 工具集除了标准库 tkinter，还有功能强大的 wxPython、PyGObject、kivy、PyQt、PySide 等，或者也可以利用有关插件和其他语言混合编程以便充分利用其他语言的 GUI。本章以标准库 tkinter 为例介绍 Python 的 GUI 应用开发。

9.1 tkinter 基础

9.1.1 tkinter 常用组件

Python 标准库 tkinter 是对 Tcl/Tk 的进一步封装，是一套完整的 GUI 开发模块的组合或套件，这些模块共同提供了强大的跨平台 GUI 编程功能，所有的源码文件位于 Python 安装目录 lib\tkinter 文件夹中。例如，tkinter 套件中的 ttk 模块提供了 Combobox、Progressbar 和 Treeview 等组件，scrolledtext 模块提供了带滚动条的文本框，messagebox、commondialog、dialog、colorchooser、simpledialog、filedialog 等模块提供了各种对话框，font 模块提供了与字体有关的对象。另外，借助扩展库 tkcalendar 还可以实现日期选择组件。

表 9-1 中列出了 tkinter 常用的组件，ttk、scrolledtext 等其他模块中包含的组件可以查阅相关资料或参考 9.2 节的例题，在作者的微信公众号"Python 小屋"中分享了更多案例。

表 9-1 tkinter 常用的组件

组 件 名 称	说　　明
Button	按钮
Canvas	画布，用于绘制直线、椭圆、多边形等各种图形
Checkbutton	复选框
Entry	单行文本框
Frame	框架，可作为其他组件的容器，常用于对组件进行分组
Label	标签，常用来显示单行文本
Listbox	列表框
Menu	菜单
Message	多行文本标签
Radiobutton	单选按钮，同一组中的单选按钮任何时刻只能有一个处于选中状态
Scrollbar	滚动条
Text	多行文本框
Toplevel	常用于创建新的窗口

本节以 Button 组件和 Entry 组件为例介绍组件的创建和用法，其他组件可以结合 9.2

节的例题和官方文档理解和使用，不再逐个介绍每个组件的详细参数、属性和方法。

下面的代码用来创建并返回一个按钮组件：

```
buttonImportXueshengXinxi = tkinter.Button(root, text='导入学生信息',
                                command=buttonImportXueshengXinxiClick)
```

其中，Button 类创建按钮组件时的参数 root 用来说明这次创建的按钮是要放置在 root 这个应用程序主界面上（也可以是 Toplevel 组件）的；参数 text 用来设置按钮上显示的文本；参数 command 用来指定单击该按钮时要执行的可调用对象；buttonImportXueshengXinxiClick 是一个函数的名字，也就是单击按钮之后会执行该函数中的代码。除此之外，常用的参数还有 background（背景色）、bitmap（按钮上显示的位图）、borderwidth（边框宽度）、cursor（鼠标形状）、font（字体）、foreground（前景色）、justify（文本对齐方式）、height（高度）、width（宽度）等。

按钮创建成功后使用变量 buttonImportXueshengXinxi 保存，然后可以通过这个变量访问按钮属性或调用按钮的方法。例如，按钮组件的 pack() 和 place() 方法用来把按钮组件放置到界面上，下面的代码使用 place() 方法把按钮放置到界面上距离左边界和上边界均为 20 像素的位置并设置宽度和高度：

```
buttonImportXueshengXinxi.place(x=20, y=20, height=30, width=100)
```

下面的代码用来创建一个单行文本框 Entry 组件并放置到界面上：

```
xuehao = tkinter.StringVar(root)
entryXuehao = tkinter.Entry(root, width=150, textvariable=xuehao)
entryXuehao.place(x=100, y=5, width=150, height=20)
```

其中，参数 textvariable 用来指定一个 tkinter 字符串变量 xuehao。当用户在界面上修改了单行文本框的内容以后，在代码中可以通过变量 xuehao 获取最新的内容。同样，如果在代码中修改了变量 xuehao 的值，也会立刻把最新的内容显示到界面上的单行文本框中。除了使用 place() 或 pack() 方法把组件放置到界面上，Entry 组件还支持使用 get() 方法获取单行文本的内容，delete(first, last) 方法删除单行文本框中下标介于 [first, last) 的文本，insert(index, string) 方法在指定位置插入字符串，等等。

9.1.2　tkinter 应用程序开发基本流程

使用 tkinter 开发 GUI 程序的基本流程如下。

（1）编写通用代码，如数据库操作、程序中需要多次调用的函数，这需要在编写代码之前对软件的功能进行分析和整理，最终提炼通用部分的代码封装成类或函数。

（2）搭建界面，放置菜单、标签、按钮、文本框、列表框、组合框等组件，设置组件属性。这个任务可以完全通过编写代码来完成，这也是 9.2 节所有案例采用的方式。这种方式设计界面的好处是代码简练，但是需要程序员记住各种组件的属性，难度较大。如果不喜欢编写代码设计界面，也可以借助 PAGE 软件协助完成，但是生成的界面代码比较啰唆，可以再根据需要进行简化。PAGE 软件的界面如图 9-1 所示。

（3）编写组件的事件处理代码。在上一步中只是搭建了一个软件界面，当用户操作这些界面上的组件时，还需要相应的代码来响应这些操作并完成预定动作和任务。例如，当单

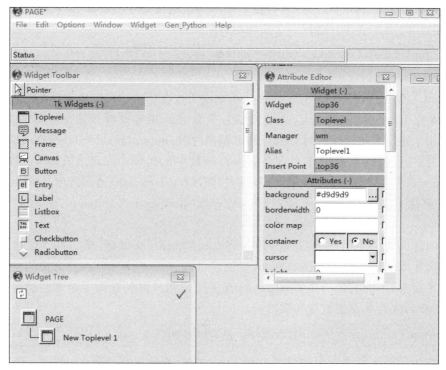

图 9-1　PAGE 软件的界面

击"登录"按钮时,代码应检查输入的用户名和密码是否正确然后决定是否允许登录;单击"保存"按钮时,代码应检查输入的数据是否合法,如果通过检查则把这些数据写入数据库或文件中;单击"发送"按钮时,代码应检查数据是否合法并通过 Socket 或其他方式把这些数据发送给一个或多个接收方;单击"删除"按钮时,代码应删除数据库或文件中符合条件的数据然后更新界面上显示的数据。除了完成预定功能之外,还应该有必要的反馈信息提示用户刚才的操作是成功还是失败,所有这些都应该在软件的设计环节定义好。

(4) 启动消息主循环,启动应用程序。

可以把下面的代码作为 tkinter 应用程序开发框架,然后在需要的位置增加代码。

```
#根据需要导入需要的模块
import tkinter
import tkinter.ttk
import tkinter.messagebox
import tkinter.simpledialog

#这里编写通用代码,或单独放置到另外的模块中再导入

#创建 tkinter 应用程序主窗口
root = tkinter.Tk()
#此处编写设置窗口属性的代码

#此处编写创建窗口上各种组件的代码,以及按钮、组合框等交互式组件的事件响应与处理代码

#启动消息主循环,启动应用程序
root.mainloop()
```

9.2　tkinter 应用案例精选

9.2.1　用户登录界面

用户输入用户名/账号和密码后,系统进行验证,通过验证才可以进行后续的操作。一般而言,用户密码都是经过安全哈希算法加密之后存储到数据库中的,并不直接保存明文。

例 9-1　tkinter 实现用户登录界面。本例主要演示如何使用 tkinter 创建应用程序窗口,以及文本框、按钮和简单消息框等组件的用法,运行界面如图 9-2~图 9-4 所示。

9.2.1

```python
import os
import os.path
import tkinter
import tkinter.messagebox

path = os.getenv('temp')
filename = os.path.join(path, 'info.txt')

root = tkinter.Tk()                              #创建应用程序窗口
root['height'] = 140                             #定义窗口大小
root['width'] = 200
#在窗口上创建标签组件
labelName = tkinter.Label(root, text='User Name:', justify=tkinter.RIGHT,
                          anchor='e', width=80)
labelName.place(x=10, y=5, width=80, height=20)
#创建字符串变量和文本框组件,同时设置关联的变量
varName = tkinter.StringVar(root, value='')
entryName = tkinter.Entry(root, width=80, textvariable=varName)
entryName.place(x=100, y=5, width=80, height=20)

labelPwd = tkinter.Label(root, text='User Pwd:', justify=tkinter.RIGHT,
                         anchor='e', width=80)
labelPwd.place(x=10, y=30, width=80, height=20)
varPwd = tkinter.StringVar(root, value='')       #创建密码文本框
entryPwd = tkinter.Entry(root, show='*', width=80, textvariable=varPwd)
entryPwd.place(x=100, y=30, width=80, height=20)

#尝试自动填写用户名和密码
try:
    with open(filename) as fp:
        n, p = fp.read().strip().split(',')
        varName.set(n)
        varPwd.set(p)
except:
    pass

#创建 Remember me 复选框
rememberMe = tkinter.IntVar(root, value=1)
#选中时变量值为 1,未选中时变量值为 0,默认选中
```

```
checkRemember = tkinter.Checkbutton(root, text='Remember me? ',
                                    variable=rememberMe, onvalue=1, offvalue=0)
checkRemember.place(x=30, y=70, width=120, height=20)

#登录按钮事件处理函数
def login():
    name = entryName.get()                                  #获取用户名和密码
    pwd = entryPwd.get()
    if name == 'admin' and pwd == '123456':
        tkinter.messagebox.showinfo(title='恭喜', message='登录成功!')
        if rememberMe.get() == 1:
                                                            #把登录成功的信息写入临时文件
            with open(filename, 'w') as fp:
                fp.write(','.join((name,pwd)))
        else:
            try:
                os.remove(filename)
            except:
                pass
    else:
        tkinter.messagebox.showerror('警告', message='用户名或密码错误')
#创建按钮组件,同时设置按钮事件处理函数
buttonOk = tkinter.Button(root, text='Login', command=login)
buttonOk.place(x=30, y=100, width=50, height=20)

#取消按钮事件处理函数
def cancel():
    varName.set('')                                         #清空用户输入的用户名和密码
    varPwd.set('')
buttonCancel = tkinter.Button(root, text='Cancel', command=cancel)
buttonCancel.place(x=90, y=100, width=50, height=20)

root.mainloop()                                             #启动消息循环
```

图 9-2　用户登录界面

图 9-3　密码正确

图 9-4　密码错误

9.2.2　选择类组件应用

下面的案例创建了一个包含文本框、单选按钮、复选框、组合框、按钮和列表框等组件的 GUI 应用程序,运行后输入学生姓名并选择年级、班级、性别以及是否班长等信息后,单击 Add 按钮可将该学生信息添加到列表框中。在列表框中选择一项后单击 DeleteSelection 按钮可将其从列表框中删除,没有选择任何项而直接单击该按钮则提示 No Selection。

例 9-2　tkinter 文本框单选按钮、复选框、组合框、按钮和列表框综合运用案例。

```python
import tkinter
import tkinter.ttk
import tkinter.messagebox

root = tkinter.Tk()
root.title('Selection widgets——by Dong Fuguo')          #窗口标题
root['height'] = 400                                     #定义窗口大小
root['width'] = 320
labelName = tkinter.Label(root, text='Name:',            #创建标签
                          justify=tkinter.RIGHT, width=50)
labelName.place(x=10, y=5, width=50, height=20)          #将标签放到窗口上
varName = tkinter.StringVar(value='')                    #与姓名关联的变量
entryName = tkinter.Entry(root, width=120,               #创建文本框
                          textvariable=varName)          #同时设置关联的变量
entryName.place(x=70, y=5, width=120, height=20)
labelGrade = tkinter.Label(root, text='Grade:', justify=tkinter.RIGHT, width=50)
labelGrade.place(x=10, y=40, width=50, height=20)
studentClasses = {'1':['1', '2', '3', '4'],              #模拟学生所在年级
                  '2':['1', '2'],                        #字典键为年级
                  '3':['1', '2', '3']}                   #字典值为班级
comboGrade = tkinter.ttk.Combobox(root,                  #学生年级组合框
                                  values=tuple(studentClasses.keys()), width=50)
comboGrade.place(x=70, y=40, width=50, height=20)
def comboChange(event):                                  #事件处理函数
    grade = comboGrade.get()
    if grade:                                            #动态改变组合框可选项
        comboClass["values"] = studentClasses.get(grade)
    else:
        comboClass.set([])
comboGrade.bind('<<ComboboxSelected>>', comboChange)     #绑定事件处理函数
labelClass = tkinter.Label(root, text='Class:', justify=tkinter.RIGHT, width=50)
labelClass.place(x=130, y=40, width=50, height=20)
comboClass = tkinter.ttk.Combobox(root, width=50)        #学生班级组合框
comboClass.place(x=190, y=40, width=50, height=20)
labelSex = tkinter.Label(root, text='Sex:', justify=tkinter.RIGHT, width=50)
labelSex.place(x=10, y=70, width=50, height=20)

sex = tkinter.IntVar(value=1)               #与性别关联的变量,1表示男;0表示女,默认为男
radioMan = tkinter.Radiobutton(root,                     #单选按钮,男
                               variable=sex, value=1, text='Man')
radioMan.place(x=70, y=70, width=50, height=20)
radioWoman = tkinter.Radiobutton(root, variable=sex, value=0, text='Woman')
radioWoman.place(x=130, y=70, width=70, height=20)

monitor = tkinter.IntVar(value=0)           #与是否班长关联的变量,默认不是班长
checkMonitor = tkinter.Checkbutton(root, text='Is Monitor?', variable=monitor,
                                   onvalue=1,            #选中时变量值为1
                                   offvalue=0)           #未选中时变量值为0
checkMonitor.place(x=20, y=100, width=100, height=20)
```

```python
def addInformation():                                    #按钮事件处理函数
    result = 'Name:' +entryName.get()
    result = result +';Grade:' +comboGrade.get()
    result = result +';Class:' +comboClass.get()
    result = result +';Sex:' +('Man' if sex.get() else 'Woman')
    result = result +';Monitor:' +('Yes' if monitor.get() else 'No')
    listboxStudents.insert(0, result)
buttonAdd = tkinter.Button(root, text='Add', width=40, command=addInformation)
buttonAdd.place(x=130, y=100, width=40, height=20)

def deleteSelection():
    selection = listboxStudents.curselection()
    if not selection:
        tkinter.messagebox.showinfo(title='Information', message='No Selection')
    else:
        listboxStudents.delete(selection)
buttonDelete = tkinter.Button(root, text='DeleteSelection',
                              width=100, command=deleteSelection)
buttonDelete.place(x=180, y=100, width=100, height=20)
listboxStudents = tkinter.Listbox(root, width=300)       #创建列表框组件
listboxStudents.place(x=10, y=130, width=300, height=200)

root.mainloop()
```

将上面的代码保存为 tkinter_selection.pyw 文件,运行后效果如图 9-5 所示。

图 9-5　程序运行效果

9.2.3　简单文本编辑器

下面的案例通过设计一个文本编辑器演示了菜单、文本框、文件对话框等组件的用法,实现了打开文件、保存文件、另存文件,以及文本的复制、剪切、粘贴和查找等功能。

例 9-3　使用 tkinter 实现文本编辑器。

```
import tkinter
import tkinter.filedialog
import tkinter.messagebox
import tkinter.scrolledtext
```

```python
import tkinter.simpledialog

app = tkinter.Tk()
app.title('My Notepad——by Dong Fuguo')
app['width'] = 800
app['height'] = 600

textChanged = tkinter.IntVar(value=0)
filename = ''                                          #当前文件名

menu = tkinter.Menu(app)                               #创建菜单
submenu = tkinter.Menu(menu, tearoff=0)                #File 子菜单
def Open():
    global filename
    #如果内容已改变,先保存
    if textChanged.get():
        yesno = tkinter.messagebox.askyesno(title='Save or not?',
                                            message='Do you want to save?')
        if yesno == tkinter.YES:
            Save()
    filename = tkinter.filedialog.askopenfilename(title='Open file',
                                            filetypes=[('Text files', '*.txt')])
    if filename:
        #清空内容,位置 0.0 是 lineNumber.Column 的表示方法,表示行号和列号
        txtContent.delete(0.0, tkinter.END)
        fp = open(filename, 'r')                       #可根据需要指定编码格式
        txtContent.insert(tkinter.INSERT, fp.read())
        fp.close()
        textChanged.set(0)                             #标记为尚未修改
#创建 Open 菜单并绑定菜单事件处理函数
submenu.add_command(label='Open', command=Open)

def Save():
    global filename
    #如果是第一次保存新建文件,则打开"另存为"窗口
    if not filename:
        SaveAs()
    #如果内容发生改变,则保存文件,可使用 with 关键字改写文件操作的代码
    elif textChanged.get():
        fp = open(filename, 'w')
        fp.write(txtContent.get(0.0, tkinter.END))
        fp.close()
        textChanged.set(0)
submenu.add_command(label='Save', command=Save)

def SaveAs():
    global filename
    #打开"另存为"窗口
    newfilename = tkinter.filedialog.asksaveasfilename(title='Save As',
                                            initialdir=r'c:\\',
                                            initialfile='new.txt')
    #如果指定了文件名,则保存文件,可使用 with 关键字改写
    if newfilename:
```

```python
            fp = open(newfilename, 'w')
            fp.write(txtContent.get(0.0, tkinter.END))
            fp.close()
            filename = newfilename
            textChanged.set(0)
submenu.add_command(label='Save As', command=SaveAs)
submenu.add_separator()                              #添加分隔线
def Close():
    global filename
    Save()
    txtContent.delete(0.0, tkinter.END)
    filename = ''                                    #置空文件名
submenu.add_command(label='Close', command=Close)
#将子菜单关联到主菜单上
menu.add_cascade(label='File', menu=submenu)

#Edit 子菜单
submenu = tkinter.Menu(menu, tearoff=0)
def Undo():
    txtContent['undo'] = True                        #启用 undo 标志
    try:
        txtContent.edit_undo()                       #撤销最后一次操作
    except Exception as e:
        pass
submenu.add_command(label='Undo', command=Undo)

def Redo():
    txtContent['undo'] = True
    try:
        txtContent.edit_redo()
    except Exception as e:
        pass
submenu.add_command(label='Redo', command=Redo)
submenu.add_separator()

def Copy():
    txtContent.clipboard_clear()
    txtContent.clipboard_append(txtContent.selection_get())
submenu.add_command(label='Copy', command=Copy)

def Cut():
    Copy()
    #删除所选内容
    txtContent.delete(tkinter.SEL_FIRST, tkinter.SEL_LAST)
submenu.add_command(label='Cut', command=Cut)

def Paste():
    #如果没有选中内容则直接粘贴到鼠标位置,如果有所选内容则先删除再粘贴
    try:
        txtContent.insert(tkinter.SEL_FIRST, txtContent.clipboard_get())
        txtContent.delete(tkinter.SEL_FIRST, tkinter.SEL_LAST)
        #如果粘贴成功就结束本函数,以免异常处理结构执行完成之后再次粘贴
        return
```

```python
        except Exception as e:
            pass
        txtContent.insert(tkinter.INSERT, txtContent.clipboard_get())
    submenu.add_command(label='Paste', command=Paste)
    submenu.add_separator()

    def Search():
        #获取要查找的内容
        textToSearch = tkinter.simpledialog.askstring(title='Search',
                                                      prompt='What to search?')
        start = txtContent.search(textToSearch, 0.0, tkinter.END)
        if start:
            tkinter.messagebox.showinfo(title='Found', message='Ok')
    submenu.add_command(label='Search', command=Search)
    menu.add_cascade(label='Edit', menu=submenu)

    #Help 子菜单
    submenu = tkinter.Menu(menu, tearoff=0)
    def About():
        tkinter.messagebox.showinfo(title='About', message='Author:Dong Fuguo')
    submenu.add_command(label='About', command=About)
    menu.add_cascade(label='Help', menu=submenu)
    #将创建的菜单关联到应用程序窗口
    app.config(menu=menu)

    #创建文本编辑组件,并自动适应窗口大小
    txtContent = tkinter.scrolledtext.ScrolledText(app, wrap=tkinter.WORD)
    txtContent.pack(fill=tkinter.BOTH, expand=tkinter.YES)
    def KeyPress(event):
        textChanged.set(1)
    txtContent.bind('<KeyPress>', KeyPress)


    app.mainloop()
```

运行结果如图 9-6 所示。

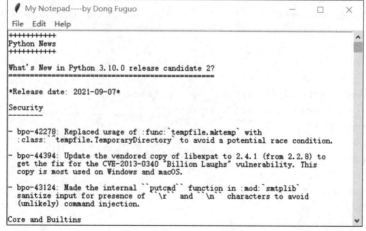

图 9-6 简单文本编辑器

9.2.4 简单画图程序

下面的程序实现了简单的画图功能,包括曲线、直线、矩形、文本的绘制,前景色和背景色的选取和设置,图片文件的打开与显示,以及橡皮擦功能。主要使用了 Canvas 和 Menu 组件,还用到颜色选择对话框,并演示了鼠标事件处理函数的运用。

例 9-4 使用 tkinter 实现画图程序。

```python
import os
import tkinter
import tkinter.simpledialog
import tkinter.colorchooser
import tkinter.filedialog
from PIL import Image
from PIL import ImageGrab


root = tkinter.Tk()
root.title('My Paint——by Dong Fuguo')
root['width'] = 800
root['height'] = 600

#控制是否允许画图的变量,1为允许,0为不允许
canDraw = tkinter.IntVar(value=0)

#控制画图类型的变量,1为曲线,2为直线,3为矩形,4为文本,5为橡皮
what = tkinter.IntVar(value=1)
X = tkinter.IntVar(value=0)                  #记录鼠标位置的变量
Y = tkinter.IntVar(value=0)
foreColor = '#000000'                        #前景色、背景色
backColor = '#FFFFFF'
image = tkinter.PhotoImage()                 #创建画布,设置尺寸和背景色
canvas = tkinter.Canvas(root, bg='white', width=800, height=600)
canvas.create_image(800, 600, image=image)

def onLeftButtonDown(event):                 #按下鼠标左键,允许画图,记录鼠标按下的位置
    canDraw.set(1)
    X.set(event.x)
    Y.set(event.y)
    if what.get() == 4:
        canvas.create_text(event.x, event.y, text=text)
canvas.bind('<Button-1>', onLeftButtonDown)

lastDraw = 0                                 #记录最后绘制图形的 id
def onLeftButtonMove(event):                 #按住鼠标左键移动,画图
    global lastDraw
    if canDraw.get() == 0:
        return
    if what.get() == 1:
        #使用当前选择的前景色绘制曲线
        canvas.create_line(X.get(), Y.get(), event.x, event.y, fill=foreColor)
```

```python
                X.set(event.x)
                Y.set(event.y)
        elif what.get() == 2:
            #绘制直线,先删除刚刚画过的直线,再画一条新的直线
            try:
                canvas.delete(lastDraw)
            except Exception as e:
                pass
            lastDraw = canvas.create_line(X.get(), Y.get(), event.x, event.y,
                                          fill=foreColor)
        elif what.get() == 3:
            #绘制矩形,先删除刚刚画过的矩形,再画一个新的矩形
            try:
                canvas.delete(lastDraw)
            except Exception as e:
                pass
            lastDraw = canvas.create_rectangle(X.get(), Y.get(), event.x, event.y,
                                               fill=backColor, outline=foreColor)
        elif what.get() == 5:
            #橡皮,使用背景色填充10×10的矩形区域,相当于擦除图像
            canvas.create_rectangle(event.x-5, event.y-5, event.x+5, event.y+5,
                                    outline=backColor, fill=backColor)
canvas.bind('<B1-Motion>', onLeftButtonMove)

def onLeftButtonUp(event):                       #鼠标左键抬起,不允许画图
    if what.get() == 2:
        #多绘制一条直线
        canvas.create_line(X.get(), Y.get(), event.x, event.y, fill=foreColor)
    elif what.get() == 3:
        #多绘制一个矩形
        canvas.create_rectangle(X.get(), Y.get(), event.x, event.y,
                                fill=backColor, outline=foreColor)
    canDraw.set(0)
    global lastDraw
    lastDraw = 0                                 #防止切换图形时误删上次绘制的图形
canvas.bind('<ButtonRelease-1>', onLeftButtonUp)

menu = tkinter.Menu(root, tearoff=0)             #创建菜单
def Open():                                      #打开图片文件
    filename = tkinter.filedialog.askopenfilename(title='Open Image',
                                                  filetypes=[('image', '*.png *.gif')])
    if filename:
        global image
        image = tkinter.PhotoImage(file=filename)
        canvas.create_image(80, 80, image=image)
menu.add_command(label='Open', command=Open)

def Save():
    #获取客户区域位置和尺寸,并截图保存,扩展库Pillow见15.2节
    left = int(root.winfo_rootx())
    top = int(root.winfo_rooty())
    width = root.winfo_width()
    height = root.winfo_height()
```

```python
        im = ImageGrab.grab((left, top, left+width, top+height))

        #保存绘制的图片
        filename = tkinter.filedialog.asksaveasfilename(title='保存图片',
                                        filetypes=[('图片文件','*.png')])
        if not filename:
            return
        if not filename.endswith('.png'):
            filename = filename + '.png'
        im.save(filename)
menu.add_command(label='Save', command=Save)

def Clear():                                            #添加菜单,清除所有图像
    for item in canvas.find_all():
        canvas.delete(item)
menu.add_command(label='Clear', command=Clear)
menu.add_separator()                                    #添加分隔线
#创建子菜单,用于选择绘图类型
menuType = tkinter.Menu(menu, tearoff=0)
def drawCurve():
    what.set(1)
menuType.add_command(label='Curve', command=drawCurve)
def drawLine():
    what.set(2)
menuType.add_command(label='Line', command=drawLine)
def drawRectangle():
    what.set(3)
menuType.add_command(label='Rectangle', command=drawRectangle)
def drawText():
    global text
    text = tkinter.simpledialog.askstring(title='Input what you want to draw',
                                    prompt='')
    what.set(4)
menuType.add_command(label='Text', command=drawText)
menuType.add_separator()
def chooseForeColor():                                  #选择前景色
    global foreColor
    foreColor = tkinter.colorchooser.askcolor()[1]
menuType.add_command(label='Choose Foreground Color', command=chooseForeColor)
def chooseBackColor():                                  #选择背景色
    global backColor
    backColor = tkinter.colorchooser.askcolor()[1]
menuType.add_command(label='Choose Background Color', command=chooseBackColor)
def onErase():                                          #橡皮
    what.set(5)
menuType.add_command(label='Erase', command=onErase)
menu.add_cascade(label='Type', menu=menuType)

def onRightButtonUp(event):                             #鼠标右键抬起,弹出菜单
    menu.post(event.x_root, event.y_root)
```

```
canvas.bind('<ButtonRelease-3>', onRightButtonUp)
canvas.pack(fill=tkinter.BOTH, expand=tkinter.YES)
root.mainloop()                                              #启动应用程序
```

程序运行结果如图 9-7 所示。

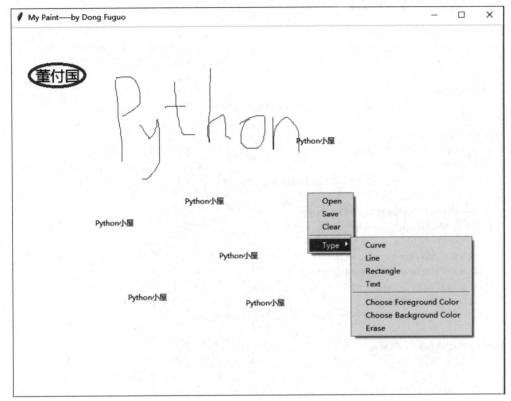

图 9-7　简单画图程序运行结果

9.2.5　电子时钟

下面的案例实现了电子时钟，使用 Label 组件实时显示当前日期和时间，涉及的知识主要有多线程（见第 13 章）、无标题栏、半透明、顶端显示、可拖动窗体的设计。

例 9-5　使用 tkinter 实现电子时钟。

```
import time
import tkinter
import datetime
import threading

app = tkinter.Tk()
app.overrideredirect(True)                    #不显示标题栏
app.attributes('-alpha', 0.9)                 #半透明
app.attributes('-topmost', 1)                 #总是在顶端
app.geometry('110x25+100+100')                #初始大小与位置,110 后面是小写字母 x
labelDateTime = tkinter.Label(app)
labelDateTime.pack(fill=tkinter.BOTH, expand=tkinter.YES)
```

```python
labelDateTime.configure(bg='gray')

X = tkinter.IntVar(value=0)                              #记录鼠标左键按下的位置
Y = tkinter.IntVar(value=0)
canMove = tkinter.IntVar(value=0)                        #窗口是否可拖动
still = tkinter.IntVar(value=1)                          #是否仍在运行

def onLeftButtonDown(event):
    app.attributes('-alpha', 0.4)                        #开始拖动时增加透明度
    X.set(event.x)                                       #鼠标左键按下,记录当前位置
    Y.set(event.y)
    canMove.set(1)                                       #标记窗口可拖动
labelDateTime.bind('<Button-1>', onLeftButtonDown)

def onLeftButtonUp(event):
    app.attributes('-alpha', 0.9)                        #停止拖动时恢复透明度
    canMove.set(0)                                       #鼠标左键抬起,标记窗口不可拖动
labelDateTime.bind('<ButtonRelease-1>', onLeftButtonUp)

def onLeftButtonMove(event):
    if canMove.get() == 0:
        return
    newX = app.winfo_x() +(event.x-X.get())
    newY = app.winfo_y() +(event.y-Y.get())
    g = '110x25+' +str(newX) +'+' +str(newY)
    app.geometry(g)                                      #修改窗口的位置
labelDateTime.bind('<B1-Motion>', onLeftButtonMove)

def onRightButtonDown(event):
    still.set(0)                                         #停止更新时间
    t.join(0.2)                                          #等待子线程结束
    app.destroy()                                        #关闭窗口
labelDateTime.bind('<Button-3>', onRightButtonDown)

def nowDateTime():
    while still.get() == 1:
        now = datetime.datetime.now()                    #获取当前日期和时间
        s = str(now)[:19]
        labelDateTime['text'] = s                        #显示当前时间
        time.sleep(0.2)
t = threading.Thread(target=nowDateTime)                 #创建用于更新时间的线程
t.daemon = True                                          #设置为守护线程,见 13.2.2 节
t.start()                                                #启动线程

app.mainloop()
```

程序运行界面截图如图 9-8 所示。

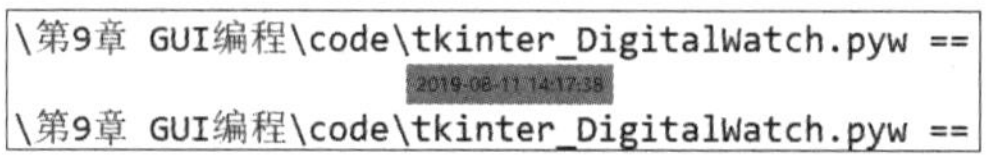

\第9章 GUI编程\code\tkinter_DigitalWatch.pyw ==
2019-06-11 14:17:18
\第9章 GUI编程\code\tkinter_DigitalWatch.pyw ==

图 9-8 电子时钟程序运行界面截图

9.2.6 简易计算器

下面的案例实现了一个简单的计算器程序,涉及的内容有动态组件生成、只读输入框组件等,大家可以在此基础之上尝试进行扩展和增加更多功能。

例 9-6 编写简易计算器。

```python
import re
import tkinter
import tkinter.messagebox

root = tkinter.Tk()
#设置窗口大小和位置,300 和 270 之间是小写字母 x,不是乘号
root.geometry('300x270+400+100')
root.resizable(False, False)                    #不允许改变窗口大小
root.title('简易计算器——董付国')                  #设置窗口标题

#放置用于显示表达式和计算结果的文本框,并设置为只读
contentVar = tkinter.StringVar(root, '')
contentEntry = tkinter.Entry(root, textvariable=contentVar)
contentEntry['state'] = 'readonly'
contentEntry.place(x=10, y=10, width=280, height=20)

#按钮通用代码
def buttonClick(btn):
    content = contentVar.get()
    if content.startswith('.'):                 #如果已有内容是以小数点开头的,则前面加 0
        content = '0' + content
    #根据不同按钮做出相应的处理
    if btn in '0123456789':
        content += btn
    elif btn == '.':
        lastPart = re.split(r'\+|-|\*|/', content)[-1]
        if '.' in lastPart:
            tkinter.messagebox.showerror('错误', '小数点太多了')
            return
        else:
            content += btn
    elif btn == 'C':
        content = ''
    elif btn == '=':
        try:
            #对输入的表达式求值
            content = str(eval(content))
        except:
            tkinter.messagebox.showerror('错误', '表达式错误')
            return
    elif btn in operators:
        if content.endswith(operators):
            tkinter.messagebox.showerror('错误', '不允许存在连续运算符')
            return
        content += btn
```

```python
        elif btn == 'Sqrt':
            n = content.split('.')
            if all(map(lambda x: x.isdigit(), n)):
                content = eval(content) ** 0.5
            else:
                tkinter.messagebox.showerror('错误', '表达式错误')
                return
        contentVar.set(content)

#放置 Clear 按钮和"="按钮
btnClear = tkinter.Button(root, text='Clear', command=lambda:buttonClick('C'))
btnClear.place(x=40, y=40, width=80, height=20)
btnCompute = tkinter.Button(root, text='=', command=lambda:buttonClick('='))
btnCompute.place(x=170, y=40, width=80, height=20)

#放置 10 个数字、小数点和计算平方根的按钮
digits = list('0123456789.') +['Sqrt']
index = 0
for row in range(4):
    for col in range(3):
        d = digits[index]
        index += 1
        btnDigit = tkinter.Button(root, text=d,
                                  command=lambda x=d:buttonClick(x))
        btnDigit.place(x=20+col*70, y=80+row*50, width=50, height=20)

#放置运算符按钮
operators = ('+', '-', '*', '/', '**', '//')
for index, operator in enumerate(operators):
    btnOperator = tkinter.Button(root, text=operator,
                                 command=lambda x=operator:buttonClick(x))
    btnOperator.place(x=230, y=80+index*30, width=50, height=20)

root.mainloop()
```

程序运行界面如图 9-9 所示。

图 9-9　简易计算器程序运行界面

9.2.7　桌面放大镜

下面的案例实现了计算机桌面放大镜功能，运行程序后，会放大鼠标当前位置附近的内容并在鼠标上方进行显示，涉及的内容主要是截图、置顶显示和鼠标事件的响应与处理，其中截图功能用到的 Pillow 扩展库可以查阅本书第 15 章以了解更多用法。

例 9-7　编写计算机桌面放大镜程序。

```python
import tkinter
from PIL import ImageGrab, ImageTk

#创建应用程序主窗口,铺满整个屏幕,并删除标题栏
root = tkinter.Tk()
screenWidth = root.winfo_screenwidth()
screenHeight = root.winfo_screenheight()
root.geometry(f'{screenWidth}x{screenHeight}+0+0')
root.overrideredirect(True)
root.resizable(False, False)
#创建画布,显示全屏截图,以便后面在全屏截图上进行区域截图并进行放大
canvas = tkinter.Canvas(root, bg='white', width=screenWidth, height=screenHeight)
image = ImageTk.PhotoImage(ImageGrab.grab())
canvas.create_image(screenWidth//2, screenHeight//2, image=image)

def onMouseRightClick(event):                          #右键退出桌面放大器程序
    root.destroy()
canvas.bind('<Button-3>', onMouseRightClick)

#截图窗口半径
radius = 20
def onMouseMove(event):
    global lastIm, subIm
    try:
        canvas.delete(lastIm)
    except:
        pass
    x, y = event.x, event.y                            #获取鼠标位置
    #二次截图,放大 3 倍,在鼠标当前位置左上方显示
    subIm = ImageGrab.grab((x-radius, y-radius, x+radius, y+radius))
    subIm = subIm.resize((radius*6, radius*6))
    subIm = ImageTk.PhotoImage(subIm)
    lastIm = canvas.create_image(x-0, y-70, image=subIm)
    #canvas.update()
canvas.bind('<Motion>', onMouseMove)                   #绑定鼠标移动事件处理函数
#把画布对象 canvas 放置到窗体上,两个方向都填满窗口,且自适应窗口大小的改变
canvas.pack(fill=tkinter.BOTH, expand=tkinter.YES)

root.mainloop()
```

程序运行界面如图 9-10 所示。

9.2.8　抽奖程序

下面的案例实现了抽奖程序，使用时可以修改嘉宾名单，然后单击"开始"和"停"按钮来

图 9-10 桌面放大镜程序运行界面

控制界面上名单的滚动实现抽奖功能,涉及的模块主要有多线程、tkinter、time、random 等。本程序也可以用于上课时的随机提问,把嘉宾名单替换为学生名单即可。

例 9-8 编写抽奖程序。

```
import tkinter
import tkinter.messagebox
import time
import random
import threading
import itertools

root = tkinter.Tk()
root.title('随机提问')
root.geometry('260x180+400+300')
root.resizable(False, False)

#关闭程序时执行的函数代码,停止滚动显示学生名单
def closeWindow():
    root.flag = False
    time.sleep(0.1)
    root.destroy()
root.protocol('WM_DELETE_WINDOW', closeWindow)

#模拟学生名单,可以改写代码从数据库或文件中读取学生名单
students = ['张三','李四','王五','赵六','周七','钱八']
#用于控制是否滚动显示学生名单
root.flag = False

def switch():
    root.flag = True
    #随机打乱学生名单
```

```python
        t = students[:]
        random.shuffle(t)
        t = itertools.cycle(t)
        while root.flag:
            #滚动显示
            lbFirst['text'] = lbSecond['text']
            lbSecond['text'] = lbThird['text']
            lbThird['text'] = next(t)
            #数字可以修改,控制滚动速度
            time.sleep(0.1)

    def btnStartClick():
        #每次单击"开始"按钮启动新线程,并禁用"开始"按钮,启用"停"按钮
        t = threading.Thread(target=switch)
        t.start()
        btnStart['state'] = 'disabled'
        btnStop['state'] = 'normal'
    btnStart = tkinter.Button(root, text='开始', command=btnStartClick)
    btnStart.place(x=30, y=10, width=80, height=20)

    def btnStopClick():
        #单击"停"按钮结束滚动显示,弹窗提示中奖名单,修改按钮状态
        root.flag = False
        time.sleep(0.3)
        tkinter.messagebox.showinfo('恭喜', '本次中奖: ' +lbSecond['text'])
        btnStart['state'] = 'normal'
        btnStop['state'] = 'disabled'
    btnStop = tkinter.Button(root, text='停', command=btnStopClick)
    btnStop['state'] = 'disabled'
    btnStop.place(x=150, y=10, width=80, height=20)

    #用于滚动显示学生名单的 3 个 Label 组件
    #可以根据需要添加 Label 组件的数量,但要修改上面的线程函数代码
    lbFirst = tkinter.Label(root, text='')
    lbFirst.place(x=80, y=60, width=100, height=20)
    #红色 Label 组件,表示中奖名单
    lbSecond = tkinter.Label(root, text='')
    lbSecond['fg'] = 'red'
    lbSecond.place(x=80, y=90, width=100, height=20)
    lbThird = tkinter.Label(root, text='')
    lbThird.place(x=80, y=120, width=100, height=20)

    #启动 tkinter 主程序
    root.mainloop()
```

程序运行界面如图 9-11 所示。

9.2.9 猜数游戏

下面的案例实现了 GUI 版猜数游戏,每局游戏开始之前都可以设置猜数的范围和次数,退出游戏时能够提示战绩,主要演示组件状态的控制方法和输入对话框的用法。

例 9-9 编写 GUI 版猜数游戏。

9.2.9

图 9-11 抽奖程序运行界面

```
import random
import tkinter
from tkinter.simpledialog import askinteger
from tkinter.messagebox import showerror, showinfo

root = tkinter.Tk()
root.title('猜数游戏——by 董付国')
root.geometry('280x80+400+300')
root.resizable(False, False)

varNumber = tkinter.StringVar(root, value='0')          #用户实际猜的数
totalTimes = tkinter.IntVar(root, value=0)              #每局游戏允许猜的次数
already = tkinter.IntVar(root, value=0)                 #已猜次数
currentNumber = tkinter.IntVar(root, value=0)           #当前生成的随机数,也就是要猜的数
times = tkinter.IntVar(root, value=0)                   #玩家玩游戏的总局数
right = tkinter.IntVar(root, value=0)                   #玩家猜对的总次数
lb = tkinter.Label(root, text='请输入一个整数: ')
lb.place(x=10, y=10, width=100, height=20)
#用户猜数并输入数字的文本框
entryNumber = tkinter.Entry(root, width=140, textvariable=varNumber)
entryNumber.place(x=110, y=10, width=140, height=20)
#默认禁用,只有开始游戏以后才允许输入
entryNumber['state'] = 'disabled'

#按钮单击事件处理函数,设置并开始一局新游戏,或提交用户猜测的数
def buttonClick():
    if button['text'] == 'Start Game':
        #每次游戏时允许用户自定义数值范围,玩家必须输入正确的数
        #猜数范围的最小数值
        while True:
            try:
                start = askinteger('允许的最小整数', '最小数(必须大于 0)',
                                    initialvalue=1)
                if start != None:
```

```python
                assert start >0
                break
        except:
            pass
#猜数范围的最大数值
while True:
    try:
        end = askinteger('允许的最大整数', '最大数(必须大于 10)',
                        initialvalue=11)
        if end != None:
            assert end>10 and end>start
            break
    except:
        pass

#在用户自定义的数值范围内生成要猜的随机数
currentNumber.set(random.randint(start, end))
#用户自定义一共允许猜几次,玩家必须输入正整数
while True:
    try:
        t = askinteger('最多允许猜几次？', '总次数(必须大于 0)',
                        initialvalue=3)
        if t != None:
            assert t >0
            totalTimes.set(t)
            break
    except:
        pass
        already.set(0)                              #已猜次数初始化为 0
    button['text'] = '剩余次数: ' + str(t)
    varNumber.set('0')                              #把文本框初始化为 0
    entryNumber['state'] = 'normal'                 #启用文本框,允许用户开始输入整数
    times.set(times.get()+1)                        #玩游戏的局数加 1
else:
    total = totalTimes.get()                        #获取本局一共允许猜几次
    current = currentNumber.get()                   #本局游戏的正确答案
    try:
        x = int(varNumber.get())                    #玩家本次猜的数
    except:
        showerror('抱歉', '必须输入整数')
        return
    if x == current:                                #猜对了
        showinfo('恭喜', '猜对了')
        button['text'] = 'Start Game'
        entryNumber['state'] = 'disabled'           #禁用文本框
        right.set(right.get()+1)                    #猜对的次数加 1
    else:
        #本局游戏已猜次数加 1
        already.set(already.get()+1)
        if x > current:
```

```
                showerror('抱歉', '猜的数太大了')
            else:
                showerror('抱歉', '猜的数太小了')
        #本局可猜次数用完了
        if already.get() == total:
            showerror('抱歉', '游戏结束了,正确的数是: ' +
                      str(currentNumber.get()))
            button['text'] = 'Start Game'
            entryNumber['state'] = 'disabled'
        else:
            button['text'] = '剩余次数: ' + str(total-already.get())

#在窗口上创建按钮,并设置事件处理函数
button = tkinter.Button(root, text='Start Game', command=buttonClick)
button.place(x=10, y=40, width=250, height=20)

#关闭程序时提示战绩
def closeWindow():
    message = '共玩游戏 {0} 次,猜对 {1} 次!\n 欢迎下次再玩!'
    message = message.format(times.get(), right.get())
    showinfo('战绩', message)
    root.destroy()
root.protocol('WM_DELETE_WINDOW', closeWindow)

root.mainloop()
```

程序运行界面如图 9-12 所示。

9.2.10 图片查看器程序

下面的案例实现了图片查看器程序,可以通过"上一张"和"下一张"按钮来查看当前文件夹中的所有图片文件,可以改写代码实现查看指定文件夹中图片文件的功能。

图 9-12 猜数游戏程序运行界面

例 9-10 编写图片查看器程序。

```
import os
import tkinter
import tkinter.messagebox
from PIL import Image, ImageTk

root = tkinter.Tk()
root.geometry('430x650+40+30')
root.resizable(False, False)
root.title('使用 Label 显示图片')

#获取当前文件夹中所有图片文件列表
suffix = ('.jpg', '.bmp', '.png')
pics = [p for p in os.listdir('.') if p.endswith(suffix)]

current = -1
def changePic(flag):
    '''flag = -1 表示上一个,flag = 1 表示下一个'''
    global current
```

```python
            new = current + flag
            if new < 0:
                tkinter.messagebox.showerror('', '这已经是第一张图片了')
            elif new >= len(pics):
                tkinter.messagebox.showerror('', '这已经是最后一张图片了')
            else:
                pic = pics[new]                        #获取要切换的图片文件名
                im = Image.open(pic)                   #创建 Image 对象并进行缩放
                w, h = im.size
                #这里假设用来显示图片的 Label 组件尺寸为 400×600
                if w > 400:
                    h = int(h*400/w)
                    w = 400
                if h > 600:
                    w = int(w*600/h)
                    h = 600
                im = im.resize((w,h))
                im1 = ImageTk.PhotoImage(im)           #创建 PhotoImage 对象,并设置 Label 组件图片
                lbPic['image'] = im1
                lbPic.image = im1

                current = new

        def btnPreClick():                             #"上一张"按钮
            changePic(-1)
        btnPre = tkinter.Button(root, text='上一张', command=btnPreClick)
        btnPre.place(x=100, y=20, width=80, height=30)
        def btnNextClick():                            #"下一张"按钮
            changePic(1)
        btnNext = tkinter.Button(root, text='下一张', command=btnNextClick)
        btnNext.place(x=230, y=20, width=80, height=30)
        #用来显示图片的 Label 组件
        lbPic = tkinter.Label(root, text='test', width=400, height=600)
        changePic(1)
        lbPic.place(x=10, y=50, width=400, height=600)

        root.mainloop()
```

程序运行界面如图 9-13 所示。

9.2.11 在 tkinter 应用程序中使用日历选择组件

tkinter 本身没有提供日历选择组件,可以安装扩展库 tkcalendar 来实现,本节案例演示了这个用法。

例 9-11 编写 tkinter 应用程序,运行程序后单击按钮弹出子窗口,在子窗口中实现日历选择组件,并实时在主窗口中显示选择的日期。请自行运行程序并观察效果。

图 9-13 图片查看器程序运行界面

```
import tkinter
import tkinter.messagebox
from datetime import date
from tkcalendar import Calendar

root = tkinter.Tk()
root.title('应用程序主界面')
root.geometry('400x300+600+400')

def show_calendar():
    #创建子窗口
    child = tkinter.Toplevel(root)
    child.geometry('240x180+400+200')
    child.title('选择日期')
    child.resizable(False, False)
    #创建日历选择组件
    cld = Calendar(child, selectmode='day', date_pattern='y-mm-dd',
                   mindate=date(1970,1,1), maxdate=date(2060,1,1),
                   background='gray', foreground='blue', bordercolor='black',
                   weekendbackground='green', weekendforeground='purple',
                   firstweekday='monday', weekenddays=[6,7],
                   year=2023, month=9, day=2)
    #实时在主窗口中显示当前选择的日期
    child.bind('<Button-1>', lambda e:txt.config(text=cld.get_date()))
    cld.pack()
```

```
        #打开子窗口
        child.mainloop()
button = tkinter.Button(root, text='选择日期', command= show_calendar)
button.place(x=10, y=20, width=80, height=20)
txt = tkinter.Label(root, text='没有选择日期')
txt.place(x=10, y=50, width=80, height=20)
root.mainloop()
```

本 章 小 结

（1） Python 标准库 tkinter 是对 Tcl/Tk 的进一步封装，与 tkinter.ttk 和 tkinter.tix 以及其他几个模块共同提供了强大的跨平台 GUI 编程的功能，IDLE 就是用 tkinter 开发的。

（2） 可以使用 `tkinter.StringVar()` 创建与特定组件关联的字符串变量，使用 `tkinter.IntVar()` 创建与特定组件关联的整型变量，使用 `tkinter.BooleanVar()` 创建布尔变量。

（3） 组合框组件比较常用的事件是 `<<ComboboxSelected>>`，表示选择项发生了改变。

（4） tkinter.filedialog 可用于显示打开文件和另存文件对话框。

（5） tkinter.colorchooser 可用于显示颜色选择器对话框。

（6） tkinter 中 Tk 类的对象可通过调用 `attributes('-alpha', 0.9)` 类似的语句对透明度进行设置。

习　题

1. 设计一个窗体，并放置一个按钮，单击按钮后弹出颜色对话框，关闭颜色对话框后提示选中的颜色。

2. 设计一个窗体，并放置一个按钮，按钮默认文本为"开始"，单击按钮后文本变为"结束"，再次单击后变为"开始"，循环切换。

3. 设计一个窗体，模拟 QQ 登录界面，当用户输入账号 123456 和密码 654321 时提示正确，否则提示错误。

4. 查阅资料，编写程序实现屏幕取色功能，使用鼠标单击屏幕任意位置，返回颜色值。

5. 查阅资料，编写程序实现倒计时启用的按钮。

6. 查阅资料，编写程序实现倒计时自动关闭的窗口。

7. 查阅资料，编写程序实现个人密码管理器，要求把密码加密后存储在 SQLite 数据库中，可以自行设计加密和解密算法。SQLite 数据库操作见第 14 章。

8. 查阅资料，编写程序生成能自动跳转到指定网页的二维码，要求网页 URL 在界面上输入。

9. 查阅资料，改写例 9-8，从 Excel 文件中读取学生名单，并语音读出中奖名单。

第 10 章 网络程序设计

Socket 是计算机之间进行网络通信的一套程序接口,最初由 Berkeley 大学研发,目前已经成为网络编程的标准,可以实现跨平台的数据传输。Socket 是网络通信的基础,相当于在发送端和接收端之间建立了一个管道来实现数据和命令的相互传递。Python 提供了 socket 模块,对 Socket 进行了封装,支持 Socket 接口的访问,大幅简化了程序的开发步骤,提高了开发效率。Python 还提供了 urllib 等大量模块可以对网页内容进行读取和处理,在此基础上结合多线程编程(见第 13 章)以及其他有关模块可以快速开发网页爬虫之类的应用。可以使用 Python 语言编写 CGI 程序,也可以把 Python 代码嵌入网页中运行。借助于 web2py、Flask、Django 框架,可以快速开发网站应用。本章将重点介绍 socket 模块和爬虫的应用开发,更多内容可参考作者另一本书《Python 网络程序设计》(微课版)。

10.1 计算机网络基础知识

1. 网络体系结构

目前较为主流的网络体系结构是 ISO/OSI 参考模型和 TCP/IP 协议族。这两种体系结构都采用了分层设计和实现的方式。例如,ISO/OSI 参考模型从上而下划分为应用层、表示层、会话层、传输层、网络层、数据链路层和物理层,TCP/IP 则将网络划分为应用层、传输层、网络层、数据链路层和物理层。分层设计的好处是,各层可以独立设计和实现,只要保证相邻层之间的调用规范和调用接口不变,就可以方便、灵活地改变某层的内部实现以进行优化或完成其他需求。

2. 网络协议

网络协议是计算机网络中为进行数据交换而建立的规则、标准或约定的集合。网络协议的三要素分别为语法、语义和时序。简单地讲,可以这么理解,语法规定数据和指令的格式,语义表示要做什么,时序规定各种事件出现的顺序。

(1) 语法。语法规定了用户数据与控制信息的结构与格式。

(2) 语义。语义用来解释控制信息每部分的含义,规定了需要发出何种控制信息,以及需要完成的动作和做出什么样的响应。

(3) 时序。时序是对事件发生顺序的详细说明,也可称为"同步"。

3. 应用层协议

应用层协议直接与最终用户进行交互,定义了运行在不同终端系统上的应用进程如何相互传递报文。下面简单列出几种常见的应用层协议。

(1) DNS。域名服务,用于实现域名与 IP 地址的转换。

(2) FTP。文件传输协议,可以通过网络在不同平台之间实现文件的传输。

(3) HTTP。超文本传输协议。

(4) SMTP。简单邮件传输协议。

（5）TELNET。远程上机协议。

4. 传输层协议

在传输层主要运行着 TCP 和 UDP 两个协议，TCP 是面向连接的、具有质量保证的可靠传输协议，但开销较大且只能单播；UDP 是尽最大能力传输的无连接协议，开销小且支持组播和广播，但存在丢包和乱序的可能，常用于视频点播（Video On Demand，VOD）之类的应用。TCP 和 UDP 并没有优劣之分，仅仅是适用场合有所不同。在传输层，使用端口号来标识和区分具体的应用层进程，每当创建一个应用层网络进程时系统就会自动分配一个端口号与之关联，是实现网络上端到端通信的重要基础。

5. IP 地址

IP 运行于网络体系结构的网络层，是网络互连的重要基础。IP 地址（32 位或 128 位二进制数）用来标识网络上的主机，在公开网络上或同一个局域网内部，每台主机都必须使用不同的 IP 地址；由于网络地址转换（Network Address Translation，NAT）和代理服务器等技术的广泛应用，不同内网之间的主机 IP 地址（第一组数字为 192 或 10）可以相同并且可以互不影响地正常工作。IP 地址与端口号共同来标识网络上特定主机上的特定应用进程，俗称 Socket 或套接字。

6. MAC 地址

MAC 地址也称网卡地址或物理地址，是一个 48 位的二进制数，用于标识不同的网卡物理地址。本机 IP 地址和 MAC 地址可以在命令提示符窗口中使用 ipconfig/all 命令查看，如图 10-1 所示。

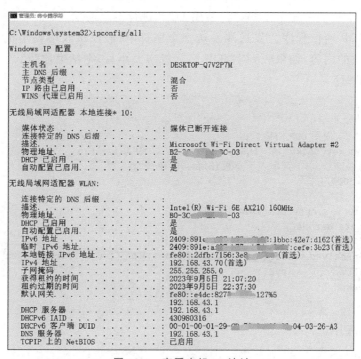

图 10-1 查看本机 IP 地址

10.2 UDP 和 TCP 编程基础

如 10.1 节所述,UDP 和 TCP 是网络体系结构中传输层运行的两大重要协议,TCP 适用于对效率要求相对低,对准确性要求相对高的场合,如文件传输、电子邮件等;UDP 适用于对效率要求相对高,对准确性要求相对低的场合,如视频点播、网络语音通话等。在 Python 中,主要使用 socket 模块来支持 TCP 和 UDP 编程。

10.2.1 UDP 编程

UDP 属于无连接协议,不需要首先建立连接(也可以建立连接,见作者另一本教材《Python 网络程序设计》(微课版)),可以直接向接收方发送信息。UDP 编程经常用到的 socket 模块函数和套接字对象方法有 3 个。

(1) socket([family[,type[,proto]]]):socket 模块函数,创建一个 Socket 对象,其中 family 为 socket.AF_INET 表示 IPv4,socket.AF_INET6 表示 IPv6;type 为 SOCK_STREAM 表示 TCP,SOCK_DGRAM 表示 UDP。

(2) sendto(string,address):套接字对象方法,把 string 指定的内容(字节串)发送给 address 指定的地址,其中 address 是一个包含接收方主机 IP 地址和应用进程端口号的元组,格式为(IP 地址,端口号)。

(3) recvfrom(bufsize[,flags]):套接字对象方法,用于接收数据(字节串)。

下面通过一个示例来简单了解如何使用 UDP 进行网络通信。

10.2.1

例 10-1 UDP 通信程序。发送端发送一个字符串,假设接收端在本机 5000 端口进行监听,并显示接收的内容,如果收到字符串 bye(忽略大小写)则结束监听。

(1) 接收端程序。

```
import socket

#使用 IPv4 协议,使用 UDP 传输数据
s = socket.socket(socket.AF_INET, socket.SOCK_DGRAM)
#绑定端口和端口号,空字符串表示本机任何可用 IP 地址
s.bind(('', 5000))
while True:
    data, addr = s.recvfrom(1024)
    data = data.decode()
    #显示接收到的内容,addr 为发送端的套接字地址
    print('received message:{0} from PORT {1[1]} on {1[0]}'.format(data, addr))
    if data.lower() == 'bye':
        break
s.close()
```

(2) 发送端程序。

```
import sys
import socket

s = socket.socket(socket.AF_INET, socket.SOCK_DGRAM)
```

```
#假设 192.168.0.103 是接收端机器的 IP 地址,假设双方均使用 UTF-8 编写格式
s.sendto(sys.argv[1].encode() , ('192.168.0.103', 5000))
s.close()
```

在上面的发送端程序中假设接收端主机 IP 地址为 192.168.0.103,可能与你的计算机配置并不一样。可以在命令提示符环境中使用命令 ipconfig/all 查看本机 IP 地址,如图 10-1 所示,然后对发送端代码中的 IP 地址做相应修改。如果对命令提示符不熟悉,也可以使用下面的代码来获取本机 IP 地址和 MAC 地址。

```
import socket
import uuid

ip = socket.gethostbyname(socket.gethostname())
node = uuid.getnode()
macHex = uuid.UUID(int=node).hex[-12:]
mac = []
for i in range(len(macHex))[::2]:
    mac.append(macHex[i:i+2])
mac = ':'.join(mac)
print('IP:', ip)
print('MAC:', mac)
```

将例题中两段代码分别保存为 receiver.py 和 sender.py,然后启动一个命令提示符环境并运行接收端程序,这时接收端程序处于阻塞状态,接下来再启动一个新的命令提示符环境并运行发送端程序,此时会看到接收端程序继续运行并显示接收到的内容以及发送端程序所在计算机 IP 地址和占用的端口号。当发送端发送字符串 Bye 后,接收端程序结束,此后再次运行发送端程序时接收端没有任何反应,但发送端程序也并不报错。这正是 UDP 协议的特点,即"尽最大努力传输",并不保证非常好的服务质量。UDP 通信程序运行结果如图 10-2 所示,代码同样适用于 Python 3.9 及更高版本。

图 10-2　UDP 通信程序运行结果

10.2.2 TCP 编程

TCP 一般用于要求可靠数据传输的场合。编写 TCP 程序时经常需要用到的 socket 模块的函数或套接字对象方法主要有 8 个。

(1) socket(): socket 模块函数,详见 10.2.1 节。
(2) connect(address): 套接字对象方法,连接远程计算机。
(3) send(bytes[,flags]): 套接字对象方法,发送数据(字节串)。
(4) recv(bufsize[,flags]): 套接字对象方法,接收数据(字节串)。
(5) bind(address): 套接字对象方法,绑定地址。
(6) listen(backlog): 套接字对象方法,开始监听,等待客户端连接。
(7) accept(): 套接字对象方法,响应客户端的请求。
(8) sendall(data[,flags]): 套接字对象方法,发送全部数据(字节串)。

例 10-2 TCP 通信程序。使用 TCP 进行通信需要首先在客户端和服务端之间建立连接,并且要在通信结束后关闭连接以释放资源。TCP 能够提供比 UDP 更好的服务质量,通信可靠性有本质上的提高。下面的代码简单模拟了机器人聊天软件原理,服务端提前建立好字典,然后根据接收到的内容自动回复。

(1) 服务端程序。

例 10-2

```python
import socket
from sys import exit
from struct import pack, unpack
from os.path import commonprefix

#缓冲区大小,或者说一次能够接收的最大字节串长度
BUFFER_SIZE = 9012

words = {'how are you?': 'Fine,thank you.', 'how old are you?': '38',
         'what is your name?': 'Dong FuGuo',
         "what's your name?": 'Dong FuGuo',
         'where do you work?': 'University', 'bye': 'Bye'}

#空字符串表示本地所有 IP 地址
#如果需要绑定到本地特定的 IP 地址,可以明确指定,例如'192.168.9.1'
HOST, PORT = '', 50007
#创建 TCP 套接字,绑定 socket 地址
sock_server = socket.socket(socket.AF_INET, socket.SOCK_STREAM)
sock_server.bind((HOST, PORT))
#声明自己为服务端套接字,开始监听,准备接收一个客户端连接
#参数表示拒绝新连接之前处于等待状态的客户端数量
sock_server.listen(10)
print('Listening on port:', PORT)

#阻塞,成功接收一个客户端连接请求之后返回新的套接字和对方地址
try:
    conn, addr = sock_server.accept()
except:
    #接收客户端连接失败,服务器故障,直接退出
    exit()
```

```python
print('Connected by', addr)
#开始聊天,使用新套接字收发信息
while True:
    #接收一个整数打包后的字节串,表示对方本次发送的实际字节串长度
    int_bytes = b''
    #在 struct 序列化规则中,整数被打包为长度为 4 的字节串
    rest = 4
    #在高并发网络服务器中,无法保证能够一次接收完 4 字节,使用循环更可靠
    #使用 TCP 通信时,必须保证接收方恰好收完发送方的数据,不能多,也不能少
    while rest > 0:
        #接收数据时自动分配缓冲区
        temp = conn.recv(rest)
        #收到空字节串,表示对方套接字已关闭
        if not temp:
            break
        int_bytes = int_bytes + temp
        rest = rest - len(temp)
    #前面的 while 循环没有接收到数据或者没有收够 4 字节
    #表示对方已结束通信或者网络故障
    if rest > 0:
        break

    #rest 表示接下来需要接收的字节串长度,unpack()的结果是一个元组
    rest = unpack('i', int_bytes)[0]
    data = b''
    while rest > 0:
        #要接收的字节串长度可能非常大,限制一次最多接收 BUFFER_SIZE 字节
        temp = conn.recv(min(rest, BUFFER_SIZE))
        data = data + temp
        rest = rest - len(temp)
    #接收数据不完整,套接字可能损坏
    if rest > 0:
        break

    #删除字符串中可能存在的连续多个空格
    data = ' '.join(data.decode().split())
    print('Received message:', data)
    #尽量猜测对方要表达的真正意思
    m = 0
    key = ''
    for k in words.keys():
        #与某个"键"有超过 70%的共同前缀,认为对方就是想问这个问题
        if len(commonprefix([k, data])) > len(k) * 0.7:
            key = k
            break
        #使用选择法,选择一个重合度较高(也就是共同单词最多)的"键"
        length = len(set(data.split())&set(k.split()))
        if length > m:
            m = length
            key = k
    #选择合适的信息进行回复
```

```
        reply = words.get(key, 'Sorry.').encode()
        #发送数据时自动确定缓冲区长度
        conn.sendall(pack('i', len(reply)) + reply)

conn.close()
sock_server.close()
```

（2）客户端程序。

```
import socket
from sys import exit
from struct import pack, unpack

#服务端主机 IP 地址和端口号
#如果服务端和客户端不在同一台计算机,需要自己修改变量 HOST 的值
HOST, PORT = '127.0.0.1', 50007
sock = socket.socket(socket.AF_INET, socket.SOCK_STREAM)
#设置超时时间,避免服务器不存在时客户端长时间等待或 GUI 无响应
sock.settimeout(0.3)
try:
    #连接服务器,成功后设置当前套接字为阻塞模式
    sock.connect((HOST, PORT))
    sock.settimeout(None)
except Exception as e:
    print('Server not found or not open')
    exit()

while True:
    msg = input('Input the content you want to send:').encode()
    #发送数据
    sock.sendall(pack('i', len(msg))+msg)
    #从服务端接收数据
    #客户端的实现可以比服务端简单一些,不需要循环接收来保证数据完整
    length = unpack('i', sock.recv(4))[0]
    data = sock.recv(length).decode()
    print('Received:', data)
    if msg.lower() == b'bye':
        break
#关闭连接
sock.close()
```

将上面代码分别保存为 chatserver.py 和 chatclient.py 文件,然后启动一个命令提示符环境并运行服务端程序,服务端开始监听;启动一个新的命令提示符环境并运行客户端程序,服务端提示连接已建立;在客户端输入要发送的信息后,服务端会根据提前建立的字典来自动回复。服务端每次都在固定的端口监听,而客户端每次建立连接时可能会使用不同的端口(如果需要也可以使用 bind()方法绑定本地地址),TCP 通信程序运行结果如图 10-3 所示。代码同样适用于 Python 3.9 及更高版本,本书配套课件中提供了增强版的服务端程序。

第10章 网络程序设计

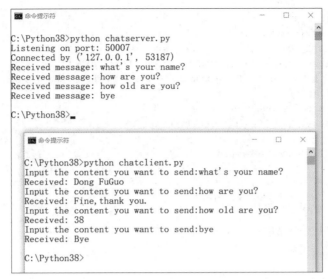

图10-3 TCP通信程序运行结果

10.3 网络编程案例精选

10.3.1 网络嗅探器

嗅探器程序可以检测本机所在局域网内的网络流量和数据包收发情况，对于网络管理具有重要作用，属于系统运维内容之一。为了实现网络流量嗅探，需要将网卡设置为混杂模式，并且运行嗅探器程序的用户账号需要拥有系统管理员权限。

例10-3 网络嗅探器程序。下面的代码运行60s，然后输出本机所在局域网内非本机发出的数据包，并统计不同主机发出的数据包数量。关于多线程的知识可参考第13章。

```
import time
import socket
import threading

activeDegree = dict()
flag = True
def main():
    global activeDegree, flag
    #the public network interface
    HOST = socket.gethostbyname(socket.gethostname())
    #create a raw socket and bind it to the public interface
    s = socket.socket(socket.AF_INET, socket.SOCK_RAW, socket.IPPROTO_IP)
    s.bind((HOST, 0))
    #include IP headers
    s.setsockopt(socket.IPPROTO_IP, socket.IP_HDRINCL, 1)
    #receive all packages
    s.ioctl(socket.SIO_RCVALL, socket.RCVALL_ON)
    #receive a package every loop
```

```
            while flag:
                c = s.recvfrom(65535)
                host = c[1][0]
                activeDegree[host] = activeDegree.get(host, 0) +1
                if c[1][0] != '10.2.1.8':  #suppose 10.2.1.8 is the IP address of current host
                    print(c)
            #disabled promiscuous mode
            s.ioctl(socket.SIO_RCVALL, socket.RCVALL_OFF)
            s.close()
t = threading.Thread(target=main)
t.start()
time.sleep(60)
flag = False
t.join()
for item in activeDegree.items():
    print(item)
```

10.3.2 多进程端口扫描器

在网络安全和黑客领域,端口扫描是经常用到的技术,可以探测指定主机上是否开放了特定端口,进一步判断主机上是否运行某些重要的网络服务,最终判断是否存在潜在的安全漏洞,从一定意义上讲也属于系统运维的范畴。

例 10-4　端口扫描器程序。下面代码模拟了端口扫描器的工作原理,并采用多进程技术提高扫描速度,关于多进程编程可参考第 13 章。

```
import socket
import multiprocessing

def ports(ports_service):
    #获取常用端口对应的服务名称
    for port in list(range(1, 100)) + [143, 145, 113, 443, 445, 3389, 8080]:
        try:
            ports_service[port] = socket.getservbyport(port)
        except socket.error:
            pass

def ports_scan(host, ports_service):
    ports_open = []
    try:
        sock = socket.socket(socket.AF_INET, socket.SOCK_STREAM)
        #超时时间的不同会影响扫描结果的精确度
        sock.settimeout(0.01)
    except socket.error:
        print('socket creation error')
        sys.exit()
    for port in ports_service:
        try:
            sock.connect((host,port))             #尝试连接指定端口
            ports_open.append(port)               #记录打开的端口
            sock.close()
        except socket.error:
```

```
            pass
    return ports_open

if __name__ == '__main__':
    m = multiprocessing.Manager()
    ports_service = dict()
    results = dict()
    ports(ports_service)
    #创建进程池,允许最多8个进程同时运行
    pool = multiprocessing.Pool(processes=8)
    net = '10.2.1.'
    for host_number in map(str, range(8, 10)):
        host = net + host_number
        #创建一个新进程,同时记录其运行结果
        results[host] = pool.apply_async(ports_scan, (host, ports_service))
        print('starting'+host + '…')
    pool.close()                              #关闭进程池,close()必须在 join()之前调用
    pool.join()                               #等待进程池中的进程全部执行结束

    for host in results:                      #输出扫描结果
        print('='*30)
        print(host,'.'*10)
        for port in results[host].get():
            print(port, ':', ports_service[port])
```

10.3.3 查看本机所有联网程序信息

Python 扩展库 psutil 是非常成熟的系统运维库,其中也提供了网络管理所需要的部分功能,例如查看本机联网程序,本节通过一个实际案例演示这个功能的用法。

例 10-5 查看本机所有联网程序信息。代码中 psutil 模块的 net_connections()函数用来查看联网进程 ID,Process 类可以根据进程 ID 查找对应的应用程序文件路径,但尝试获取系统进程信息时会引发异常,在代码中使用异常处理结构跳过了这些系统进程。

```
from os.path import basename
from psutil import net_connections, Process

for conn in net_connections('all'):
    laddr, raddr, status, pid = conn[3:]
    if not raddr:
        continue
    try:
        filename = basename(Process(pid).exe())
    except:
        pass
    else:
        msg = '''\n 程序文件名: {}
本地地址: {}
远程地址: {}
连接状态: {}'''.format(filename, laddr, raddr, status)
        print(msg)
```

10.3.4　查看局域网内 IP 地址与 MAC 地址的对应关系

在本节案例中使用 ARP 获取局域网内所有计算机的 IP 地址与 MAC 地址，首先使用系统命令 arp 获取 ARP 表并生成文本文件，然后从文件中读取和解析这些信息。

例 10-6　查看局域网内 IP 地址与 MAC 地址的对应关系。

```python
import os
from socket import gethostbyname, gethostname

#获取本机 IP 地址
host = gethostbyname(gethostname())
#获取 ARP 表,建议使用记事本打开文件 temp.txt 查看内容,方便理解代码
os.system('arp -a >temp.txt')
with open('temp.txt') as fp:
    for line in fp:
        line = line.split()[:2]
        if line and line[0].startswith(host[:4]) and (not line[0].endswith('255')):
            print(':'.join(line))
```

10.3.5　查看本机网络流量

在本节案例中，使用系统运维扩展库 psutil 提供的函数 net_io_counters()查看本机收发数据情况，该函数返回值前两项分别为上行数据和下行数据的数量（单位是字节），根据单位时间内收发数据情况可以计算本机网络流量。

例 10-7　编写本机网络流量监视器。

```python
import time
import psutil

def main():
    #初始流量情况
    traffic_io = psutil.net_io_counters()[:2]
    while True:
        #0.5s 之后再次获取流量情况
        time.sleep(0.5)
        traffic_ioNew = psutil.net_io_counters()[:2]
        diff = traffic_ioNew[0]-traffic_io[0], traffic_ioNew[1]-traffic_io[1]
        #记录新的流量情况,以便下次比较和计算
        traffic_io = traffic_ioNew
        #乘以 2 是因为 0.5s 查看一次,除以 1024 是为了把单位变成 KB
        diff = tuple(map(lambda x: x*2/1024, diff))
        message = '↑{0[0]:#f} kbytes/s\t↓{0[1]:#f} kbytes/s'.format(diff)
        print(message)
```

10.3.6

10.3.6　局域网内服务器自动发现

在本节案例中，服务器使用 UDP 定时在局域网内群发信息，客户端监听指定的端口接收服务器发送的信息，收到后则输出服务器 IP 地址。在实际应用中，可以修改代码实现客户端收到信息后自动修正服务器 IP 地址，然后进行必要的操作。

例 10-8　使用 UDP 广播实现局域网内服务器自动发现。

（1）服务端程序。

```python
import socket
import time

def sendServerIP():
    sock = socket.socket(socket.AF_INET, socket.SOCK_DGRAM)
    while True:
        IP = socket.gethostbyname(socket.gethostname())
        #255 表示广播地址
        IP = IP[:IP.rindex('.')] + '.255'
        #发送广播信息
        sock.sendto('ServerIP'.encode(), (IP, 5050))
        time.sleep(1)

print('Server started…')
sendServerIP()
```

（2）客户端程序。

```python
import socket
import time

def findServer():
    sock = socket.socket(socket.AF_INET, socket.SOCK_DGRAM)
    sock.bind(('', 5050))
    while True:
        #接收信息,没有消息到达时处于阻塞状态
        data, addr = sock.recvfrom(1024)
        #检查是否为服务器广播的信息
        if data.decode() == 'ServerIP':
            print(addr[0])
        #休息 1s,其实没有必要,可以安全地删除下一行代码
        time.sleep(1)
findServer()
```

10.3.7　多线程＋Socket 实现素数远程查询

在本节案例中,服务器程序使用一个线程不断搜索并存储素数,使用另一个线程响应客户端程序查询一个正整数是否为素数的请求。

例 10-9　多线程＋Socket 实现素数远程查询,代码解释可扫描二维码观看微课。

（1）服务端程序。

例 10-9

```python
import socket
from threading import Thread

primes = {2, 3, 5}
def getPrimes():
    def isPrime(n):
        m = n % 6
        if m!=1 and m!=5:
```

```
                return False
        else:
            for i in range(3, int(n**0.5)+1, 2):
                if n%i == 0:
                    return False
                else:
                    return True
    num = 7
    while True:
        if isPrime(num):
            primes.add(num)
        num = num + 2

def recieveNumber():
    sock = socket.socket(socket.AF_INET, socket.SOCK_STREAM)
    sock.bind(('', 5005))
    sock.listen(1)
    while True:
        conn, _ = sock.accept()
        data = int(conn.recv(1024))
        if data > max(primes):
            conn.sendall(b'too big, wait a moment.')
        elif data in primes:
            conn.sendall(b'yes')
        else:
            conn.sendall(b'no')

Thread(target=getPrimes).start()
Thread(target=recieveNumber).start()
```

（2）客户端程序。

```
import socket

while True:
    data = input('Input an integer(q to quit):').strip()
    if not data:
        continue
    if data == 'q':
        break
    if data.isdigit():
        sock = socket.socket(socket.AF_INET, socket.SOCK_STREAM)
        try:
            sock.connect(('127.0.0.1', 5005))
        except:
            print('server not found')
            exit(0)
        sock.sendall(data.encode())
        print(sock.recv(100).decode())
        sock.close()
```

10.3.8　建立和使用 TCP 长连接

一般来说，客户端连接服务端之后每次通信只发送少量数据，完成会话之后立刻断开连接释放资源，需要时再次发起连接请求，这样可以减轻服务端压力。但建立连接和释放连接本身也是需要时间的，如果频繁创建连接和释放连接反而会浪费服务器资源，所以有些场合中需要长时间保持连接。默认情况下，如果对方长时间没有收发数据，TCP 连接会自动断开，以免服务端长期存在半开放连接浪费资源。如果需要长时间保持连接，需要显式设置套接字为长连接模式。

例 10-10　编写程序，使用 TCP 协议进行通信，并且使得 TCP 连接能够长时间保持存活。

（1）服务端程序。

```python
import socket
from struct import unpack

sockServer = socket.socket(socket.AF_INET, socket.SOCK_STREAM)
sockServer.bind(('', 6666))
sockServer.listen(1)

conn, addr = sockServer.accept()
#为客户端套接字开启长连接
#保活设置只需要在一端启用就可以,不需要在服务端和客户端都设置
#可以注释掉下面两行代码再运行,对比运行结果,理解保活机制的作用
conn.setsockopt(socket.SOL_SOCKET, socket.SO_KEEPALIVE, True)
conn.ioctl(socket.SIO_KEEPALIVE_VALS,
          (1,             #开启保活机制
           60 * 1000,     #60s 后如果对方还没有反应,开始探测连接是否存在
           30 * 1000)     #30s 探测一次,默认探测 10 次,失败则断开
          )
while True:
    #这里没有考虑高并发网络服务器,假设一次可以接收完数据
    #接收 4 字节,解包为整数,表示对方要发送的字节串长度
    data_length = conn.recv(4)
    if not data_length:
        conn.close()
        break
    data_length = unpack('i', data_length)[0]
    data = conn.recv(data_length).decode()
    print(data)
```

（2）客户端程序。

```python
import socket
from time import sleep
from struct import pack
from datetime import datetime

sockClient = socket.socket(socket.AF_INET, socket.SOCK_STREAM)

try:
    #实际运行时建议使用两台计算机进行测试
    #一台计算机运行服务端,另一台计算机运行客户端
    #并把下面代码中的'127.0.0.1'修改为服务端计算机的 IP 地址
    sockClient.connect(('127.0.0.1', 6666))
except:
    print('服务器不存在。')
```

```
        exit()

for i in range(5):
    msg = str(datetime.now())[:19]
    print(msg)
    msg = msg.encode()
    sockClient.sendall(pack('i', len(msg)))
    sockClient.sendall(msg)
    # 每隔 6min 发送一次数据
    sleep(360)

sockClient.close()
```

使用联网的两台计算机测试程序,一台计算机运行服务端程序,另一台计算机运行客户端程序,并把客户端程序中的本地回环地址 127.0.0.1 修改为服务端程序所在计算机的真实 IP 地址。删除或注释掉服务端开启保活机制的两行代码之后再运行,观察两种运行结果的区别。可以看出,如果没有开启保活机制,无法保持 TCP 长连接。如果客户端长时间不操作,再次发送数据时会因为连接已断开而失败。更多关于套接字编程的案例见《Python 网络程序设计》(微课版)(董付国,清华大学出版社)。

10.4 网页内容读取与网页爬虫

10.4.1

10.4.1 网页内容读取与域名处理基础知识

Python 3.x 标准库 urllib 支持读取网页内容所需要的相关功能,主要包含 urllib.request、urllib.response、urllib.parse、urllib.robotparser 和 urllib.error 这几个模块。

下面的代码演示了如何读取并显示指定网页的内容。

```
>>> import urllib.request
>>> fp = urllib.request.urlopen(r'http://www.python.org')
>>> print(fp.read(100))
>>> print(fp.read(100).decode())
>>> fp.close()
```

下面的代码演示了如何使用 GET 方式读取并显示指定 URL 的内容。

```
from urllib.request import urlopen
from urllib.parse import urlencode

params = urlencode({'keyword':'董付国', 'keytm':'8D3538239084926C8C'})
url = f'http://www.tup.tsinghua.edu.cn/booksCenter/booklist.html?{params}'
with urlopen(url) as fp:
    content = fp.read().decode('utf8')
print(content)
```

下面的代码演示了如何使用 POST 方式提交参数并读取指定页面内容。

```
from urllib.parse import urlencode, quote
from urllib.request import urlopen, Request
```

```
#查询普通高校招生专业(专业类)选考科目要求,以北京大学为例
#完整代码见《Python 网络程序设计》(微课版)(董付国,清华大学出版社)
url = r'https://xkkm.sdzk.cn/xkkm/queryXxInfor'
data = urlencode({'dm': '10001', 'mc': quote('北京大学'),
                  'yzm':'ok', 'nf':'2024'}).encode('ascii')
headers = {'user-agent': 'Chrome/99.0.4844.84', 'Cookie': ''}
req = Request(url, data=data, headers=headers)
with urlopen(req) as fp:
    content = fp.read().decode()
print(content)
```

有些服务器设置了反爬机制,如果发现不是来自浏览器的正常请求就拒绝响应,此时可以伪造请求的头部,假冒浏览器向服务器发起请求,下面的代码演示了这个用法,其中 dstUrl 表示要爬取的目标网页地址,可自行定义。

```
>>> import urllib.request
>>> headers = {'User-Agent':'Mozilla/5.0 (Windows NT 6.1; Win64; x64) AppleWebKit/537.36 (KHTML, like Gecko) Chrome/62.0.3202.62 Safari/537.36'}
>>> req = urllib.request.Request(url=dstUrl, headers=headers)
>>> with urllib.request.urlopen(req) as fp:
    content = fp.read().decode()
```

在下载文件时,可能会遇到服务器开启了防盗链设置,使得无法从网站外部直接访问文件。此时可以修改代码伪造请求的头部,假装是从网站内部正常访问和下载文件。下面的代码演示了这个用法,其中的要点是 headers 字典中的 Referer 信息。

```
from re import findall
from urllib.parse import urljoin
from urllib.request import urlopen, Request

url = r'http://jwc.sdtbu.edu.cn/info/2002/5418.htm'
headers = {'User-Agent':'Mozilla/5.0 (Windows NT 6.1; Win64; x64)
           AppleWebKit/537.36(KHTML, like Gecko) Chrome/62.0.3202.62 Safari/537.36',
           'Referer': url}                          #不加这一项会有防盗链提示

req = Request(url=url, headers=headers)
with urlopen(req) as fp:                            #读取网页源代码
    content = fp.read().decode()
pattern = r'附件【<a href="(.+?)"><span>(.+?)</span>'
for fileUrl, fileName in findall(pattern, content): #提取要下载的文件地址
    fileUrl = urljoin(url, fileUrl)
    req = Request(url=fileUrl, headers=headers)
    with urlopen(req) as fp1:                       #下载文件
        with open(fileName, 'wb') as fp2:
            fp2.write(fp1.read())
```

10.4.2　网页爬虫实战

网页爬虫用来模拟人类浏览网页时的行为并在互联网上自动采集感兴趣的文本、图片和文件,结合数据分析技术可以得到更深层次的信息。虽然有 requests、scrapy、bs4 等非常成熟的爬虫和页面内容解析框架,但标准库 urllib 和 re 仍是学习和编写爬虫程序的重要基

础,本节重点演示这两个标准库在编写爬虫程序中的应用,要多爬虫案例见《Python 网络程序设计》(微课版)(董付国,清华大学出版社)。虽然目标网页的源代码分析是编写网络爬虫的第一步,也是最重要的一步,但为节约篇幅,省去了对网页源代码的分析过程,可使用浏览器打开例题中涉及的网页然后自行分析,或者观看微课视频。

例 10-11 编写多进程爬虫程序,爬取中国工程院院士文字简介和图片,多进程编程的内容见本书第 13 章。

例 10-11

```python
import re
import os
import os.path
from time import sleep
from urllib.parse import urljoin
from urllib.request import urlopen
from multiprocessing import Pool

dstDir = 'YuanShi'
if not os.path.isdir(dstDir):
    os.mkdir(dstDir)

url = r'http://www.cae.cn/cae/html/main/col48/column_48_1.html'
with urlopen(url) as fp:                                    #读取网页源代码
    content = fp.read().decode()

pattern = r'<li class="name_list"><a href="(.+?)" target="_blank">(.+?)</a></li>'
result = re.findall(pattern, content)                       #提取每位院士的链接地址

def crawlEveryUrl(item):
    perUrl, name = item
    perUrl = urljoin(url, perUrl)
    name = os.path.join(dstDir, name)
    print(perUrl)
    try:
        with urlopen(perUrl) as fp:
            content = fp.read().decode()
    except:
        print('出错了,1s 后自动重试…')
        sleep(1)
        crawlEveryUrl(item)
        return

    pattern = r'<img src="(.+?)" style=.*?/>'
    imgUrls = re.findall(pattern, content)                  #提取图片地址
    if imgUrls:
        imgUrl = urljoin(url, imgUrls[0])
        try:
            with urlopen(imgUrl) as fp1:                    #下载图片
                with open(name+'.jpg', 'wb') as fp2:
```

```
                    fp2.write(fp1.read())
            except:
                pass
        
        pattern = r'<p>(.+?)</p>'
        intro = re.findall(pattern, content, re.M)          #提取院士的简介,写入本地文件
        if intro:
            intro = '\n'.join(intro)
            intro = re.sub('( )|( )|(<a href.*?</a>)', '', intro)
            with open(name+'.txt', 'w', encoding='utf8') as fp:
                fp.write(intro)

if __name__ == '__main__':
    with Pool(10) as p:                                     #使用10个进程同时采集,见13.4节
        p.map(crawlEveryUrl, result)
```

例 10-12 编写爬虫程序,爬取微信公众号"Python 小屋"所有历史文章中的图片,每篇文章的图片单独存放于当前文件夹下以文章标题为名的一个子文件夹中,该文章中的所有图片存放于子文件夹中并分别命名为 0.png、1.png、2.png 等。

例 10-12

```
from os import mkdir
from os.path import isdir
from re import findall, sub
from urllib.request import urlopen

#公众号"Python 小屋"历史文章清单,可以用手机进入公众号,打开菜单"最新资源"→"历史文章"
#复制链接,再使用计算机端浏览器打开该链接
startUrl = r'https://mp.weixin.qq.com/s/u9FeqoBaA3Mr0fPCUMbpqA'

#读取网页源代码,提取每篇文章的 URL
pattern = r'<p>.*?<a href="(.+?)" target="_blank".+?data-linktype="2">(.+?)</a>.*?</p>'
with urlopen(startUrl) as fp:
    content = fp.read().decode()
for perUrl, title in findall(pattern, content):
    #删除文章标题中可能的 HTML 代码
    title = sub('<.+?>', '', title)
    print('正在爬取文章: ', title)
    with urlopen(perUrl) as fp:
        content = fp.read().decode()
    if not isdir(title):
        mkdir(title)
    #爬取网页中的图片地址
    pattern = r'<img .*? data-type="png".*?data-src="(.+?)"'
    for index, picUrl in enumerate(findall(pattern, content)):
        with open(title+'\\{}.png'.format(index), 'wb') as fpPic:
            fpPic.write(urlopen(picUrl).read())
```

本 章 小 结

(1) IP 地址和端口号共同来标识网络上特定主机上的特定应用进程,称为 Socket。
(2) TCP 适用于对效率要求较低,对准确性要求较高的场合,是面向连接、具有服务质

量保证的可靠传输协议。UDP属于无连接协议,可能会发生丢包、乱序或其他错误,适用于视频点播、网络语音通信之类的场合。

（3）标准库 urllib 中的模块提供了大量对象和函数支持网页内容读取,结合 re 模块以及多线程和多进程编程可以轻松实现网络爬虫程序。

（4）UDP 的特点是"尽最大努力传输",不保证非常好的服务质量。

（5）为了实现网络流量嗅探,需要将网卡设置为混杂模式,并且运行嗅探器程序的用户账号需要拥有系统管理员权限。

（6）端口扫描技术可以探测指定主机上是否开放了某些端口,进一步判断主机上是否运行某些重要的网络服务,最终判断是否存在潜在的安全漏洞。

（7）TCP 属于端到端的协议,无法实现一对多的群发功能,而 UDP 可以。

（8）Python 扩展库 psutil 提供了很多网络管理和进程管理的功能。

（9）标准库 urllib.request 的函数 urlopen() 可以像打开文件一样打开远程的 URL,打开成功之后可以使用返回对象的 read() 方法读取目标网页中的内容,返回字节串。

（10）有些服务器设置了反爬机制,如果发现不是来自浏览器的正常请求就拒绝响应,此时可以伪造请求的头部,假冒浏览器向服务器发起请求。

（11）在下载文件时,可能会遇到服务器开启了防盗链设置,使得无法从网站外部直接访问文件,从而导致无法使用爬虫下载文件。此时可以修改代码伪造请求的头部,假装是从网站内部正常访问和下载文件。

习　题

1. 简单解释 TCP 和 UDP 的区别。

2. 同学之间合作编写 UDP 通信程序,分别编写发送端和接收端代码,发送端发送一个字符串"Hello world!"。假设接收端在计算机的 5000 端口进行接收,并显示接收到的内容。

3. 简单介绍 socket 模块中用于 TCP 编程的常用函数和对象方法。

4. 编写代码读取搜狐网页首页源代码,要求使用标准库 urllib,并把结果写入文件 sohu.txt。

5. 查阅资料,编写程序模拟 FTP 通信,要求使用 TCP。

6. 查阅资料,编写程序模拟端口映射原理,要求使用 TCP。

7. 编写程序,爬取微信公众号"Python 小屋",每篇文章生成一个 Word 文档。

8. 编写程序,爬取 www.sdtbu.edu.cn 网站上的所有新闻,每个新闻生成一个子文件夹,网页上的文本存放于一个文本文件,网页上的图片单独存放于该文件夹中。

9. 查阅资料,使用 scrapy 编写爬虫项目,爬取 www.weather.com.cn/shandong 各城市未来 7 天的天气情况。

第 11 章 安卓平台的 Python 编程

11.1 QPython 简介

SL4A 是 Scripting Layer for Android 的缩写，与 Android Scripting Environment（ASE）含义相同。SL4A 将脚本语言引入 Android 平台，允许用户直接在 Android 设备上编写和运行脚本程序，目前支持 Python、Perl、JavaScript、Tcl 等多种脚本语言。

QPython 扩展了 SL4A 框架，可以访问网络、蓝牙、GPS 等安卓特性，并且支持 Bottle 和 Django 作为 Web 开发框架。QPython 主要有两个版本，其中 QPython 支持 Python 2.7.2，QPython3 支持 Python 3.6.6，并且设计更加开放和灵活。在 QPython 中调用安卓的 SL4A 接口，需要导入 androidhelper 库。

本章以 QPython 和 QPython3 为例来介绍安卓平台的 Python 编程，极个别扩展库暂时还不支持 QPython3。打开 QPython3 主界面，单击"终端"按钮进入交互式开发模式，如图 11-1 和图 11-2 所示，更多详细用法以及另一个手机端开发环境 Pydroid3 的相关内容可以到微信公众号"Python 小屋"发送消息"安卓"查看。

图 11-1 列表推导式与内置函数用法　　　图 11-2 基本表达式和 math 模块用法

在 QPython 或 QPython3 主界面中选择"程序"或"编辑器"，可以创建和运行 Python 程序，QPython 和 QPython3 都是用 QEdit 作为 Python 程序编辑器，支持语法高亮显示。图 11-3 和图 11-4 演示了程序开发模式的用法。QPython 和 QPython3 默认开发模式是 console，即控制台应用程序，如果开发图形用户界面则需要在开头部分加上一行"＃qpy:kivy"，如果开发 Web App 则需要在程序开头部分加上"＃qpy:webapp:(App标题)"和"＃qpy:(Web服务地址:端口号)/(Web服务路径)"。

在 QPython 和 QPython3 中可以使用 pip 工具来管理 Python 扩展库，pip 支持的命令主要有：①bundle，创建包含多个包的 pybundles；②freeze，显示所有已安装的包；③help，显示可用命令；④install，安装包；⑤search，搜索 PyPi；⑥uninstall，卸载包；⑦unzip，解压缩单个包；⑧zip，压缩单个包。

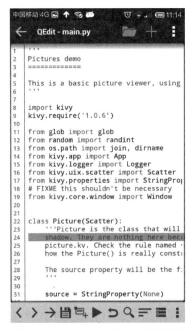

图 11-3　QPython3 程序开发界面　　　图 11-4　QPython3 新建项目或文件界面

11.2　安卓应用开发案例

例 11-1　在屏幕上显示字符串。

```
import android

d = android.Android()
d.makeToast('Hello, Python in Android')
```

例 11-2　获取用户输入并在屏幕上显示字符串。

```
import androidhelper

droid = androidhelper.Android()
line = droid.dialogGetInput()
s = 'Hello %s' %line.result
droid.makeToast(s)
```

例 11-3　扫描并显示条形码。

```
import android

droid = android.Android()
code = droid.scanBarcode()
droid.makeToast(code[1]['extras']['SCAN_RESULT'])
```

例 11-4　打开手机摄像头并保存照片。

```
import sl4a
import os
```

```
droid = sl4a.Android()
#获取已拍照片数量
t = [x for x in os.listdir('/sdcard') if x.startswith('pic') and x.endswith('.jpg')]
n = len(t) + 1
#打开摄像头并保存拍到的照片
droid.cameraInteractiveCapturePicture('/sdcard/pic'+str(n)+'.jpg')
```

例 11-5 获取 GPS 和罗盘信息。

```
import android
import time
from math import radians

droid = android.Android()
droid.startSensingTimed(1, 250)
droid.startLocating()
while True:
    gpsdata = droid.readLocation().result
    s6data = droid.sensorsReadOrientation().result
    if len(gpsdata) > 0:
        print(gpsdata['gps']['bearing'])          #取得 GPS 导向(bearing)(角度)
    if len(s6data) > 0:
        print(s6data[0])                          #取得罗盘方位角(azimuth)(弧度)
    time.sleep(0.5)
droid.stopLocating()
droid.stopSensing()
```

例 11-6 编写 Web App。

```
#qpy:3
#qpy:webapp:Hello QPython
#qpy://localhost:8080/hello
from bottle import route, run

@route('/hello')
def hello():
    return "Hello World!"
run(host='localhost', port=8080)
```

例 11-7 使用 kivy 包生成按钮。

```
#-*-coding:utf8;-*-
#qpy:2
#qpy:kivy
from kivy.app import App
from kivy.lang import Builder

kv = '''
FloatLayout:
    Button:
        text: 'hello world'
        size_hint: None, None
        pos_hint: {'center_x': .5, 'center_y': .5}
        canvas.before:
            PushMatrix
            Rotate:
```

```
            angle: 45
            origin: self.center
    canvas.after:
        PopMatrix
'''
class RotationApp(App):
    def build(self):
        return Builder.load_string(kv)
RotationApp().run()
```

例 11-8　使用 kivy 包绘制可改变大小和形状的椭圆。

```
#-*-coding:utf8;-*-
#qpy:2
#qpy:kivy
from kivy.app import App
from kivy.lang import Builder

kv = '''
BoxLayout:
    orientation: 'vertical'
    BoxLayout:
        size_hint_y: None
        height: sp(100)
        BoxLayout:
            orientation: 'vertical'
            Slider:
                id: e1
                min: -360
                max: 360
            Label:
                text: 'angle_start={}'.format(e1.value)
        BoxLayout:
            orientation: 'vertical'
            Slider:
                id: e2
                min: -360
                max: 360
                value: 360
            Label:
                text: 'angle_end={}'.format(e2.value)
    BoxLayout:
        size_hint_y: None
        height: sp(100)
        BoxLayout:
            orientation: 'vertical'
            Slider:
                id: wm
                min: 0
                max: 2
                value: 1
```

```
            Label:
                text: 'Width mult.={}'.format(wm.value)
        BoxLayout:
            orientation: 'vertical'
            Slider:
                id: hm
                min: 0
                max: 2
                value: 1
            Label:
                text: 'Height mult.={}'.format(hm.value)
        Button:
            text: 'Reset ratios'
            on_press: wm.value=1; hm.value=1
    FloatLayout:
        canvas:
            Color:
                rgb: 1, 1, 1
            Ellipse:
                pos: 100, 100
                size: 200 *wm.value, 201 *hm.value
                source: 'data/logo/kivy-icon-512.png'
                angle_start: e1.value
                angle_end: e2.value
'''
class CircleApp(App):
    def build(self):
        return Builder.load_string(kv)
CircleApp().run()
```

例 11-9 生成二维码。

Python 扩展库 qrcode 提供了生成二维码的功能，可以使用 pip 工具安装，生成的二维码可使用手机微信扫描并识别其中的信息。本例代码为计算机端 Python 程序。

```
import qrcode

qr = qrcode.QRCode(version=10, box_size=10, border=4,
                error_correction=qrcode.constants.ERROR_CORRECT_L)
qr.add_data('http://user.qzone.qq.com/306467355/blog/1439803492')
qr.make(fit=True)
img = qr.make_image()
img.save('D:\Python_dfg.png')
```

最后，如果需要将自己的 Python 程序以及所有依赖包打包成为 APK 文件以便直接在安卓平台上运行，可以使用 Python for Android 或者 Buildozer，本书不再赘述。

本 章 小 结

（1）SL4A 将脚本语言引入 Android 平台，允许用户编辑和执行脚本，直接在 Android 设备上运行解释器。

（2）QPython 和 QPython3 扩展了 SL4A 框架，可以访问网络、蓝牙、GPS 等安卓特性，并且支持 Bottle 和 Django 作为 Web 开发框架。

（3）QPython 和 QPython3 默认的开发模式是 console，即控制台应用程序，如果开发图形界面则需要在开头部分加上一行"`#qpy:kivy`"，如果开发 Web App 则需要在程序开头部分加上"`#qpy:webapp:(App标题)`"和"`#qpy:(Web服务地址:端口号)/(Web服务路径)`"。

（4）除了 QPython3 之外，Pydroid3 也是一款非常不错的手机端 Python 开发环境，可以微信关注公众号"Python 小屋"，发送消息"安卓"了解相关的安装与使用方法。

习　题

查阅资料，了解 androidhelper、android、sl4a 3 个包的异同。

第 12 章　Windows 系统编程

　　Python 是一门功能强大的通用编程语言,它可以把其他语言编写的程序黏合在一起,可以很容易地调用外部程序,以及调用其他语言编写的动态链接库中的代码,甚至还可以将 Python 程序打包为.exe 可执行程序以便在没有安装 Python 的 Windows 系统中运行。在本章中通过大量示例来介绍 Windows 平台的混合编程技术及底层编程技术,有些内容可能需要读者对 Windows 平台有较深层的了解,可以查阅《Windows 内核原理与实现》《Windows 核心编程》《深入解析 Windows 操作系统》或其他相关书籍。

12.1　注册表编程

　　对于 Windows 操作系统,注册表无疑是非常重要的组成部分,Windows 将几乎所有软硬件系统配置信息都保存在注册表中。通过读取注册表中的数据,可以获取 Windows 平台的相应信息,例如,已安装的服务和程序列表、开机自动运行的程序列表、文件类型与程序的关联关系等;通过修改注册表中的数据,可以对 Windows 系统进行详细配置。

　　Windows 注册表有如下 5 个根键。

　　(1) HKEY_LOCAL_MACHINE(HKLM)。

　　(2) HKEY_CURRENT_CONFIG(HKCC)。

　　(3) HKEY_CLASSES_ROOT(HKCR)。

　　(4) HKEY_USERS(HKU)。

　　(5) HKEY_CURRENT_USER(HKCU)。

　　单击"开始"→"运行"命令,弹出"运行"对话框,在对话框中输入 regedit.exe 并按 Enter 键,可以打开"注册表编辑器"窗口,如图 12-1 所示,在"注册表编辑器"窗口中可以对注册表的键和值进行增加、删除、修改、查询等操作。

　　在注册表中,值可以为数值、字符串等多种类型,详细类型如表 12-1 所示。

表 12-1　注册表中值的类型

类 型 名	说　　明
REG_NONE	没有类型
REG_SZ	字符串类型
REG_EXPAND_SZ	一个可扩展的字符串值,其中可以包含环境变量
REG_BINARY	二进制类型
REG_DWORD / REG_DWORD_LITTLE_ENDIAN	DWORD 类型,用于存储 32 位无符号整数,即 0~4 294 967 295 的整数,以 little-endian 格式存储

续表

类 型 名	说　　明
REG_DWORD_BIG_ENDIAN	DWORD 类型，用于存储 32 位无符号整数，即 0~4 294 967 295 的整数，以 big-endian 格式存储
REG_LINK	到其他注册表键的链接，指定根键或到目标键的路径
REG_MULTI_SZ	一个多字符串值，指定一个非空字符串的排序列表
REG_RESOURCE_LIST	资源列表，用于枚举即插即用硬件及其配置
REG_FULL_RESOURCE_DESCRIPTOR	资源标识符，用于枚举即插即用硬件及其配置
REG_RESOURCE_REQUIREMENTS_LIST	资源需求列表，用于枚举即插即用硬件及其配置
REG_QWORD / REG_QWORD_LITTLE_ENDIAN	QWORD 类型，用于存储 64 位无符号整数，以 little-endian 格式存储或未指定存储格式

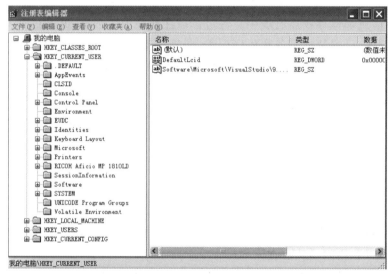

图 12-1　注册表编辑器

对于注册表编程，可以使用 Pywin32 中的 win32api 模块和 win32con 模块，其中 win32api 模块封装了 Windows API 函数，提供了非常友好的接口。该模块中常用的注册表操作函数有 8 个。

（1）`RegOpenKey()/RegOpenKeyEx()`：打开注册表键。

（2）`RegCloseKey()`：关闭注册表键。

（3）`RegQueryValue()/RegQueryValueEx()`：读取项值。

（4）`RegSetValue()/RegSetValueEx()`：设置项值。

（5）`RegCreateKey()/RegCreateKeyEx()`：添加项。

（6）`RegDeleteKey()`：删除项。

（7）`RegEnumKey()`：枚举子键。

(8) `RegDeletetValue()`：删除值。

例如，下面的代码用于查询注册表并输出本机安装的 IE 浏览器软件版本信息：

```python
import win32api
import win32con

key = win32api.RegOpenKey(win32con.HKEY_LOCAL_MACHINE,
                          'SOFTWARE\Microsoft\Internet Explorer',
                          0, win32con.KEY_ALL_ACCESS)
print(win32api.RegQueryValue(key, ''))              #返回默认值
print(win32api.RegQueryValueEx(key, 'Version'))     #返回指定键的值
#返回项的子项数目、项值数目，以及最后一次修改时间
print(win32api.RegQueryInfoKey(key))
win32api.RegCloseKey(key)
```

例 12-1 检查随系统启动而启动的程序列表。

```python
from win32api import *
from win32con import *

def GetValues(fullname):
    name = fullname.split('\\', 1)
    try:
        if name[0] == 'HKEY_LOCAL_MACHINE':
            key = RegOpenKey(HKEY_LOCAL_MACHINE, name[1], 0, KEY_READ)
        elif name[0] == 'HKEY_CURRENT_USER':
            key = RegOpenKey(HKEY_CURRENT_USER, name[1], 0, KEY_READ)
        elif name[0] == 'HKEY_CURRENT_ROOT':
            key = RegOpenKey(HKEY_CURRENT_ROOT, name[1], 0, KEY_READ)
        elif name[0] == 'HKEY_CURRENT_CONFIG':
            key = RegOpenKey(HKEY_CURRENT_CONFIG, name[1], 0, KEY_READ)
        elif name[0] == 'HKEY_USERS':
            key = RegOpenKey(HKEY_USERS, name[1], 0, KEY_READ)
        else:
            print('Error, no key named ', name[0])
            return
        info = RegQueryInfoKey(key)
        for i in range(0, info[1]):
            ValueName = RegEnumValue(key, i)
            print(str.ljust(ValueName[0], 20), ValueName[1])
        RegCloseKey(key)
    except BaseException as e:
        print('Sth is wrong')
        print(e)
if __name__ == '__main__':
    KeyNames = ['HKEY_LOCAL_MACHINE\\SOFTWARE\\Microsoft\\Windows\\
                CurrentVersion\\Run',
                'HKEY_LOCAL_MACHINE\\SOFTWARE\\Microsoft\\Windows\\
                CurrentVersion\\RunOnce',
                'HKEY_LOCAL_MACHINE\\SOFTWARE\\Microsoft\\Windows\\
                CurrentVersion\\RunOnceEx',
```

```
            'HKEY_CURRENT_USER\\Software\\Microsoft\\Windows\\
             CurrentVersion\\Run',
            'HKEY_CURRENT_USER\\Software\\Microsoft\\Windows\\
             CurrentVersion\\RunOnce']
for KeyName in KeyNames:
    print(KeyName)
    GetValues(KeyName)
```

操作 Windows 注册表的另一种常用方式是使用 Python 模块 winreg，该模块提供了 OpenKey()、DeleteKey()、DeleteValue()、CreateKey()、SetValue()、QueryValueEx()、EnumValue() 和 EnumKey() 等大量用于注册表访问和操作的方法。下面的代码演示了使用模块 winreg 枚举注册表值的用法。

例 12-2　枚举注册表。

```
import winreg
key = winreg.OpenKey(winreg.HKEY_CURRENT_USER,
                r'Software\Microsoft\Windows\CurrentVersion\Explorer')
try:
    i = 0
    while True:
        Name, Value, Type = winreg.EnumValue(key, i)
        print(Name, Value, Type, sep=':')
        i += 1
except WindowsError:
    pass
print('='*20)
Name = 'FaultTime'
Value, Type = winreg.QueryValueEx(key, Name)
print(Name, Value)
```

12.2　创建可执行文件

把 Python 程序打包成二进制可执行文件再发布，可以脱离开发环境独立运行，同时也可以对源码进行保密。比较常用的打包工具有 pyinstaller、Nuitka、cx_Freeze、py2exe 和 py2app，其中 py2app 仅适用于在 macOS X 平台打包 Python 程序。本节简单介绍 pyinstaller 打包工具的使用。

首先，进入命令提示符环境并切换至 Python 安装目录的 scripts 子文件夹，然后执行 pip install pyinstaller 命令安装扩展库，安装成功之后就可以使用 pyinstaller 工具了。例如，已有 Python 程序文件 CheckAndViewAutoRunsInSystem.py，在命令提示符环境执行命令

```
pyinstaller CheckAndViewAutoRunsInSystem.py
```

即可把该程序以及所依赖的扩展库和系统文件打包，生成文件夹 dist\CheckAndViewAutoRunsInSystem，文件夹中的内容如图 12-2 所示，其中箭头所指即为可执行文件，其他文件是运行程序所需要的依赖文件。

如果想把程序打包为一个可以独立运行的可执行文件，可以使用下面的命令：

```
 _bz2.pyd                               2023/6/19 14:35    Python Extensio...    47 KB
 _hashlib.pyd                           2023/6/19 14:35    Python Extensio...    25 KB
 _lzma.pyd                              2023/6/19 14:35    Python Extensio...    82 KB
 _socket.pyd                            2023/6/19 14:35    Python Extensio...    39 KB
 _ssl.pyd                               2023/6/19 14:35    Python Extensio...    51 KB
 base_library.zip                       2023/6/19 14:34    WinRAR ZIP 压缩...    761 KB
 CheckAndViewAutoRunsInSystem.exe       2023/6/19 14:36    应用程序             1,418 KB
 CheckAndViewAutoRunsInSystem.exe...    2023/6/19 14:36    MANIFEST 文件         2 KB
 libcrypto-1_1-x64.dll                  2023/6/19 14:35    应用程序扩展          739 KB
 libssl-1_1-x64.dll                     2023/6/19 14:35    应用程序扩展          156 KB
 pyexpat.pyd                            2023/6/19 14:35    Python Extensio...    76 KB
 python37.dll                           2023/6/19 14:35    应用程序扩展        1,218 KB
 select.pyd                             2023/6/19 14:35    Python Extensio...    22 KB
 unicodedata.pyd                        2023/6/19 14:35    Python Extensio...   275 KB
 VCRUNTIME140.dll                       2022/6/23 15:00    应用程序扩展           87 KB
```

图 12-2　pyinstaller 打包结果

```
pyinstaller -F -w CheckAndViewAutoRunsInSystem.py
```

安装扩展库 tinyaes 后执行 pyinstaller 命令时加选项 --key 可以进行代码混淆来对抗反编译。关于 pyinstaller 工具的更多用法，可以在命令提示符环境执行下面的命令进行查看：

```
pyinstaller -h
```

使用 pyinstaller 打包 Python 程序时，如果遇到由于 vcruntime140.dll 文件有问题而无法执行的情况，一般是因为该文件版本不对。此时可以使用本书资源中的文件 vcruntime140.dll 替换自己机器上 Python 安装目录下的同名文件，或者从网上下载后进行替换。

12.3　调用外部程序

1. 使用 os 模块的函数调用外部程序

```
>>> import os
>>> os.system('notepad.exe')
>>> os.system('notepad C:\\dir.txt')
```

使用上面的 system() 函数也可以调用 Windows 系统命令，如 dir、xcopy 等，但是有一个缺点，不论启动什么程序都会先启动一个控制台窗口，然后再打开被调程序，如图 12-3 所示。

也可以使用 os 模块的 popen() 函数来打开外部程序，这样不会出现命令提示符窗口。

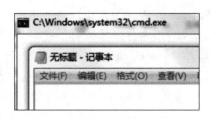

图 12-3　使用 os 模块的 system() 方法启动记事本

```
>>> os.popen(r'C:\windows\notepad.exe')
<open file 'C:\\windows\\notepad.exe', mode 'r' at 0x012BEF98>
```

或者，还可以使用 os 模块的 startfile() 函数来打开外部程序或文件，系统将自动关联相应的程序来打开或执行文件，传入 URL 时会自动打开默认浏览器。

```
>>>os.startfile(r'C:\windows\notepad.exe')
>>>os.startfile(r'wxIsPrime.py')
```

2. 使用 win32api 模块调用 ShellExecute() 函数来启动外部程序

```
>>> import win32api
>>>win32api.ShellExecute(0, 'open', 'notepad.exe', '', '', 0)     #0表示后台运行程序
>>>win32api.ShellExecute(0, 'open', 'notepad.exe', '', '', 1)     #1表示前台运行程序
>>>win32api.ShellExecute(0,'open','notepad.exe','C:\\dir.txt', '', 1)
                                                                  #传递参数打开指定文件
>>>win32api.ShellExecute(0, 'open', 'www.python.org',',' '', 1)  #打开网址
>>>win32api.ShellExecute(0, 'open',r'C:\dir.txt', '', '', 1)     #相当于双击文件
```

使用这种方式运行程序或者打开文件时，不会像 os 模块的 `system()` 函数那样先打开一个命令提示符窗口，并且系统将根据文件类型自动关联相应程序并打开文件，类似于在资源管理器中双击打开文件或单击打开超链接。例如，如果打开的是记事本文件，则会自动使用记事本程序打开；如果指定的是个域名，则会自动使用默认浏览器打开该网址；如果打开的是可执行文件，则会自动打开并运行该程序文件。

3. 通过创建进程来启动外部程序

```
>>> import win32process
>>> handle = win32process.CreateProcess(r'C:\windows\notepad.exe','', None,None,0,
          win32process.CREATE_NO_WINDOW, None, None, win32process.STARTUPINFO())
                                                  #打开记事本程序
>>>win32process.TerminateProcess(handle[0], 0)    #关闭刚才打开的程序
>>> handle = win32process.CreateProcess(r'C:\windows\notepad.exe','', None, None, 0,
          win32process.CREATE_NO_WINDOW,None,None,win32process.STARTUPINFO())
>>> import win32event
>>>win32event.WaitForSingleObject(handle[0], -1) #需要手动关闭记事本
```

4. 通过 ctypes 来调用动态链接库代码

ctypes 是 Python 处理动态链接库的标准模块，提供了与 C 语言兼容的数据类型，允许在 Python 程序中调用动态链接库或共享库中的代码，从而支持 Python 与其他编程语言的混合编程，充分发挥各自的优势，大幅提高开发效率和运行效率。另外，NumPy 模块也提供了一个函数 `numpy.ctypeslib.load_library()`，用来打开指定的动态链接库并返回一个 ctypes 对象，通过该对象可以访问动态链接库中的函数。或者，使用 weave 扩展库也可以方便地将 C++ 程序以字符串的形式嵌入 Python 程序中。

ctypes 提供了 3 种方法调用动态链接库：cdll、windll 和 oledll，它们的不同之处在于函数调用时的参数传递方式和返回时栈的平衡方式。cdll 加载的库导出的函数必须使用标准的 cdecl 调用约定（函数的参数从右往左依次压入栈内，在函数执行完成后，由函数的调用者负责函数的栈帧平衡），windll 方法加载的库导出的函数必须使用 stdcall 调用约定（Win32 API 的原生约定），oledll 方法与 windll 方法类似，不过假设函数返回一个 HRESULT 错误代码。

下面的代码调用 Windows 动态链接库 user32.dll 中的 **MessageBoxA()** 函数来显示对话框：

```
>>> import ctypes                                 #通过 ctypes 可以调用动态链接库中的函数
>>> user32 = ctypes.windll.LoadLibrary('user32.dll')
>>> user32.MessageBoxA(0, b'Hello world!', b'Python ctypes', 0)
```

或者使用下面更为简洁的形式：

```
>>> import ctypes
>>> ctypes.windll.user32.MessageBoxA(0, b'Hello world!', b'Python ctypes', 0)
```

下面的代码调用标准 C 函数库 msvcrt 中的 printf() 函数来输出文本：

```
import ctypes
msvcrt = ctypes.cdll.LoadLibrary('msvcrt')
printf = msvcrt.wprintf
printf('Hello world!')
```

或者使用下面的形式：

```
import ctypes
ctypes.cdll.msvcrt.wprintf('Hello world!')
```

该程序需要在命令提示符环境中而不是在 IDLE 中执行，在 IDLE 中运行输出的是字符数量而不是字符串。假设将上面的代码保存为 useprintfthroughctypes.py 文件，然后在命令提示符中运行结果如图 12-4 所示，代码同样适用于 Python 3.8 及更高版本。

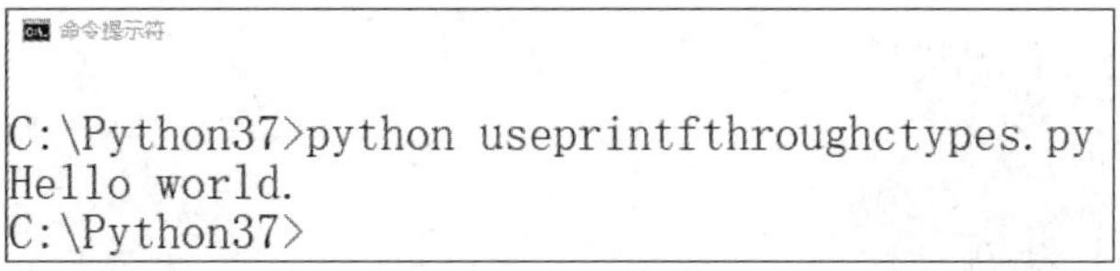

```
C:\Python37>python useprintfthroughctypes.py
Hello world.
C:\Python37>
```

图 12-4　使用 ctypes 库调用 C 语言的 printf 函数

下面的代码可以用来清空系统剪切板：

```
ctypes.windll.user32.OpenClipboard(None)
ctypes.windll.user32.EmptyClipboard()
ctypes.windll.user32.CloseClipboard()
```

ctypes 提供了与 C 语言兼容的数据类型，但在 Python 中使用 C 语言的结构体时，需要用类来改写。表 12-2 给出了基本类型的对应关系，关于结构体改写的内容可以通过后面给出的示例代码了解大概思路，更多面向对象程序设计的内容见第 6 章。

表 12-2　基本类型的对应关系

ctypes type	C type	Python type
c_bool	_Bool	bool (1)
c_char	char	1- character bytes object
c_wchar	wchar_t	1- character string
c_byte	char	int
c_ubyte	unsigned char	int
c_short	short	int
c_ushort	unsigned short	int
c_int	int	int
c_uint	unsigned int	int

续表

ctypes type	C type	Python type
c_long	long	int
c_ulong	unsigned long	int
c_longlong	__int64 or long long	int
c_ulonglong	unsigned __int64 or unsigned long long	int
c_float	float	float
c_double	double	float
c_longdouble	long double	float
c_char_p	char *(NUL terminated)	bytes object or None
c_wchar_p	wchar_t *(NUL terminated)	string or None
c_void_p	void *	int or None

例 12-3 枚举进程列表。

```
#EnumProcess.py
from ctypes.wintypes import *
from ctypes import *
import collections

kernel32 = windll.kernel32

class tagPROCESSENTRY32(Structure):                  #定义结构体
    _fields_ = [('dwSize',               DWORD),
                ('cntUsage',             DWORD),
                ('th32ProcessID',        DWORD),
                ('th32DefaultHeapID',    POINTER(ULONG)),
                ('th32ModuleID',         DWORD),
                ('cntThreads',           DWORD),
                ('th32ParentProcessID',  DWORD),
                ('pcPriClassBase',       LONG),
                ('dwFlags',              DWORD),
                ('szExeFile',            c_char*260)]

def enumProcess():
    hSnapshot = kernel32.CreateToolhelp32Snapshot(15, 0)
    fProcessEntry32 = tagPROCESSENTRY32()
    processClass = collections.namedtuple('processInfo', 'proessName processID')
    processSet = []
    if hSnapshot:
        fProcessEntry32.dwSize = sizeof(fProcessEntry32)
        listloop = kernel32.Process32First(hSnapshot, byref(fProcessEntry32))
        while listloop:
            processName = (fProcessEntry32.szExeFile)
            processID = fProcessEntry32.th32ProcessID
            processSet.append(processClass(processName, processID))
            listloop = kernel32.Process32Next(hSnapshot, byref(fProcessEntry32))
    return processSet
```

```
for i in enumProcess():
    print(i.processName, i.processID)
```

12.4 创建窗口

例 12-4 编写程序,调用 Windows 底层 API 函数来创建窗口并构建消息循环。

```
import win32gui
from win32con import *

def WndProc(hwnd, msg, wParam, lParam):
    if msg == WM_PAINT:
        hdc, ps = win32gui.BeginPaint(hwnd)
        rect = win32gui.GetClientRect(hwnd)
        win32gui.DrawText(hdc, 'GUI Python', len('GUI Python'), rect,
                        DT_SINGLELINE|DT_CENTER|DT_VCENTER)
        win32gui.EndPaint(hwnd, ps)
    if msg == WM_DESTROY:
        win32gui.PostQuitMessage(0)
    return win32gui.DefWindowProc(hwnd, msg, wParam, lParam)

wc = win32gui.WNDCLASS()
wc.hbrBackground = COLOR_BTNFACE + 1
wc.hCursor = win32gui.LoadCursor(0, IDC_ARROW)
wc.hIcon = win32gui.LoadIcon(0, IDI_APPLICATION)
wc.lpszClassName = 'Python on Windows'
wc.lpfnWndProc = WndProc

reg = win32gui.RegisterClass(wc)
hwnd = win32gui.CreateWindow(
    reg, 'Python', WS_OVERLAPPEDWINDOW, CW_USEDEFAULT, CW_USEDEFAULT,
    CW_USEDEFAULT, CW_USEDEFAULT, 0, 0, 0, None)
win32gui.ShowWindow(hwnd, SW_SHOWNORMAL)
win32gui.UpdateWindow(hwnd)
win32gui.PumpMessages()
```

例 12-5 编写程序,使用 MFC 创建窗口,并创建菜单。

```
import win32ui
import win32api
from win32con import *
from pywin.mfc import window

class MyWnd(window.Wnd):
    def __init__(self):
        window.Wnd.__init__(self, win32ui.CreateWnd())
        self._obj_.CreateWindowEx(WS_EX_CLIENTEDGE,
                        win32ui.RegisterWndClass(0, 0, COLOR_WINDOW +1),
                        'MFC GUI', WS_OVERLAPPEDWINDOW,
                        (10,10,800,500), None, 0, None)
        self.HookMessage(self.OnRClick, WM_RBUTTONDOWN)
```

```python
        submenu = win32ui.CreateMenu()
        menu = win32ui.CreateMenu()
        submenu.AppendMenu(MF_STRING, 1051, '&Open')
        submenu.AppendMenu(MF_STRING, 1052, '&Close')
        submenu.AppendMenu(MF_STRING, 1053, '&Save')
        menu.AppendMenu(MF_STRING | MF_POPUP, submenu.GetHandle(), '&File')

        submenu = win32ui.CreateMenu()
        submenu.AppendMenu(MF_STRING, 1054, '&Copy')
        submenu.AppendMenu(MF_STRING, 1055, '&Paste')
        submenu.AppendMenu(MF_SEPARATOR, 1056, None)
        submenu.AppendMenu(MF_STRING, 1057, 'C&ut')
        menu.AppendMenu(MF_STRING | MF_POPUP, submenu.GetHandle(), '&Edit')

        submenu = win32ui.CreateMenu()
        submenu.AppendMenu(MF_STRING, 1058, 'Tools')
        submenu.AppendMenu(MF_STRING | MF_GRAYED, 1059, 'Settings')
        m = win32ui.CreateMenu()
        m.AppendMenu(MF_STRING | MF_POPUP | MF_CHECKED, submenu.GetHandle(), 'Option')
        menu.AppendMenu(MF_STRING | MF_POPUP, m.GetHandle(), '&Other')

        self._obj_.SetMenu(menu)
        self.HookCommand(self.MenuClick, 1051)
        self.HookCommand(self.MenuClick, 1052)
        self.HookCommand(self.MenuClick, 1053)
        self.HookCommand(self.MenuClick, 1054)
        self.HookCommand(self.MenuClick, 1060)

    def OnRClick(self,param):
        submenu = win32ui.CreatePopupMenu()
        submenu.AppendMenu(MF_STRING, 1060, 'Copy')
        submenu.AppendMenu(MF_STRING, 1061, 'Paste')
        submenu.AppendMenu(MF_SEPARATOR, 1062, None)
        submenu.AppendMenu(MF_STRING, 1063, 'Cut')
        submenu.TrackPopupMenu(param[5], TPM_LEFTALIGN | TPM_LEFTBUTTON | TPM_
                               RIGHTBUTTON, self)

    def MenuClick(self, lParam, wParam):
        if lParam == 1051:
            self.MessageBox('Open', 'Python', MB_OK)
        elif lParam == 1053:
            self.MessageBox('Save', 'Python', MB_OK)
        elif lParam == 1052:
            self.OnClose()
        elif lParam == 1060 or lParam == 1054:
            self.MessageBox('Copy', 'Python', MB_OK)

    def OnClose(self):
        self.EndModalLoop(0)

    def OnPaint(self):
        dc, ps = self.BeginPaint()
        dc.DrawText('MFC GUI', self.GetClientRect(),
                    DT_SINGLELINE | DT_CENTER | DT_VCENTER)
        self.EndPaint(ps)
```

```python
w = MyWnd()
w.ShowWindow()
w.UpdateWindow()
w.RunModalLoop(1)
```

例 12-6 编写程序，创建 MFC 窗口并响应按钮消息。

```python
import win32ui
import win32con
from pywin.mfc import dialog

class MyDialog(dialog.Dialog):
    def OnInitDialog(self):
        dialog.Dialog.OnInitDialog(self)
        self.HookCommand(self.OnButton1, 1051)
        self.HookCommand(self.OnButton2, 1052)
    def OnButton1(self, wPrarm, lParam):
        win32ui.MessageBox('Button1', 'Python', win32con.MB_OK)
        #self.EndDialog(1)
    def OnButton2(self, wParam, lParam):
        text = self.GetDlgItemText(1054)
        win32ui.MessageBox(text, 'Python', win32con.MB_OK)
        #self.EndDialog(1)

style = win32con.DS_MODALFRAME | win32con.WS_POPUP | win32con.WS_VISIBLE \
        | win32con.WS_CAPTION | win32con.WS_SYSMENU | win32con.DS_SETFONT
childstyle = win32con.WS_CHILD | win32con.WS_VISIBLE
buttonstyle = win32con.WS_TABSTOP | childstyle

di = ['Python', (0,0,300,180), style, None, (8,'MS Sans Serif')]
Button1 = (['Button', 'Button1', 1051, (80, 150, 50, 14),
           buttonstyle | win32con.BS_PUSHBUTTON])
Button2 = (['Button', 'Button2', 1052, (160, 150, 50, 14),
           buttonstyle | win32con.BS_PUSHBUTTON])
Static = (['Static', 'Python Dialog', 1053, (130, 50, 60, 14), childstyle])
Edit = (['Edit', '', 1054, (130, 80, 60, 14),
        childstyle | win32con.ES_LEFT | win32con.WS_BORDER | win32con.WS_TABSTOP])

init = [di, Button1, Button2, Stadic, Edit,]
mydialog = MyDialog(init)
mydialog.DoModal()
```

12.5 判断 Windows 操作系统的版本

某些情况下，程序可能依赖于特定版本操作系统中的功能或者希望程序在不同版本的操作系统中有不同的表现，因此能够在程序运行时获知操作系统的版本就变得非常有必要。Python 支持使用多种不同的方法来获取操作系统的版本信息，请自行运行下面的代码。

```
>>> import os
>>> print(os.popen('ver').read())
>>> import sys
```

```
>>> print(sys.getwindowsversion())
>>> import platform
>>> print(platform.platform())
```

Windows 管理规范（Windows Management Instrumentation，WMI）是 Windows 的一项核心技术，它以公共信息模型对象管理器（Common Information Model Object Manager，CIMOM）为基础，是一个描述 Windows 操作系统构成单元的对象数据库。WMI 是 Windows 的核心组件，通过编写 WMI 脚本和应用程序可以获取计算机系统、软件和硬件信息，还可以对计算机进行管理，如关机、重新启动计算机等。

```
>>> import wmi
>>> wmiShell = wmi.WMI()
>>> print(wmiShell.Win32_OperatingSystem()[0].Caption)
```

还可以通过 os.system('ver')语句来查看 Windows 操作系统的版本，但需要编写程序并在命令提示符环境中运行，在 IDLE 中运行无法直接查看结果。

12.6 系统运维

系统运维涉及的内容非常广泛，包括文件系统、数据库、用户账号的维护，任务调度与分配，CPU、内存、网络带宽、硬盘空间、IP 地址等资源的分配与运行状态监测等。第 7 章关于文件夹增量备份、文件夹大小计算、删除指定类型文件和第 10 章的网络嗅探器、端口扫描器、网络流量查看器的案例都属于系统运维范畴。Python 标准库 os 提供了大量可用于系统运维的函数，如表 12-3 所示。另外，Python 标准库 sys、platform 以及扩展库 psutil 等也提供了很多支持系统运维的功能。

表 12-3 Python 标准库 os 中常用系统运维的函数

函 数 名	功 能 说 明	函 数 名	功 能 说 明
getcwd()	获取当前工作目录	kill()	结束进程
chdir()	改变当前工作目录	scandir()	遍历指定文件夹
get_exec_path()	返回可执行文件搜索路径列表	cpu_count()	查看处理器数量
getlogin()	获取当前登录的用户名	getpid()	查看当前进程 ID
listdir()	列出指定文件夹中的所有文件和子文件夹	system()、startfile()	启动外部程序或打开指定文件
mkdir()、makedirs()	创建文件夹	getppid()	查看父进程 ID
remove()、rmdir()、removedirs()	删除文件、文件夹	rename()、renames()	重命名文件

12.6.1 Python 扩展库 psutil

跨平台的 Python 扩展库 psutil 可以用来查询进程或 CPU、内存、硬盘及网络等系统资源占用率等信息，常用于系统运行状态检测和维护，可以使用 pip 工具安装该库。

(1) 查看 CPU 信息。

```
>>> psutil.cpu_count()                    #查看 CPU 的核数
>>> psutil.cpu_count(logical=False)       #查看物理 CPU 的个数
>>> psutil.cpu_percent()                  #查看 CPU 的使用率
>>> psutil.cpu_percent(percpu=True)       #查看每个 CPU 的使用率
>>> psutil.cpu_times()                    #查看 CPU 时间分配情况
```

(2) 查看开机时间。

```
>>> import datetime
>>> t = psutil.boot_time()
>>> datetime.datetime.fromtimestamp(t).strftime('%Y-%m-%d %H:%M:%S')
```

(3) 查看内存信息。

```
>>> virtual_memory = psutil.virtual_memory()
>>> virtual_memory.total/1024/1024/1024    #内存总大小
>>> virtual_memory.used/1024/1024/1024     #已使用内存
>>> virtual_memory.free/1024/1024/1024     #空闲内存
>>> virtual_memory.percent                 #内存使用率
```

(4) 查看磁盘信息。

```
>>> psutil.disk_partitions()                    #查看所有分区信息
>>> psutil.disk_usage('c:\\')                   #查看指定分区的磁盘空间情况
>>> psutil.disk_io_counters(perdisk=True)       #查看硬盘读写操作情况
```

(5) 查看网络流量与收发包信息。

```
>>> psutil.net_io_counters()
```

(6) 查看当前登录用户信息。

```
>>> psutil.users()
```

(7) 查看进程信息。进程有关的概念和应用见第 13 章。

```
>>> psutil.pids()                    #查看当前所有进程 ID
>>> p = psutil.Process(4204)         #获取指定 ID 的进程
>>> p.name()                         #进程名
>>> p.username()                     #查看创建该进程的用户名
>>> p.cmdline()                      #查看该进程对应的 EXE 文件
>>> p.cwd()                          #查看该进程的工作目录
>>> p.exe()                          #进程对应的可执行文件名
>>> p.cpu_affinity()                 #该进程 CPU 占用情况(运行在哪个 CPU 上)
>>> p.num_threads()                  #该进程包含的线程数量
>>> p.threads()                      #该进程所有线程对象
>>> p.status()                       #进程状态
>>> p.is_running()                   #进程是否正在运行
>>> p.suspend()                      #挂起
>>> p.resume()                       #恢复运行
>>> p.kill()                         #结束进程
```

(8) 检查记事本程序是否在运行,如果在运行则返回记事本程序对应的进程 ID。

```
>>>for pid in psutil.pids():
    try:
        p = psutil.Process(pid)
        if os.path.basename(p.exe()) == 'notepad.exe':    #需要先导入 os.path
            print(id)
    except:
        pass
```

12.6.2 使用 Pywin32 实现事件查看器

Windows 系统会对运行过程中发生的很多事情进行记录,通过事件查看器可以查看系统日志,常用于计算机取证、事后调查、责任定位以及攻击向量分析等,对于服务器管理和运行维护具有重要意义。Windows 7 操作系统中打开事件查看器的步骤:单击"开始"菜单→右击"计算机"→单击"管理"菜单→单击"事件查看器",界面如图 12-5 所示。Windows 10 与此类似,不再赘述。

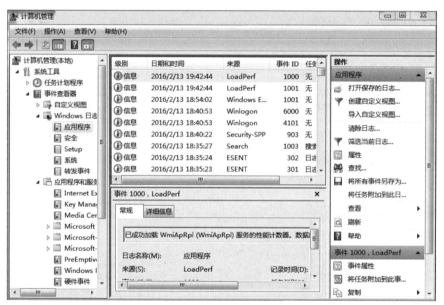

图 12-5 Windows 7 操作系统事件查看器

例 12-7 编写程序,实现多线程事件查看器。

```
import re
import sys
import time
import traceback
import threading
import winerror
import win32con
import win32evtlog
import win32security
import win32evtlogutil

def date2sec(evt_date):
```

```python
    '''把类似于'6/26/23 15:54:09'格式的日期时间字符串转换为纪元时间以来经过的秒数'''
    #把日期和时间分开
    the_date, the_time = evt_date.split()
    (month, day, year) = map(int, the_date.split('/'))
    (hour, minute, second) = map(int, the_time.split(':'))
    if 70 < year < 100:
        year = year + 1900
    elif year < 50:
        year = year + 2000
    tup = (year, month, day, hour, minute, second, 0, 0, 0)
    seconds = time.mktime(tup)
    return seconds

def main(computer='.', logtype='System', interval=480):
    flags = win32evtlog.EVENTLOG_BACKWARDS_READ|\
            win32evtlog.EVENTLOG_SEQUENTIAL_READ
    evt_dict = {win32con.EVENTLOG_AUDIT_FAILURE:'审核失败事件',
                win32con.EVENTLOG_AUDIT_SUCCESS:'审核成功事件',
                win32con.EVENTLOG_INFORMATION_TYPE:'通知事件',
                win32con.EVENTLOG_WARNING_TYPE:'警告事件',
                win32con.EVENTLOG_ERROR_TYPE:'错误事件'}
    begin_sec = time.time()
    begin_time = time.strftime('%H:%M:%S', time.localtime(begin_sec))
    try:
        #打开日志
        hand = win32evtlog.OpenEventLog(computer,logtype)
    except:
        print('无法打开"{0}"服务器上的"{1}"日志'.format(computer, logtype))
        return
    print(logtype, 'events found in the last {0} hours before {1}'.format(interval/60/60,
                                                                          begin_time))
    events = 1
    while events:
        events = win32evtlog.ReadEventLog(hand, flags, 0)
        for ev_obj in events:
            try:
                the_time = ev_obj.TimeGenerated.Format('%D %H:%M:%S')
                seconds = date2sec(the_time)
                #只查看指定时间段内的日志
                if seconds < begin_sec-interval:
                    break
                computer = ev_obj.ComputerName
                cat = str(ev_obj.EventCategory)
                src = str(ev_obj.SourceName)
                record = str(ev_obj.RecordNumber)
                evt_id = str(winerror.HRESULT_CODE(ev_obj.EventID))
                evt_type = evt_dict[ev_obj.EventType]
                msg = win32evtlogutil.SafeFormatMessage(ev_obj, logtype)
                print(':'.join((the_time, computer, src, cat, record,
                                evt_id, evt_type, msg)))
                print('='*20)
                if seconds < begin_sec-interval:
                    break
```

```
            except:
                pass
        win32evtlog.CloseEventLog(hand)

t3 = threading.Thread(target=main, args=('.', 'Application', 5400))
t3.start()
t3.join()
```

12.6.3 切换用户登录身份

大型服务器可能会有多个不同权限的管理员账号,有时可能会需要在多个账号之间切换,下面的代码使用 Pywin32 实现了临时登录为另一个账号的功能。

例 12-8 临时登录为另一个账号。

首先在系统中创建一个账号 ddddd 并设置密码为 123456,创建文件夹 D:\test_ddd。

```
import os
import win32con
import win32api
import win32security

class Impersonate:
    def __init__(self, loginName, password):
        self.domain = 'WORKGROUP'
        self.loginName = loginName
        self.password = password

    def logon(self):
        self.handel = win32security.LogonUser(self.loginName, self.domain,
                                self.password, in32con.LOGON32_LOGON_INTERACTIVE,
                                win32con.LOGON32_PROVIDER_DEFAULT)
        #登录另一个账号
        win32security.ImpersonateLoggedOnUser(self.handel)

    def logoff(self):
        #切换至本来的用户名
        win32security.RevertToSelf()
        print('OK. I am back '+win32api.GetUserName())
        #关闭句柄
        self.handel.Close()

print('Origionally I am '+win32api.GetUserName())
#要模仿的用户名和密码
a = Impersonate('ddddd','123456')
try:
    #以别人身份登录
    a.logon()
    #显示当前的登录用户名
    print('Now I become '+win32api.GetUserName())
    os.mkdir(r'D:\test_ddd\ddd')
```

```
        #注销并切换至本来的用户身份
        a.logoff()
except:
    print("Denied.Now I will become an administrator and try again")
    a.logoff()
    os.mkdir(r'D:\test_ddd\administrator')
```

运行上面的代码,会发现在文件夹 D:\test_ddd 中创建了文件夹 ddd。接下来删除刚刚创建的 ddd 文件夹,并设置文件夹 D:\test_ddd 的权限,拒绝账号 ddd 的任何操作。主要步骤:右击 D:\test_ddd 文件夹,在属性对话框中选择"安全"选项卡,然后单击"编辑"按钮,依次单击"添加"→"高级"→"立即查找"命令,选择刚刚创建的 ddddd 账号,然后设置权限为全部拒绝,如图 12-6 所示。再次运行上面的代码,会发现在 D:\test_ddd 文件夹中创建了子文件夹 administrator。

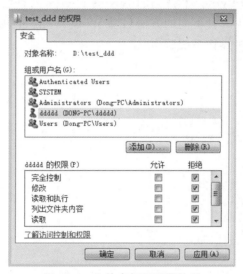

图 12-6　文件夹权限配置结果

本 章 小 结

(1) Pywin32 封装了 Windows 底层几乎所有 API 函数。

(2) 使用 ctypes 模块可以调用其他语言编写的动态链接库或共享库文件。

(3) 使用 pyinstaller 可以方便地将 Python 程序打包为可执行程序。

(4) 可以通过 os 模块的 `system()`、`popen()` 和 `startfile()` 等函数调用外部程序或打开磁盘上的文件,也可以使用 win32api 模块的 `ShellExecute()` 函数或 win32process 模块的 `CreateProcess()` 函数实现这一目的。

(5) ctypes 提供了 3 种调用动态链接库文件的方法,分别为 cdll、windll 和 oledll,它们的不同之处在于调用函数时的参数传递方式和函数返回时的栈平衡方式。

(6) 在 Windows 平台上,可以通过 sys 模块的 `getwindowsversion()`、platform 模块的 `platform()` 及 wmi 模块等多种方法来动态检测系统版本。

(7) 系统运维涉及的内容非常广泛,包括文件系统、数据库、用户账号的维护,任务调度与分配，CPU、内存、网络带宽、硬盘空间、IP 地址等资源的分配与运行监测，等等。

（8）Python 标准库 os、sys、platform 提供了大量可用于系统运维的函数。

（9）跨平台的 Python 扩展库 psutil 可以用来查询进程或 CPU、内存、硬盘以及网络等系统资源占用率等信息，常用于系统运行状态检测和维护。

（10）Windows 系统会对运行过程中发生的很多事情进行记录，通过事件查看器可以查看系统记录的所有事件，常用于计算机取证、事后调查、责任定位以及攻击向量分析等，对于服务器的管理和运行维护具有非常重要的意义。

习　题

1. 查阅相关资料，解释注册表几大根键的用途。
2. 选择一个编写好的 Python 程序，使用 pyinstaller 将其打包为.exe 可执行文件。
3. 编写代码，使用至少 3 种不同的方法启动 Windows 自带的计算器程序。
4. 编写代码，检测本机操作系统版本。

第 13 章 多线程与多进程编程

由于硬件技术的飞速发展,早期的多核、多处理器等高端技术已经走进了普通家庭,再加上内存、主频、硬盘等各种硬件配置的飞速提高,大幅提高了普通 PC 的运算速度和数据处理能力。在多核、多处理器平台上,每个核可以运行一个线程,多个线程同时运行并相互协作,从而达到高速处理任务的目的。但由于全局解释锁(Global Interpreter Lock,GIL)的存在,Python 多线程编程并不能充分利用 CPU 资源,好在官方技术团队一直在努力,Python 3.12 已经支持编译安装时选择不使用 GIL。

然而,即使是高端服务器或工作站甚至集群系统,处理器和核的数量总是有限的,如果线程的数量多于核的数量,就必然需要调度。在调度时,处理器分配一个很短的时间片,所有线程根据具体的调度算法轮流获得该时间片。当时间片用完以后,即使该线程还没有执行完也要退出处理器并等待下次调度。由于处理器中寄存器的数量有限,而不同的线程很可能需要使用到相同的一组寄存器来保存中间计算结果或当前状态。因此,在调度线程时必须做好上下文保存和恢复工作,以保证该线程下次被调度进处理器后能够继续上次的工作。虽然这些工作并不需要 Python 程序员操心,但是我们必须清楚的一件事是,并不是使用的线程数量越多越好,如果线程太多,线程调度带来的开销可能会比线程实际执行的开销还大,这样使用多线程就失去本来的意义了。

13.1 threading 模块

threading 模块是 Python 支持多线程编程的重要模块,其中的常用函数如表 13-1 所示。

表 13-1 threading 模块常用函数

函 数	功 能 说 明
active_count()	返回当前处于 alive 状态的 Thread 对象数量
current_thread()	返回当前 Thread 对象
enumerate()	返回当前处于 alive 状态的所有 Thread 对象列表,其中包含守护线程、主线程以及 current_thread()函数创建的伪线程对象,不包含已结束和未启动的线程
get_ident()	返回当前线程的线程标识符。线程标识符是一个非负整数,并无特殊含义,只是用来标识线程,该整数可能会被循环利用
main_thread()	返回主线程对象,即启动 Python 解释器的线程对象
stack_size([size])	返回创建线程时使用的栈的大小,如果指定 size 参数,则用来指定后续创建的线程使用的栈大小,size 必须是 0(表示使用系统默认值)或大于 32KB 的正整数

在 threading 模块中还提供了 Thread、Timer、Lock、RLock、Condition、Event、Semaphore、Barrier 等类，其中前两个是用来创建线程对象的类（Thread 类用得更多一些，见 13.2 节），后面几个类用来支持线程同步（见 13.3 节）。

13.2　Thread 对象

Thread 类支持使用两种方法来创建线程：一种方法是使用一个可调用对象作参数创建对象；另一种方法是继承 Thread 类并在派生类中重写 __init__() 和 run() 方法。创建线程对象以后，可以调用其 start() 方法来启动，该方法自动调用该类对象的 run() 方法，此时该线程处于 alive 状态，直至线程的 run() 方法运行结束。Thread 对象成员如表 13-2 所示。

表 13-2　Thread 对象成员

成　　员	说　　明
start()	自动调用 run() 方法，启动线程，执行线程代码
run()	线程代码，用于实现线程的功能与业务逻辑，可以在子类中重写该方法来自定义线程的行为
__init__(self,group=None,target=None,name=None, 　　　　args=(),kwargs=None,*,daemon=None)	构造方法
name	用于读取或设置线程的名字
ident	线程标识，非 0 数字或 None（线程未被启动时）
is_alive()	测试线程是否处于 alive 状态
daemon	布尔值，表示线程是否为守护线程
join(timeout=None)	等待线程结束或超时返回

13.2.1　Thread 对象中的方法

（1）join([timeout])：阻塞当前线程，等待被调线程结束或超时返回后再继续执行当前线程的后续代码，参数 timeout 用于指定最长等待时间，单位为秒。

例 13-1　创建多线程。

例 13-1

```
import time
import threading

def func1(x, y):                                    #线程函数
    for i in range(x, y):
        print(i, end=' ')
    time.sleep(10)                                  #注释掉这里试试

t1 = threading.Thread(target=func1, args=(15, 20))
t1.start()                                          #启动线程
t1.join(5)                                          #注释掉这里试试
t2 = threading.Thread(target=func1, args=(5, 10))
t2.start()
```

保存并运行上面的程序将会发现，首先输出 15~19 这 5 个整数，然后程序暂停，几秒以后又继续输出 5~9 这 5 个整数。如果将 t1.join(5) 这一行注释掉再运行，两个线程的输出将会重叠在一起，这是因为两个线程并发运行，而不是第一个结束以后再运行第二个。

（2）is_alive()：测试线程是否处于运行状态。

例 13-2 查看线程状态。

```
import time
import threading

def func1(x, y):
    for i in range(x, y):
        print(i)

t1 = threading.Thread(target=func1, args=(15, 20))
t1.start()
t1.join(5)          #注释掉这里试试
t2 = threading.Thread(target=func1, args=(5, 10))
t2.start()
t2.join()           #注释掉这里试试
print('t1:', t1.is_alive())
print('t2:', t2.is_alive())
```

运行上面的程序会发现，最后两个的输出都是 False，即两个线程都执行完了。如果将 t1.join(5) 这一行注释掉会发现最后两行的输出结果没有变化，这是因为 t2.join() 这一行代码会阻塞当前程序直至线程 t2 运行结束，此时线程 t1 基本也运行结束了。如果将线程 t1 的参数范围增大则会发现，倒数第二行的输出结果很可能会变为 True。

13.2.2 Thread 对象中的 daemon 属性

在程序运行过程中有一个主线程，若在主线程中创建子线程，当主线程结束时根据子线程 daemon 属性值的不同可能会发生下面两种情况之一：①当某子线程的 daemon 属性为 False 时，主线程结束时进程会检测该子线程是否结束，如果该子线程尚未完成，则进程会等待它完成后再退出；②当某子线程的 daemon 属性为 True 时，主线程运行结束时进程不对该子线程进行检查而直接退出，同时所有 daemon 值为 True 的子线程将随主线程一起结束，不论是否运行完成。daemon 属性的值默认为 False，如果需要修改，必须在调用 start() 方法启动线程之前修改。

以上论述不适用于 IDLE 中的交互模式或程序运行模式，因为在该环境中的主线程只有在退出 IDLE 时才终止。

例 13-3 线程对象 daemon 属性的作用。

例 13-3

```
import time
import threading

class MyThread(threading.Thread):
    def __init__(self, num, threadname):
        threading.Thread.__init__(self, name=threadname)
        self.num = num
```

```
        def run(self):
            time.sleep(self.num)
            print(self.num)

t1 = MyThread(1, 't1')
t2 = MyThread(5, 't2')
t2.daemon = True
print(t1.daemon)
print(t2.daemon)
t1.start()
t2.start()
```

在 IDLE 中运行结果如图 13-1 所示,在命令提示符环境中运行结果如图 13-2 所示。

图 13-1　在 IDLE 中的运行结果　　图 13-2　在命令提示符环境中的运行结果

派生自 Thread 类的自定义线程类除了拥有线程类特有的 run()、start()、join()等一系列方法,也可以在线程类中定义普通方法并通过线程对象来调用。

例 13-4　调用线程对象的普通方法,面向对象程序设计的内容见第 6 章。

例 13-4

```
import time
import threading

class MyThread(threading.Thread):
    def __init__(self, threadName):
        threading.Thread.__init__(self)
        self.name = threadName
    def run(self):
        time.sleep(1)
        print('In run:', self.name)
    def output(self):                          #在线程类中定义普通方法
        print('In output:', self.name)

t = MyThread('test')                           #创建线程对象
t.start()                                      #启动线程
t.output()                                     #调用普通方法
time.sleep(2)
print('OK')
```

13.3　线程同步技术

多线程是为了充分利用硬件资源尤其是 CPU 资源来提高任务处理速度和效率的技术。将任务拆分成互相协作的多个线程同时运行,那么属于同一个任务的多个线程之间必然会有交互和同步,以便互相协作地完成任务。另外,还可以使用多线程来为用户提供很多的方便。例如,打开软件时可能需要加载大量的模块和库,这可能需要较长的时间,此时可以使用一个线程来显示一个小动画来表示当前软件正在启动,当后台线程加载完所有的模块和库之后,结束该动画的播放并打开软件主界面,这是多线程同步的一个典型应用。再如,字处理软件可以使用一个线程来接收用户键盘输入,同时使用一个后台线程来进行拼写检查以及字数统计之类的功能,并实时将结果显示在状态栏上,这无疑会极大方便用户的使用,对于提高用户体验有重要帮助。

13.3.1 Lock/RLock 对象

Lock 是比较低级的同步原语,当被锁定以后不属于特定的线程。一个锁有两种状态:locked 和 unlocked。如果锁处于 unlocked 状态,`acquire()`方法将其修改为 locked 并立即返回;如果锁已处于 locked 状态,则阻塞当前线程并等待其他线程释放锁,然后将其修改为 locked 并立即返回。`release()`方法用来将锁的状态由 locked 修改为 unlocked 并立即返回,如果锁状态本来已经是 unlocked,调用该方法会抛出异常。

可重入锁 RLock 对象也是一种常用的线程同步原语,可被同一个线程 `acquire()` 多次。当处于 locked 状态时,某线程拥有该锁;当处于 unlocked 状态时,该锁不属于任何线程。RLock 对象的 `acquire()`/`release()` 调用对可以嵌套,仅当最后一个或者最外层的 `release()` 执行结束后,锁才会被设置为 unlocked 状态。

例 13-5　使用 Lock/RLock 对象实现线程同步。

```
import time
import threading

class MyThread(threading.Thread):
    def __init__(self):                              #派生类中定义了构造方法
        threading.Thread.__init__(self)              #必须显式调基类的构造方法
    def run(self):
        global x
        lock.acquire()
        for i in range(3):
            x = x + i
        time.sleep(2)
        print(x)
        lock.release()

lock = threading.RLock()                             #也可以用 lock = threading.Lock()
tl = []
for i in range(10):
    t = MyThread()
    tl.append(t)
x = 0
for i in tl:
    i.start()
```

例 13-6　使用多线程编程技术统计指定整数范围内的素数数量。

```
import time
```

例 13-6

```python
import threading

def prime(x):
    if x < 2: return False                      #这里是为了节约篇幅
    if x in (2, 3): return True                 #实际编写时建议换行缩进
    if x%2 == 0: return False
    for i in range(3, int(x**0.5)+1, 2):
        if x%i == 0: return False
    return True

def worker(p):
    global c                                     #声明使用全局变量,见 5.5 节
    while True:
        try:
            x = next(p)                          #从迭代器对象中获取下一个整数
        except:
            break                                #获取失败说明已处理完,结束循环
        else:
            if prime(x):
                lock.acquire()                   #申请进入临界区,任意时刻只有一个线程修改数值
                c = c +1
                lock.release()

c = 0                                            #记录素数数量的变量
lock = threading.Lock()
numbers = iter(range(1000000))                   #其中每个元素被获取之后就没有了
tList = []
for i in range(10):
    t = threading.Thread(target=worker, args=(numbers,))
    tList.append(t)

start = time.time()
for t in tList: t.start()
for t in tList: t.join()
print(time.time()-start)
print("Count=", c)
```

13.3.2 Condition 对象

使用 Condition 对象可以在某些事件触发或某个条件满足后才处理数据,可以用于不同线程之间的通信或通知,以实现更高级别的同步。Condition 对象除了具有 `acquire()` 和 `release()` 方法之外,还有 `wait()`、`notify()` 和 `notify_all()` 等方法。下面通过经典生产者-消费者问题来演示 Condition 对象的用法。

例 13-7 使用 Condition 对象实现线程同步,模拟生产者-消费者问题。

```python
import threading

class Producer(threading.Thread):                #生产者线程类
    def __init__(self, threadname):
        threading.Thread.__init__(self, name=threadname)
```

```python
    def run(self):
        global x
        con.acquire()
        if x == 20:                                    #假设缓冲区满
            con.wait()                                 #生产者等待
        print('\nProducer:', end='')
        for i in range(20):                            #生产数据
            print(x, end='')
            x = x + 1
        print(x)
        con.notify()                                   #通知消费者
        con.release()

class Consumer(threading.Thread):                       #消费者线程类
    def __init__(self, threadname):
        threading.Thread.__init__(self, name=threadname)
    def run(self):
        global x
        con.acquire()
        if x == 0:                                     #假设缓冲区空,无数据可消费
            con.wait()                                 #消费者等待
        print('\nConsumer:', end='')
        for i in range(20):                            #消费数据
            print(x, end='')
            x = x - 1
        print(x)
    con.notify()                                       #通知生产者
    con.release()

con = threading.Condition()
x = 0
p = Producer('Producer')
c = Consumer('Consumer')
p.start()                                              #启动线程
c.start()
p.join()                                               #等待线程结束
c.join()
print('After Producer and Consumer all done:', x)
```

该程序的运行结果如图 13-3 所示。

```
>>> ================================ RESTART ========================
>>>
Producer: 0 1 2 3 4 5 6 7 8 9 10 11 12 13 14 15 16 17 18 19 20
Consumer: 20 19 18 17 16 15 14 13 12 11 10 9 8 7 6 5 4 3 2 1 0
After Producer and Consumer all done: 0
```

图 13-3　使用 Condition 对象实现线程同步

例 13-8　使用 Condition 对象实现线程同步,模拟生产者-消费者问题。在代码中使用列表模拟物品池,生产者往里放置东西,消费者从池中获取物品。物品池满时生产者等待,空时消费者等待。

例 13-8

```python
import threading
from random import randint
from time import sleep

class Producer(threading.Thread):
    def __init__(self, threadname):
        threading.Thread.__init__(self, name=threadname)
    def run(self):
        while True:
            sleep(1)
            con.acquire()                              #获取锁
            if len(x) == 5:                            #假设共享列表中最多能容纳 5 个元素
                print('Producer is waiting…')          #如果共享列表已满,生产者等待
                con.wait()
            else:
                r = randint(1, 1000)
                print('Produced:', r)                  #产生新元素,添加至共享列表
                x.append(r)
                con.notify()                           #唤醒等待条件的线程
            con.release()                              #释放锁

class Consumer(threading.Thread):
    def __init__(self, threadname):
        threading.Thread.__init__(self, name=threadname)
    def run(self):
        while True:
            sleep(3)
            #获取锁
            con.acquire()
            if not x:
                #物品池已空,消费者等待
                print('Consumer is waiting…')
                con.wait()
            else:
                print('Consumed:', x.pop(0))
                con.notify()
            con.release()

#创建 Condition 对象以及生产者线程和消费者线程
con = threading.Condition()
x = []
p = Producer('Producer')
c = Consumer('Consumer')
p.start()
c.start()
```

13.3.3　queue 模块

queue 模块实现了多生产者-多消费者队列,尤其适合需要在多个线程之间进行信息交换的场合,该模块的 Queue 对象实现了多线程编程所需要的所有锁语义。

```python
import time
import queue
```

```python
import threading

class Producer(threading.Thread):                    #生产者线程类
    def __init__(self, threadname):
        threading.Thread.__init__(self, name=threadname)
    def run(self):
        global myqueue                               #可增加 sleep()函数再观察效果
        myqueue.put(self.getName())                  #put()用于在队列尾部追加元素
        print(self.getName(), ' put ', self.getName(), ' to queue.')

class Consumer(threading.Thread):                    #消费者线程类
    def __init__(self, threadname):
        threading.Thread.__init__(self, name=threadname)
    def run(self):
        global myqueue                               #get()用于获取队列头部的元素
        print(self.getName(), ' get ', myqueue.get(), ' from queue.')

myqueue = queue.Queue()
plist = []
clist = []
for i in range(10):
    p = Producer('Producer'+str(i))                  #每个生产者线程的名字不同
    plist.append(p)
    c = Consumer('Consumer'+str(i))                  #每个消费者线程的名字不同
    clist.append(c)
for i in plist:
    i.start()
    i.join()                                         #join()本身就有同步的作用
for i in clist:
    i.start()
    i.join()                                         #可以删除这两行 join()再观察效果
```

13.3.4 Event 对象

Event 对象是一种常用的线程通信技术，一个线程设置 Event 对象，另一个线程等待 Event 对象。Event 对象的 `set()`方法可以设置 Event 对象内部的信号标志为真；`clear()`方法可以清除 Event 对象内部的信号标志，将其设置为假；`is_set()`方法用来判断其内部信号标志的状态；`wait()`方法在其内部信号状态为真时立即执行并返回，若 Event 对象的内部信号标志为假，`wait()`方法将一直等待至超时或内部信号状态为真。

例 13-9　使用 Event 对象实现线程同步。

```python
import threading

class myThread(threading.Thread):
    def __init__(self, threadname):
        threading.Thread.__init__(self, name=threadname)
    def run(self):
        if myEvent.is_set():
            myEvent.clear()                          #清除事件标志
            print(self.name+' is waiting…')
            myEvent.wait()                           #等待事件再次发生后继续执行
```

```
                print(self.name+' resumed…')
            else:
                print(self.name)
                myEvent.set()                           #设置事件标志

myEvent = threading.Event()
#设置初始状态,清除事件标志
myEvent.clear()

for i in range(10):
    myThread(str(i)).start()                            #创建并启动 10 个非守护线程
```

将上面的代码保存为 ThreadSynchronizationUsingEvent.py 文件并运行,每次的运行结果略有不同,还有可能发生死锁,图 13-4 是其中一次的运行结果。

图 13-4 使用 Event 对象实现线程同步

13.3.5 Semaphore 与 BoundedSemaphore

Semaphore 对象维护着一个内部计数器,调用 `acquire()`方法时该计数器减 1,调用 `release()`方法时该计数器加 1,适用于需要控制特定资源并发访问线程数量的场合。

调用 `acquire()`方法时,如果计数器已经为 0 则阻塞当前线程,直到有其他线程调用了 `release()`方法,所以计数器的值永远不会小于 0。

Semaphore 对象可以调用任意次 `release()`方法,而 BoundedSemaphore 对象可以保证计数器的值不超过特定的值。

例 13-10

例 13-10 使用 BoundedSemaphore 对象限制特定资源并发访问线程数量。

```
from time import time, sleep
from random import randrange
from threading import Thread, BoundedSemaphore

def worker(value):
    start = time()                                      #记录线程启动的时间
    with sema:
        end = time()                                    #记录获取资源访问权限的时间
        t = randrange(5)
```

```
                #第一个冒号后面是该线程等待的时间
                print(value, end-start, t, sep=':')
                sleep(2)

    sema = BoundedSemaphore(3)              #同一时刻最多允许 3 个线程访问特定资源
    for i in range(10):                     #创建并启动 10 个非守护线程
        Thread(target=worker, args=(i,)).start()
```

13.3.6 Barrier 对象

Barrier 对象常用于实现这样的线程同步：多个线程运行到某个时间点以后每个线程都需要等着其他线程准备好以后再同时进行下一步工作。类似于赛马时需要先用栅栏拦住，每个试图穿过栅栏的选手都需要明确说明自己准备好了，当所有选手都表示准备好以后，栅栏打开，所有选手同时冲出栅栏。

Barrier 对象最常用的方法是 `wait()`。线程调用该方法后会阻塞，当所有线程都调用了该方法后，会被同时释放并继续执行后面的代码。Barrier 对象的 `wait()` 方法会返回一个 0~parties-1 的整数，每个线程都会得到一个不同的整数。

例 13-11 创建一个允许 3 个线程互相等待的 Barrier 对象，每个线程做完一些准备工作后调用 Barrier 对象的 `wait()` 方法等待其他线程，当所有线程都调用了 `wait()` 方法之后，会调用指定的 action 对象，然后同时开始执行 `wait()` 之后的代码。

例 13-11

```
import time
import random
import threading

def worker(arg):
    #假设每个线程需要不同的时间来完成准备工作
    time.sleep(random.randint(1, 20))
    #假设已知任何线程的准备工作最多需要 20s
    #每个线程调用 wait()时，返回值不一样
    r = b.wait(20)
    if r == 0:
        print(arg)

def printOk():
    print('ok')

#允许 3 个线程等待，如果线程调用 wait()时没有指定超时时间，默认为 20s
b = threading.Barrier(parties=3, action=printOk, timeout=20)

#创建并启动 3 个线程，线程数量必须与创建 Barrier 对象时的 parties 参数值一致
for i in range(3):
    threading.Thread(target=worker, args=(i,)).start()
```

13.4 多进程编程

由于 GIL 的问题，Python 程序使用多线程并不能大幅提高任务吞吐量和整体处理速度，也无法实现真正的多任务并行执行，尤其是计算密集型的任务。如果有这样的功能需

求,目前还是建议使用多进程编程,Python 3.12 开始可能会逐步取消 GIL。

Python 标准库 multiprocessing 支持使用类似于 threading 的用法来创建与管理进程,不存在 GIL 问题,可以更有效地利用 CPU 资源。本节简单介绍进程创建与启动以及数据交换的内容,更多案例可以参考作者教材《Python 网络程序设计》(微课版)和《Python 程序设计开发宝典》。

13.4.1 创建与启动进程

可以通过创建 Process 对象来创建一个进程并通过调用进程对象的 start()方法来启动,然后调用 join()方法等待执行结果。多进程的程序需要在命令提示符或者 Powershell 中运行,不能在 IDLE 中直接运行。

创建进程

```
import os
from multiprocessing import Process

def f(name):
    print('module name:', __name__)
    print('parent process:', os.getppid())      #查看父进程 ID
    print('process id:', os.getpid())            #查看当前进程 ID
    print('hello', name)

if __name__ == '__main__':
    p = Process(target=f, args=('bob',))
    p.start()                                    #strart()方法会自动调用 run()方法
    p.join()                                     #等待子进程结束
```

或者使用下面的方法,创建自定义类并继承自 Process 类,然后实现 run()方法。和线程类对象一样,调用进程对象 start()方法时将自动调用 run()方法。

```
from multiprocessing import Process

class MyProcess(Process):
    def __init__(self):                          #如果在派生类中定义了构造方法
        Process.__init__(self)                   #就必须显式调用基类的构造方法
    def run(self):
        print('ok')

if __name__ == '__main__':
    MyProcess().start()
```

multiprocessing 还提供了 Pool 对象支持数据的并行操作。例如,下面的代码可以并行计算二维数组每行的平均值。

```
from statistics import mean
from multiprocessing import Pool

def f(x):
    return mean(x)

if __name__ == '__main__':
```

```
        x = [list(range(10)), list(range(20, 30)), list(range(50,60)), list(range(80, 90))]
        with Pool(5) as p:
            print(p.map(f, x))                              #可以把 f 直接改为 mean
```

下面的代码并行判断 100000000 以内的数字是否为素数,并统计素数个数。

```
from multiprocessing import Pool

def isPrime(n):
    if n < 2:
        return 0
    if n == 2:
        return 1
    if not n&1:                                             #相当于 if n%2 == 0:
        return 0
    for i in range(3, int(n**0.5)+1, 2):
        if n%i == 0:
            return 0
    return 1

if __name__ == '__main__':
    with Pool(5) as p:
        print(sum(p.map(isPrime, range(100000000))))
```

13.4.2 进程间数据交换

同一个进程中的所有线程共用地址空间,可以直接访问其他线程申请或创建的全局资源。不同进程之间的地址空间是互相隔离的,必须使用专门的方式进行数据交换和共享,本节通过几个例题演示相关用法。

例 13-12 使用 Queue 对象在进程间交换数据。

```
import multiprocessing as mp

def foo(q):
    q.put('hello world!')                                   #子进程的地址空间与主进程是隔离的
                                                            #彼此之间不能直接访问数据

if __name__ == '__main__':
    mp.set_start_method('spawn')                            #Windows 系统创建子进程的默认方式
    q = mp.Queue()
    p = mp.Process(target=foo, args=(q,))                   #需要使用参数把对象传递给子进程
    p.start()
    p.join()
    print(q.get())
```

也可以使用上下文对象 context 的 Queue 对象实现进程间的数据交换。

```
import multiprocessing as mp

def foo(q):
    q.put('hello world')

if __name__ == '__main__':
```

```python
ctx = mp.get_context('spawn')
q = ctx.Queue()
p = ctx.Process(target=foo, args=(q,))
p.start()
p.join()
print(q.get())
```

例 13-13　使用管道实现进程间数据交换。

```python
from multiprocessing import Process, Pipe

def f(conn):
    conn.send('hello world')             #向管道中发送数据
    conn.close()

if __name__ == '__main__':
    parent_conn, child_conn=Pipe()       #创建管道对象
    p = Process(target=f, args=(child_conn,))  #将管道的一方作为参数传递给子进程
    p.start()
    print(parent_conn.recv())            #通过管道的另一方获取数据
    p.join()
```

例 13-14　使用共享内存实现进程间数据交换。

```python
from multiprocessing import Process, Value, Array

def f(n, a):
    n.value = 3.1415927                  #在子进程中修改主进程中对象的值
    for i in range(len(a)):
        a[i] = a[i]*a[i]

if __name__ == '__main__':
    num = Value('d', 0.0)                #实型
    arr = Array('i', range(10))          #整型数组
    p = Process(target=f, args=(num, arr))
    p.start()
    p.join()
    print(num.value)                     #查看被子进程修改后的值
    print(arr[:])
```

例 13-15　使用 Manager 对象实现进程间数据交换。Manager 对象控制一个拥有 list、dict、Lock、RLock、Semaphore、BoundedSemaphore、Condition、Event、Barrier、Queue、Value、Array、Namespace 等对象的服务端进程，并且允许其他进程访问这些对象。

```python
from multiprocessing import Process, Manager

def f(d, l, t):
    d['name'] = 'Dong Fuguo'             #在子进程中修改主进程中的对象
    d['age'] = 38
    d['sex'] = 'Male'
    d['affiliation'] = 'YanTai'
    l.reverse()
    t.value = 3
```

```
if __name__ == '__main__':
    with Manager() as manager:
        d = manager.dict()
        l = manager.list(range(10))
        t = manager.Value('i', 0)
        p = Process(target=f, args=(d,l,t))
        p.start()
        p.join()
        for item in d.items():
            print(item)
        print(l)
        print(t.value)
```

例 13-16 使用 Manager 对象实现不同机器上的进程跨网络共享数据。

(1) 编写程序文件 multiprocessing_server.py，启动服务器进程，创建可共享的队列对象。

```
from queue import Queue
from multiprocessing.managers import BaseManager

q = Queue()
class QueueManager(BaseManager):
    pass
QueueManager.register('get_queue', callable=lambda:q)
m = QueueManager(address=('',30030), authkey=b'dongfuguo')
s = m.get_server()
s.serve_forever()
```

(2) 编写程序文件 multiprocessing_client1.py，连接服务器进程，并往共享的队列中存入一些数据。

```
from multiprocessing.managers import BaseManager

class QueueManager(BaseManager):
    pass
QueueManager.register('get_queue')
#假设服务器的 IP 地址为 10.2.1.2
m = QueueManager(address=('10.2.1.2', 30030), authkey=b'dongfuguo')
m.connect()
q = m.get_queue()
for i in range(3):
    q.put(i)
```

(3) 编写程序文件 multiprocessing_client2.py，连接服务器进程，从共享的队列对象中读取数据并输出显示。

```
from multiprocessing.managers import BaseManager

class QueueManager(BaseManager):
    pass
QueueManager.register('get_queue')
m = QueueManager(address=('10.2.1.2', 30030), authkey=b'dongfuguo')
```

```
m.connect()
q = m.get_queue()
for i in range(3):
    print(q.get())
```

13.4.3 进程同步

在需要协同工作完成大型任务时,多个进程间的同步非常重要。进程同步方法与线程同步方法类似,代码稍微改写一下即可,本节以 Lock 对象和 Event 对象为例简单演示其用法。

例 13-17 编写程序,使用 Lock 对象实现进程同步。运行程序会发现每次顺序不一样。

```
from multiprocessing import Process, Lock

def f(l, i):
    l.acquire()                          #获取锁
    try:
        print('hello world', i)
    finally:
        l.release()                      #释放锁

if __name__ == '__main__':
    lock = Lock()                        #创建锁对象
    for num in range(10):
        Process(target=f, args=(lock, num)).start()
```

例 13-18 编写程序,使用 Event 对象实现进程同步。运行程序会发现每次结果不一样,并且只有一部分进程有输出结果。

```
from multiprocessing import Process, Event

def f(e, i):
    if e.is_set():
        e.wait()
        print('hello world', i)
        e.clear()
    else:
        e.set()

if __name__ == '__main__':
    e = Event()
    for num in range(10):
        Process(target=f, args=(e,num)).start()
```

13.4.4 标准库 subprocess

标准库 subprocess 允许创建子进程,连接子进程的输入输出管道,并获得子进程的返回码,也是常用的并行执行技术之一。

该标准库提供了 run()、call()和 Popen() 3 种不同的函数用来创建子进程,其中 run()

函数会阻塞当前进程,子进程结束后返回包含返回码和其他信息的 CompletedProcess 对象;call()函数也会阻塞当前进程,子进程结束后直接得到返回码;Popen()函数创建子进程时不阻塞当前进程,直接返回得到 Popen 对象,通过该对象可以对子进程进行更多的操作和控制。

Popen 对象的 kill()和 terminate()方法可以用来结束该进程,send_signal()可以给子进程发送指定信号,wait()方法用来等待子进程运行结束,pid 属性用来表示子进程 ID,等等。

```
>>> p = subprocess.Popen('c:\\windows\\notepad.exe')    #创建并运行子进程
>>> p.pid                                                #进程 ID
2744
>>> p.kill()                                             #结束进程
```

例 13-19 编写程序,演示在一个程序中调用另一个程序并控制输入和输出的用法。

(1) 假设有一个程序 externProgram.py,内容如下:

```
x = input()
print('hello world', x)
```

(2) 编写测试程序 test.py,内容如下:

```
from subprocess import PIPE, Popen

text = '董付国'
test = Popen('python externProgram.py', stdin=PIPE, stdout=PIPE, stderr=PIPE)
test.stdin.write(text.encode())
test.stdin.close()

with open('b.txt', 'w') as result:
    result.write(test.stdout.read().decode())
```

(3) 运行程序 test.py,自动调用和执行程序 externProgram.py,在当前文件夹中创建文件 b.txt,其中内容为

```
hello world 董付国
```

(4) 把程序 externProgram.py 修改为

```
x = input()
print('董付国系列教材: ')
for item in x.split():
    print('\t', item)
```

(5) 把 test.py 代码修改为

```
from subprocess import PIPE, Popen

text = '董付国'
test = Popen('python externProgram.py', stdin=open('in.txt'),stdout=open('out.txt', 'w'))
```

(6) 创建文本文件 in.txt,内容如下:

《Python 程序设计基础》(第 3 版)《Python 程序设计》(第 4 版)《Python 可以这样学》《Python 程序设计开发宝典》《Python 程序设计实验指导书》《Python 网络程序设计》《Python 数据分析与数据可视化》

（7）然后运行程序 test.py，自动调用程序 externProgram.py 并从 in.txt 文件中读取内容再写入 out.txt 文件中，其中内容为

```
董付国系列教材：
    《Python 程序设计基础》(第 3 版)
    《Python 程序设计》(第 4 版)
    《Python 可以这样学》
    《Python 程序设计开发宝典》
    《Python 程序设计实验指导书》
    《Python 网络程序设计》
    《Python 数据分析与数据可视化》
```

在线评测（Online Judge，OJ）系统的核心功能是自动执行学生提交的代码并判断功能是否正确，下面的代码演示了在服务端执行代码的 4 种方式，结合本章的多线程多进程编程技术、第 10 章的套接字编程技术、第 14 章的数据库编程技术以及网站开发技术（本书没有涉及网站开发，请自行查阅资料），就可以开发 OJ 系统了。在作者另一本教材《Python 网络程序设计》（微课版）中例 2-23 演示了 OJ 系统更详细的实现。

例 13-20 编写程序，模拟 OJ 系统中执行客户端代码并获取其输出结果和异常信息。

```python
from time import sleep
from os import remove
from threading import Thread
from subprocess import (run, call, check_output, Popen, STDOUT, CREATE_NO_WINDOW,
                        CalledProcessError, TimeoutExpired)

#模拟学生提交的多段代码
code_blocks = (
    '''print("你好")''',
    #反斜线表示不换行
    '''\
for i in range(5):
    print(i, end=' ')
print(5)''',
    #这里故意制造了一个错误
    '''\
for i in range(10):
    if i%2 ==0:
        print(i, end=" "''',
    '''\
def func():
    return "ok"
print(func())''',
    #这里故意制造了一个无限循环
    '''\
from time import sleep

while True:
    sleep(1)''',
)
```

```python
#用于存放进程输出信息和出错信息的临时文件
file_path = 'test_result.txt'

#第一种方式
for code in code_blocks:
    try:
        call(['python', '-c', code],
             timeout=120, shell=False,
             creationflags=CREATE_NO_WINDOW,
             stdout=open(file_path,'w'),
             stderr=open(file_path,'w'))
    except TimeoutExpired:
        with open(file_path, 'w') as fp:
            fp.write('timeout')

    #读取并显示代码执行结果或出错信息
    with open(file_path) as fp:
        print(fp.read())

#删除临时文件
remove(file_path)

#第二种方式
for code in code_blocks:
    try:
        child = run(['python', '-c', code],
                    timeout=120, shell=False,
                    creationflags=CREATE_NO_WINDOW,
                    capture_output=True)
        print((child.stdout+child.stderr).decode('gbk'))
    except:
        print('timeout')

#第三种方式
def func():
    for code in code_blocks:
        try:
            print(check_output(['python', '-c', code],
                               timeout=120,
                               shell=False,         #防止命令注入
                               stderr=STDOUT,       #捕获异常信息
                               text=True,           #以字符串形式返回
                               creationflags=CREATE_NO_WINDOW))
        except CalledProcessError as e:
            #输出捕获到的异常信息
            print(e.output)
        except TimeoutExpired as e:
            print('timeout')
Thread(target=func).start()

#第四种方式
for code in code_blocks:
    p = Popen(['python', '-c', code], shell=False,
```

```
            stdout=open(file_path,'w'),
            stderr=open(file_path,'w'),
            creationflags=CREATE_NO_WINDOW)
    for _ in range(1200):
        if p.poll() is None:
            sleep(0.1)
        else:
            #子进程主动结束时提前结束循环
            break
    else:
        #子进程超时时强制结束
        p.kill()
    #读取并显示代码执行结果或出错信息
    with open(file_path) as fp:
        content = fp.read()
        if content:
            print(content)
        else:
            print('timeout')
remove(file_path)
```

本 章 小 结

（1）线程数量并不是越多越好。

（2）threading 模块是 Python 支持多线程编程的重要模块，提供了 Thread、Lock、RLock、Condition、Event、Timer、Semaphore 等大量类和对象。

（3）Thread 类支持两种方法来创建线程：一种方法是使用一个可调用对象创建线程；另一种方法是继承 Thread 类并在派生类中重写 __init__() 和 run() 方法。

（4）创建了线程对象之后，可以调用其 start() 方法来启动该线程，join() 方法用来等待线程结束或超时后返回。

（5）可以通过设置线程的 daemon 属性来决定主线程结束时是否需要等待子线程结束，daemon 属性的值默认为 False，即主线程结束时检查并等待子线程结束，如果需要修改 daemon 属性的值则必须在调用 start() 方法之前进行修改。

（6）除了 threading 模块中提供的线程同步对象之外，queue 模块也实现了多生产者-多消费者队列，尤其适合需要在多个线程之间进行信息交换的场合。

（7）派生自 Thread 类的自定义线程类首先也是一个普通类，同时还拥有线程类特有的 run()、start()、join() 等一系列方法。也就是说，可以在线程类中定义其他方法并且通过线程对象调用。

（8）Python 标准库 multiprocessing 用于创建和管理进程，不存在 GIL 的问题，可以更加有效地利用 CPU 资源。

（9）可以通过创建 Process 对象创建一个进程并通过调用进程对象的 start() 方法来启动该进程的运行。

（10）可以使用 Queue 对象、管道、共享内存、Manager 对象在进程之间交换数据。

（11）标准库 subprocess 提供了更多创建、管理进程，并与子进程通信的功能。

习 题

1. 叙述创建线程对象的几种方法。
2. 叙述 Thread 对象的常用方法语法和功能。
3. 叙述线程对象的 daemon 属性的作用和影响。
4. 解释至少 3 种线程同步方法。
5. 解释进程与线程的概念和区别。
6. 叙述进程间交换数据的常用方式。
7. 编写程序,禁用本机的记事本程序,发现启动就立即将其关闭。

第 14 章　数据库编程

毫无疑问，数据库技术的发展为各行各业都带来了很大方便，数据库不仅支持各类数据的长期保存，更重要的是支持各种跨平台、跨时空的数据查询、共享以及修改，极大方便了人类的生活和工作。金融行业、聊天系统、各类网站、办公自动化系统、各种管理信息系统等都少不了数据库技术的支持。本章主要介绍 SQLite、Access、MySQL、MS SQL Server 等几种数据库的 Python 接口，并通过几个示例来演示数据的增加、删除、修改、查询等操作。为了节省篇幅，本章并没有详细介绍数据库的原理、概念以及 SQL 语句的语法，而是假设读者已经了解或者可以通过查阅相关资料学习这部分内容。

14.1　SQLite 应用

SQLite 是内嵌在 Python 中的轻量级、基于磁盘文件的关系数据库管理系统，不需要服务器进程，也不使用任何端口号，支持使用 SQL 语句来访问数据库。该数据库使用 C 语言开发，支持大多数 SQL91 标准（部分语法略有不同），支持原子的、一致的、独立的和持久的事务，不支持外键限制；通过数据库级的独占性和共享锁定来实现独立事务，当多个线程同时访问同一个数据库并试图写入数据时，每一时刻只有一个线程可以写入数据。

SQLite 理论上支持 128TB 的单个数据库，每个数据库完全存储在单个磁盘文件中，以 B^+ 树数据结构的形式存储，一个数据库就是一个文件，通过复制即可实现数据库的备份和还原。如果需要可视化管理工具，可使用 SQLite Manager、SQLite DataBase Browser、SQLiteExpert、SQLiteStudio、Navicat for SQLite。

标准库 sqlite3 提供了访问和操作 SQLite 数据库的功能，使用该模块时，首先需要创建一个与数据库关联的 Connection 对象，例如：

```
import sqlite3
conn = sqlite3.connect('example.db')        #数据库文件存在则连接,否则创建空文件
```

成功创建 Connection 对象以后，再创建一个 Cursor 对象，然后调用 Cursor 对象的 **execute()** 方法（也可以直接使用 Connection 对象的同名方法）来执行 SQL 语句创建数据表以及查询、插入、修改或删除数据库中的数据，例如：

```
c = conn.cursor()
#创建表
c.execute('CREATE TABLE stocks(date text,trans text,symbol text,qty real, price real)')
#插入一条记录
c.execute("INSERT INTO stocks VALUES ('2024-01-05','BUY','RHAT',100,35.14)")
#提交当前事务,保存数据
conn.commit()
#关闭数据库连接
conn.close()
```

如果需要查询表中的内容，那么重新创建 Connection 对象和 Cursor 对象之后，可以使用下面的代码来查询：

```
>>> for row in c.execute('SELECT * FROM stocks ORDER BY price'):
        print(row)
```

14.1.1　Connection 对象

Connection 是 sqlite3 模块中最基本也是最重要的一个对象，其主要方法如表 14-1 所示。

表 14-1　Connection 对象的主要方法

方　　法	说　　明
execute(sql[, parameters])	执行一条 SQL 语句，可使用 parameters 指定参数
executemany(sql[, parameters])	使用不同参数执行多次 SQL 语句
cursor()	返回游标对象
commit()	提交当前事务，如果不提交，那么自上次调用 commit() 方法之后的所有修改都不会真正保存到数据库文件中
rollback()	撤销当前事务，将数据库恢复至上次调用 commit() 方法后的状态
close()	关闭数据库连接
create_function(name, num_params, func)	创建可在 SQL 语句中调用的函数，其中 name 为函数名，num_params 表示该函数可以接收的参数个数，func 表示 Python 可调用对象

Connection 对象的其他几个函数都比较容易理解，与 14.1.2 节介绍的 Cursor 对象同名方法用法类似。下面的代码演示了如何在 sqlite3 连接中创建并调用自定义函数：

```
import sqlite3
import hashlib

def md5sum(t):
    return hashlib.md5(t).hexdigest()          #计算字节串的 MD5 值(十六进制形式)

con = sqlite3.connect(':memory:')              #在内存中创建数据库
con.create_function('md5', 1, md5sum)          #第二个参数 1 表示 md5sum 函数的参数个数
cur = con.cursor()
cur.execute('SELECT md5(?)', (b'foo',))        #在 SQL 语句中调用自定义函数
print(cur.fetchone()[0])                       #fetchone()方法见 14.1.2 节
```

14.1.2　Cursor 对象

Cursor 也是 sqlite3 模块中比较重要的一个对象，该对象具有如下常用方法。

1．execute(sql[,parameters])

该方法用于执行一条 SQL 语句，下面的代码演示了该方法的用法，以及为 SQL 语句传递参数的两种形式，分别使用问号和命名变量作为占位符。

```python
import sqlite3

con = sqlite3.connect(':memory:')                    #Connection 对象支持 with 关键字,见例 14-1
cur = con.cursor()
cur.execute('CREATE TABLE people(name_last, age)')
who, age = 'Dong', 38
#使用问号作为占位符,插入数据,内存中的临时数据库不用提交事务
cur.execute('INSERT INTO people VALUES(?,?)', (who, age))
#使用命名变量作为占位符,使用字典提供数据,参数化查询可防范 SQL 注入攻击,推荐优先使用
cur.execute('SELECT * FROM people WHERE name_last=:who AND age=:age',
            {'who': who, 'age': age})
print(cur.fetchone())
```

运行结果如图 14-1 所示。

2. executemany(sql,seq_of_parameters)

该方法用来对所有给定参数执行同一条 SQL 语句,参数序列可以使用不同的方式产生,例如,下面的代码使用迭代器来产生参数序列:

```
>>>
======================= RESTART:
('Dong', 38)
>>>
```

图 14-1 运行结果

```python
import sqlite3

class IterChars:
    def __init__(self):
        self.count = ord('a')
    def __iter__(self):                              #特殊方法见 6.4 节
        return self
    def __next__(self):
        if self.count > ord('z'):
            raise StopIteration
        self.count += 1
        return(chr(self.count-1),)                   #生成下一个字符

con = sqlite3.connect(':memory:')                    #在内存中创建数据库
cur = con.cursor()
cur.execute('CREATE TABLE characters(c)')
cur.executemany('INSERT INTO characters(c) VALUES(?)', IterChars())
cur.execute('SELECT c FROM characters')              #内存数据库不用提交事务即可生效
print(cur.fetchall())                                #fetchall()方法返回所有数据行
```

下面的代码使用生成器来产生参数:

```python
import string
import sqlite3

def char_generator():                                #生成器函数,见 5.8 节
    for c in string.ascii_lowercase:
        yield (c,)
con = sqlite3.connect(':memory:')                    #在内存中创建数据库
cur = con.cursor()
cur.execute('CREATE TABLE characters(c)')            #不需要提交事务
cur.executemany('INSERT INTO characters(c) VALUES(?)', char_generator())
cur.execute('SELECT c FROM characters')
```

```python
print(cur.fetchall())
```

下面的代码则使用直接创建的序列作为 SQL 语句的参数：

```python
import sqlite3

persons = [('Hugo', 'Boss'), ('Calvin', 'Klein')]
con = sqlite3.connect(':memory:')
con.execute('CREATE TABLE person(firstname,lastname)')
#插入数据,推荐使用这种方式提交参数,避免使用字符串拼接 SQL 语句
con.executemany('INSERT INTO person(firstname,lastname) VALUES(?, ?)', persons)
for row in con.execute('SELECT firstname,lastname FROM person'):
    print(row)
print('I just deleted', con.execute('DELETE FROM person').rowcount, 'rows')
```

运行结果如图 14-2 所示。

3. fetchone()、fetchmany(size=cursor.arraysize)、fetchall()

```
('Hugo', 'Boss')
('Calvin', 'Klein')
I just deleted 2 rows
```

图 14-2　运行结果

这 3 个方法用于读取数据。假设数据库通过下面的代码创建并插入数据：

```python
import sqlite3
conn = sqlite3.connect("D:/addressBook.db")
conn.execute("INSERT INTO addressList(name,sex,phon,QQ,address) VALUES"
             "('王小丫','女','138XXXXXXXX','66735','北京市')")
conn.commit()                                    #操作磁盘上的数据库文件需要提交事务
conn.close()
```

下面的代码演示了使用 fetchall() 读取数据的方法，fetchone() 的用法见前一页代码。

```python
import sqlite3
conn = sqlite3.connect('D:/addressBook.db')
cur = conn.cursor()
cur.execute('SELECT * FROM addressList')
li = cur.fetchall()                              #读取所有查询结果,返回列表
for line in li:                                  #列表中每个元组对应一条记录
    for item in line:
        print(item, end='\t')
    print()
conn.close()
```

14.1.3　Row 对象

下面通过一个示例来演示 Row 对象的用法，假设数据库以下面的方式创建并插入数据：

```python
conn = sqlite3.connect('database.db')
c = conn.cursor()
c.execute('create table stocks(date text,trans text,symbol text,qty real,price real)')
c.execute("INSERT INTO stocks VALUES('2024-01-05','BUY','RHAT',100,35.14)")
conn.commit()
c.close()
```

那么，重新连接数据库后可以使用下面的方式来读取其中的数据：

```
>>> conn.row_factory = sqlite3.Row
>>> c = conn.cursor()
>>> c.execute('SELECT * FROM stocks')
<sqlite3.Cursor object at 0x7f4e7dd8fa80>
>>> r = c.fetchone()                           #读取一条查询结果
>>> type(r)
<class 'sqlite3.Row'>
>>> tuple(r)
('2024-01-05', 'BUY', 'RHAT', 100.0, 35.14)
>>> len(r)
5
>>> r[2]                                       #支持整数序号作下标
'RHAT'
>>> r.keys()
['date', 'trans', 'symbol', 'qty', 'price']
>>> r['qty']                                   #支持使用"键"(字段名)作下标
100.0
>>> for member in r:
        print(member)
2024-01-05
BUY
RHAT
100.0
35.14
```

例 14-1

例 14-1　生成 50 个 Excel 文件，写入随机数据，然后创建 SQLite 数据库和数据表，把 Excel 文件中的数据导入 SQLite 数据库。Excel 文件的读写操作见第 7 章，更新 SQLite 案例见作者微信公众号"Python 小屋"。

```
import sqlite3
from time import time
from os.path import isdir
from os import listdir, mkdir
from random import choice, randrange
from string import digits, ascii_letters
from openpyxl import Workbook, load_workbook

def generateRandomData():
    ''' 生成测试数据,共 50 个 Excel 文件,每个文件有 5 列随机字符串'''
    #如果不存在子文件夹 xlsxs 就创建
    if not isdir('xlsxs'):
        mkdir('xlsxs')
    global total
    characters = digits + ascii_letters
    #生成 50 个 Excel 文件
    for i in range(50):
        xlsName = 'xlsxs\\{i}.xlsx'
        #随机数,每个 XLSX 文件的行数不一样
        totalLines = randrange(10**4)
        #创建 Workbook,获取第 1 个 worksheet
        wb = Workbook()
```

```
        ws = wb.worksheets[0]
        #写入表头
        ws.append(['a', 'b', 'c', 'd', 'e'])
        #随机数据,每行5个字段,每个字段30个字符,生成器表达式可改用choices()函数
        for j in range(totalLines):
            line = [''.join((choice(characters) for ii in range(30))) for jj in range(5)]
            ws.append(line)
            total += 1
        wb.save(xlsName)

def eachXlsx(xlsxFn):
    '''针对每个XLSX文件的生成器,见5.8节'''
    #打开Excel文件,获取第1个worksheet
    wb = load_workbook(xlsxFn)
    ws = wb.worksheets[0]
    for index, row in enumerate(ws.rows):
        #忽略表头
        if index == 0:
            continue
        yield tuple(map(lambda x:x.value, row))

def xlsx2sqlite():
    '''从批量Excel文件中导入数据到SQLite数据库'''
    xlsxs = ('xlsxs\\'+fn for fn in listdir('xlsxs'))
    with sqlite3.connect('dataxlsx.db') as conn:
        cur = conn.cursor()
        for xlsx in xlsxs:
            #每次导入一个Excel文件,减少提交事务的次数,可以提高速度
            sql = 'INSERT INTO fromxlsx VALUES(?,?,?,?,?)'
            cur.executemany(sql, eachXlsx(xlsx))
            conn.commit()

total = 0                           #用于记录生成的随机数据的总行数
generateRandomData()                #生成随机数据
start = time()                      #导入数据,并测试速度
xlsx2sqlite()
delta = time() - start
print('导入用时: ', delta)
print('导入速度(条/秒): ', total/delta)
```

14.2 访问其他类型数据库

除了 SQLite 数据库以外,Python 还可以操作 Access、MS SQL Server 和 MySQL 等多种类型的数据库,本节中对几种常见的接口逐一进行简单介绍。

14.2.1 操作 Access 数据库

Python 支持使用 ADODB 和 ODBC 两种方式操作 mdb 和 accdb 格式的 Access 数据库,本节主要介绍 ADODB,关于 ODBC 的方式可以安装 Python 扩展库 pypyodbc 之后查阅相关文档。

使用 ADODB 方式操作 Access 数据库之前,需要安装 Python 扩展库 Pywin32,然后通

过下面代码中演示的方法读写数据库。

```python
from win32com import client

#建立数据库连接,分别适用于 mdb 和 accdb 文件
DSN1 = 'PROVIDER=Microsoft.Jet.OLEDB.4.0;DATA SOURCE=MyDB.mdb'
DSN2 = r'Provider=Microsoft.ACE.OLEDB.12.0;Data Source=MyDB.accdb'
conn = client.Dispatch('ADODB.Connection')
conn.Open(DSN1)

rs = client.Dispatch('ADODB.Recordset')            #打开数据表
table_name = '学生信息表'
#最后一个参数 3 表示可读、可写、可删除
rs.Open(f'[{table_name}]', conn, 1, 3)

#增加一条记录,设置各字段的值
rs.AddNew()
rs.Fields.Item(1).Value = '1001'
rs.Fields.Item(2).Value = '张三'
rs.Fields.Item(3).Value = 42
rs.Fields.Item(4).Value = '男'
rs.Update()
rs.Close()

#使用 SQL 语句增加多条记录
sql = 'INSERT INTO [学生信息表]([学号],[姓名],[年龄],[性别]) VALUES("{}","{}",{},"{}")'
information = [('1002', '李四', 43, '男'), ('1003', '王五', 44, '女')]
conn.BeginTrans()                                  #开始事务
for info in information:
    t = sql.format(*info)
    conn.Execute(t)
conn.CommitTrans()                                 #提交事务,写入数据库


#重新打开数据表
rs.Open(f'[{table_name}]', conn, 1, 3)

#查看所有记录
rs.MoveFirst()
while not rs.EOF:
    for i in range(1,rs.Fields.Count):
        print(rs.Fields.Item(i).Value, end=' ')
    print()
    rs.MoveNext()
rs.Close()

conn.Close()                                       #关闭数据库连接
```

14.2.2 操作 MS SQL Server 数据库

可以使用 Pywin32 和 pymssql 两种不同的方式来访问 MS SQL Server 数据库。先来了解一下 Pywin32 模块访问 MS SQL Server 数据库的步骤。

1. 添加引用

```
import adodbapi
adodbapi.adodbapi.verbose = False # adds details to the sample printout
import adodbapi.ado_consts as adc
```

2. 创建连接

```
Cfg = {'server':'192.168.29.86\\eclexpress','password':'xxxx','db':'pscitemp'}
constr = r"Provider=SQLOLEDB.1; Initial Catalog=%s; Data Source=%s; user ID=%s;
Password=%s; "%(Cfg['db'], Cfg['server'], 'sa', Cfg['password'])
conn = adodbapi.connect(constr)
```

3. 执行 SQL 语句

```
cur = conn.cursor()
sql = "SELECT * FROM softextBook WHERE title='{0}' AND remark3!='{1}'".format(bookName,flag)
cur.execute(sql)
data = cur.fetchall()
cur.close()
```

4. 执行存储过程

```
#假设 procName 有 3 个参数,最后一个参数传了 None
ret = cur.callproc('procName', (parm1, parm2, None))
conn.commit()
```

5. 关闭连接

```
conn.close()
```

下面的代码演示了 pymssql 模块访问 MS SQL Server 数据库的方法。

```
import pymssql
conn = pymssql.connect(host='SQL01', user='user', password='password',
                      database= 'mydatabase')
cur = conn.cursor()
cur.execute('CREATE TABLE persons(id INT, name VARCHAR(100))')
cur.executemany('INSERT INTO persons VALUES(%d, %s)', [(1, 'John Doe'), (2, 'Jane Doe')])
conn.commit()
cur.execute('SELECT *FROM persons WHERE name="John Doe"')
while (row:= cur.fetchone()):
    print("ID=%d, Name=%s" %(row[0], row[1]))
cur.execute("SELECT * FROM persons WHERE name LIKE 'J%'")
print(cur.fetchall())
conn.close()
```

14.2.3 操作 MySQL 数据库

Python 扩展库 pymysql 提供了操作 MySQL 数据库的接口,使用 pip install pymysql 安装并导入之后,可以使用内置函数 dir()和 help()了解该模块的用法,也可以查阅官方的帮助文档和扩展库源码。本节通过一段代码来了解 Python 使用 pymysql 扩展库操作 MySQL 数据库的方法。

```
import pymysql
```

```python
conn = pymysql.connect(host='127.0.0.1', user='root', password='123456',
                       database='mysql', charset='UTF8MB4')
cursor = conn.cursor()

def doSQL(sql):
    cursor.execute(sql)
    conn.commit()

doSQL('DROP DATABASE IF EXISTS onelinelearning;')               #删除数据库


doSQL('CREATE DATABASE IF NOT EXISTS onelinelearning;')         #创建数据库
doSQL('DROP TABLE IF EXISTS questions')                         #删除数据表
#创建数据表
sql = '''
CREATE TABLE IF NOT EXISTS questions(
id INT auto_increment PRIMARY KEY,
wenti CHAR(200) NOT NULL UNIQUE,
daan CHAR(50) NOT NULL
) ENGINE=innodb DEFAULT CHARSET=UTF8MB4;
'''
doSQL(sql)

doSQL('DELETE FROM questions;')                                 #删除所有数据
for i in range(10):                                             #插入数据
    sql = 'INSERT INTO questions(wenti,daan) VALUES("测试问题{0}","答案{0}");'.format(i)
    cursor.execute(sql)
conn.commit()
#修改数据
doSQL('UPDATE questions SET daan="被修改了" WHERE wenti="测试问题 6";')
doSQL('DELETE FROM questions WHERE daan="答案 8";')             #删除指定的数据
#查询并输出数据
sql = 'SELECT * FROM questions'
cursor.execute(sql)
for row in cursor.fetchall():
    print(row)
cursor.close()                                                  #关闭游标和连接
conn.close()
```

本 章 小 结

（1）SQLite 是内嵌在 Python 中的轻量级、基于磁盘文件的数据库管理系统，不需要服务器进程，支持使用 SQL 语句来访问数据库。

（2）访问和操作 SQLite 数据库时，需要首先导入 sqlite3 模块。

（3）可以使用 Pywin32 模块来操作 Access 数据库和 MS SQL Server 数据库，也可以使用 pymssql 模块来操作 MS SQL Server 数据库。

（4）可以使用 pymysql 模块操作 MySQL 数据库。

（5）对于 SQLite 数据库，删除数据时并不会释放空间，数据库文件仍然很大。如果需要释放空间压缩数据库文件大小，可以通过 Connection 对象或 Cursor 对象执行 SQL 语句

'vacuum',例如 conn.execute('vacuum')。

（6）Python 也支持操作 MongoDB 数据库，可以通过微信公众号"Python 小屋"或官方文档了解和学习。

习　题

1. 简单介绍 SQLite 数据库。
2. 使用 Python 内置函数 `dir()` 查看 Cursor 对象中的方法，并使用内置函数 `help()` 查看其用法。
3. 叙述使用 Python 操作 Access 数据库的步骤。
4. 叙述使用 Python 操作 MS SQL Server 数据库的步骤。
5. 叙述 pymysql 模块提供的数据库访问方法。
6. 结合第 9 章内容，编写 GUI 版通讯录管理系统，使用 tkinter 实现界面。
7. 结合第 9 章内容，编写 GUI 版个人密码管理系统，使用 tkinter 实现界面，并对密码进行加密存储。

第 15 章 多媒体编程

在本章中,主要介绍图形编程、图像编程、音乐编程、视频编程以及声音处理与语音识别等模块和技术,由于篇幅限制,更多案例可关注作者微信公众号"Python 小屋"学习。本章中用到的大部分模块不是默认安装的,可根据需要下载并安装相应的模块。

15.1 图形编程

计算机图形学主要研究如何使用计算机来生成具有真实感的图形,涉及的内容主要包括三维建模、图形变换、光照模型、纹理映射和阴影模型等内容,在机械制造、虚拟现实、游戏开发、漫游系统设计、产品展示、元宇宙等多个领域具有重要的应用。随着 3D 打印机的诞生,只要有模型就能够快速生成实物,这无疑会大大扩展计算机图形学的应用范围,例如,可以使用计算机图形学制作出各种可爱的模型,然后使用 3D 打印机批量生产各种食品、玩偶和饰品等。目前大部分关于计算机图形学的书籍都是基于 OpenGL 的,Python 也提供了扩展库 PyOpenGL,这极大地方便了编写图形学程序的 Python 程序员。

15.1.1 创建图形编程框架

Python 的跨平台扩展模块 PyOpenGL 封装了 OpenGL API,支持图形编程所需要的所有功能。安装扩展库 PyOpenGL 之后,使用该模块进行图形编程的步骤如下。

(1) 导入模块。

```
import sys
from OpenGL.GL import *
from OpenGL.GLU import *
from OpenGL.GLUT import *
```

(2) 使用 OpenGL 创建窗口类。

```
class MyPyOpenGLTest:
```

(3) 重写构造方法,初始化 OpenGL 环境,指定显示模式以及用于绘图的函数。

```
def __init__(self, width=640, height=480, title='MyPyOpenGLTest'.encode('gbk')):
    glutInit(sys.argv)
    glutInitDisplayMode(GLUT_RGBA | GLUT_DOUBLE | GLUT_DEPTH)
    glutInitWindowSize(width, height)
    self.window = glutCreateWindow(title)
    glutDisplayFunc(self.draw)
    glutIdleFunc(self.draw)
    self.InitGL(width, height)
```

(4) 根据特定的需要,进一步完成 OpenGL 的初始化。

```
def InitGL(self, width, height):
```

```
            glClearColor(1.0, 1.0, 1.0, 0.0)
            glClearDepth(1.0)
            glDepthFunc(GL_LESS)
            glShadeModel(GL_SMOOTH)
            glEnable(GL_POINT_SMOOTH)
            glEnable(GL_LINE_SMOOTH)
            glEnable(GL_POLYGON_SMOOTH)
            glMatrixMode(GL_PROJECTION)
            glHint(GL_POINT_SMOOTH_HINT,GL_NICEST)
            glHint(GL_LINE_SMOOTH_HINT,GL_NICEST)
            glHint(GL_POLYGON_SMOOTH_HINT,GL_FASTEST)
            glLoadIdentity()
            gluPerspective(45.0, float(width)/float(height), 0.1, 100.0)
            glMatrixMode(GL_MODELVIEW)
```

（5）定义自己的绘图方法。

```
        def draw(self):
            glClear(GL_COLOR_BUFFER_BIT | GL_DEPTH_BUFFER_BIT)
            #在这里编写绘制图形的代码
            glutSwapBuffers()
```

（6）消息主循环。

```
        def mainLoop(self):
            glutMainLoop()
```

（7）实例化窗口类，运行程序。

```
    if __name__ == '__main__':
        w = MyPyOpenGLTest()
        w.mainLoop()
```

15.1.2 绘制文字

可以使用 glutBitmapCharacter() 函数在窗口上每次绘制一个字符，可以使用循环结构来绘制多个字符。改写前面图形编程框架中的 draw() 方法，就可以实现该功能。

```
        def draw(self):
            glClear(GL_COLOR_BUFFER_BIT | GL_DEPTH_BUFFER_BIT)
            glLoadIdentity()
            glColor3f(1.0, 1.0, 1.0)
            glTranslatef(0.0, 0.0, -1.0)
            glRasterPos2f(0.0, 0.0)
            s = 'PyOpenGL is the binding layer between Python and OpenGL.'
            for ch in s:
                glutBitmapCharacter(GLUT_BITMAP_8_BY_13, ord(ch))
```

15.1.3 绘制图形

在 OpenGL 中绘制图形的代码需要放在 glBegin(mode) 和 glEnd() 这一对函数的调用之间，其中 mode 表示绘图类型，mode 的取值范围如表 15-1 所示。

表 15-1 mode 的取值范围

取值	说明	取值	说明
GL_POINTS	绘制点	GL_TRIANGLE_STRIP	绘制三角形串
GL_LINES	绘制直线	GL_TRIANGLE_FAN	绘制三角扇形
GL_LINE_STRIP	绘制连续直线,不封闭	GL_QUADS	绘制四边形
GL_LINE_LOOP	绘制封闭的连续直线	GL_QUAD_STRIP	绘制四边形串
GL_TRIANGLES	绘制三角形	GL_POLYGON	绘制多边形

例如,将前面给出的图形编程框架中的 `draw()` 方法改写成下面的代码,则可以绘制一个彩色三角形和一条彩色直线。在这段代码中,首先设置绘制模式为多边形,然后依次绘制该多边形的顶点,绘制每个顶点之前设置顶点颜色,最后修改绘制模式为直线并指定直线段的端点颜色和位置。需要注意的是,使用 `glColor3f()` 函数设置颜色之后,直到下一次使用该函数改变颜色之前,绘制的所有顶点都使用这个颜色。或者说,OpenGL 采用的是"状态机"工作方式,一旦设置了某种状态之后,除非显式修改该状态,否则该状态将一直保持。

```
def draw(self):
    glClear(GL_COLOR_BUFFER_BIT|GL_DEPTH_BUFFER_BIT)
    glLoadIdentity()
    glTranslatef(-2.0, 0.0, -8.0)
    #绘制二维图形,z坐标为0
    glBegin(GL_POLYGON)              #绘制多边形
    glColor3f(1.0, 0.0, 0.0)         #设置顶点颜色
    glVertex3f(0.0, 1.0, 0.0)        #绘制多边形顶点,内部自动插值
    glColor3f(0.0, 1.0, 0.0)
    glVertex3f(1.0, -1.0, 0.0)
    glColor3f(0.0, 0.0, 1.0)
    glVertex3f(-1.0, -1.0, 0.0)
    glEnd()
    glTranslatef(2.5, 0.0, 0.0)      #向右平移
    #绘制三维图形
    glBegin(GL_LINES)                #绘制直线,两个端点颜色不一样
    glColor3f(1.0, 0.0, 0.0)
    glVertex3f(1.0, 1.0, -1.0)
    glColor3f(0.0, 1.0, 0.0)
    glVertex3f(-1.0, -1.0, 3.0)
    glEnd()
    glutSwapBuffers()
```

上面的代码运行结果如图 15-1 所示。

15.1.4 纹理映射

在现实中,人们主要通过物体表面丰富的纹理细节来区分具有相同形状的不同物体。在三维建模时也往往通过纹理映射来简化建模的工作量,可以在保证图形具有较强真实感的前提下大幅提高渲染效率。

简单地说,纹理映射就是为物体表面进行贴图以使其呈现出特定的视觉效果。这需要首先准备好纹理,然后构建物体空间坐标和纹理坐标之间的对应关系来完成贴图。可以使

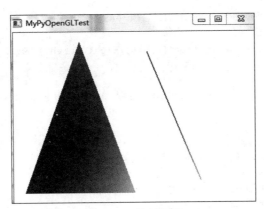

图 15-1 绘制图形

用函数来生成一些规则的纹理，如粗布纹理、棋盘纹理等；也可以将拍摄或通过网络搜索下载的图片作为纹理映射到物体表面上。进行纹理映射之前，首先要读取并设置纹理数据。在前面给出的图形编程框架中增加如下函数用来读取和设置纹理数据：

```python
    def LoadTexture(self):
        img = Image.open('sample.bmp')
        width, height = img.size
        img = img.tostring('raw', 'RGBX', 0, -1)
        glBindTexture(GL_TEXTURE_2D, glGenTextures(1))
        glPixelStorei(GL_UNPACK_ALIGNMENT, 1)
        glTexImage2D(GL_TEXTURE_2D, 0, 4, width, height, 0, GL_RGBA,
                    GL_UNSIGNED_BYTE, img)
        glTexParameterf(GL_TEXTURE_2D, GL_TEXTURE_WRAP_S, GL_CLAMP)
        glTexParameterf(GL_TEXTURE_2D, GL_TEXTURE_WRAP_T, GL_CLAMP)
        glTexParameterf(GL_TEXTURE_2D, GL_TEXTURE_WRAP_S, GL_REPEAT)
        glTexParameterf(GL_TEXTURE_2D, GL_TEXTURE_WRAP_T, GL_REPEAT)
        glTexParameterf(GL_TEXTURE_2D, GL_TEXTURE_MAG_FILTER, GL_NEAREST)
        glTexParameterf(GL_TEXTURE_2D, GL_TEXTURE_MIN_FILTER, GL_NEAREST)
        glTexEnvf(GL_TEXTURE_ENV, GL_TEXTURE_ENV_MODE, GL_DECAL)
```

然后修改图形编程框架中的初始化函数，设置纹理映射属性，并进行背面剔除，修改后的代码如下：

```python
    def InitGL(self, width, height):
        self.LoadTexture()
        glEnable(GL_TEXTURE_2D)
        glClearColor(0.0, 0.0, 0.0, 0.0)
        glClearDepth(1.0)
        glDepthFunc(GL_LESS)
        glShadeModel(GL_SMOOTH)
        glEnable(GL_CULL_FACE)
        glCullFace(GL_BACK)                            #背面剔除
        glEnable(GL_LINE_SMOOTH)
        glEnable(GL_POLYGON_SMOOTH)
        glMatrixMode(GL_PROJECTION)
        glHint(GL_LINE_SMOOTH_HINT, GL_NICEST)
```

```
glHint(GL_POLYGON_SMOOTH_HINT,GL_FASTEST)
glLoadIdentity()
gluPerspective(45.0, float(width)/float(height), 0.1, 100.0)
glMatrixMode(GL_MODELVIEW)
```

接下来,修改图形编程框架中的 draw()函数,绘制立方体盒子,并使用上面代码读取到的纹理数据进行表面映射:

```
def draw(self):
    glClear(GL_COLOR_BUFFER_BIT | GL_DEPTH_BUFFER_BIT)
    glLoadIdentity()
    glTranslate(0.0, 0.0, -9.0)
    glRotatef(self.x, 1.0, 0.0, 0.0)
    glRotatef(self.y, 0.0, 1.0, 0.0)
    glRotatef(self.z, 0.0, 0.0, 1.0)

    #依次绘制立方体的6个面并进行纹理映射
    glBegin(GL_QUADS)
    glTexCoord2f(0.0, 0.0)           #设置纹理坐标
    glVertex3f(-1.0, -1.0, 1.0)      #指定顶点位置
    glTexCoord2f(1.0, 0.0)
    glVertex3f(1.0, -1.0, 1.0)
    glTexCoord2f(1.0, 1.0)
    glVertex3f(1.0, 1.0, 1.0)
    glTexCoord2f(0.0, 1.0)
    glVertex3f(-1.0, 1.0, 1.0)
    #另外5个面的纹理映射代码见配套资源
    glEnd()
    glutSwapBuffers()
```

15.1.5 处理键盘/鼠标事件

如果需要使用键盘或鼠标来操作图形,如平移、旋转和缩放等,那么首先需要在初始化函数中指定接收键盘/鼠标事件的函数,即增加下面两行代码:

```
def __init__(self, width=640, height=480, title=b'MyPyOpenGLTest'):
    :
    glutKeyboardFunc(self.keypress)
    glutMouseFunc(self.mouse)
    :
```

然后在窗口类中增加下面的函数定义,用来接收并处理键盘/鼠标事件:

```
def mouse(self, button, mode, x, y):
    if button == GLUT_RIGHT_BUTTON and mode == GLUT_DOWN:
        print('yes')
def keypress(self, key, x, y):
    print(key)
```

15.2 图像编程

15.2.1 图像处理模块 Pillow 功能简介

PIL(Python Imaging Library)是 Python 的图像处理扩展模块,支持多种图像格式,提供了非常强大的图像处理功能。在 PIL 中主要提供 Image、ImageChops、ImageColor、ImageDraw、ImagePath、ImageFile、ImageEnhance 和 PSDraw,以及其他一些模块来支持图像的处理。在 Python 3.x 中需要安装 Pillow,然后通过 PIL 导入相应模块。

使用该扩展库时,首先需要导入它,例如:

```
>>> from PIL import Image
```

接下来,通过几个示例来简单演示一下该模块的用法。

(1) 打开图像文件。

```
>>> im = Image.open('sample.jpg')
```

(2) 显示图像。

```
>>> im.show()
```

(3) 查看图像信息。

```
>>> print(im.format)        #图像格式
>>> print(im.size)          #图像大小
```

(4) 查看图像直方图。

```
>>> im.histogram()
```

(5) 读取像素值。

```
>>> print(im.getpixel((100,50)))    #返回标量或格式为(r,g,b)的元组
```

(6) 设置像素值,通过读取和修改图像像素值可以实现图像点运算。

```
>>> im.putpixel((100,50), (128,30,120))    #第二个参数用于指定目标像素的颜色值
```

(7) 保存图像文件。

```
>>> im.save('sample1.jpg')     #可增加 quality 参数设置图像质量
```

(8) 转换图像格式。

```
>>> im.save('sample.bmp')      #通过该方法可以进行格式转换
```

(9) 图像缩放。

```
>>> im1 = im.resize((100,100))
```

(10) 旋转图像,rotate()方法支持任意角度的旋转,而 transpose()方法支持部分特殊角度的旋转,如 90°、180°、270°旋转,以及水平、垂直翻转等。

```
>>> im2 = im.rotate(90)
>>> im3 = im.transpose(Image.ROTATE_180)     #180°旋转
>>> im4 = im.transpose(Image.FLIP_LEFT_RIGHT)  #水平翻转
```

(11) 图像裁剪与粘贴。

```
>>> box = (120, 194, 220, 294)
>>> region = im.crop(box)                              #定义裁剪区域
>>> region = region.transpose(Image.ROTATE_180)
>>> im.paste(region, box)                              #粘贴
>>> im.show()
```

例如,图 15-2 是老虎的原始图像,图 15-3 是将其中一部分旋转 180°以后的结果,请注意左下角区域图像的变化。

图 15-2　原始图像

图 15-3　部分区域被旋转 180°以后的效果图

(12) 将彩色图像分离为红、绿、蓝三分量子图,分离后每个图像大小与原图像一样,但是只包含一个颜色分量。

```
>>> r, g, b = im.split()
```

(13) 图像增强。

```
>>> from PIL import ImageFilter
>>> im5 = im.filter(ImageFilter.DETAIL)
```

(14) 图像模糊。

```
>>> im6 = im.filter(ImageFilter.BLUR)
```

(15) 图像边缘提取。

```
>>> im7 = im.filter(ImageFilter.FIND_EDGES)
```

(16) 图像点运算,整体变暗或变亮。

```
>>> im8 = im.point(lambda i: i*1.3)
>>> im9 = im.point(lambda i: i*0.7)
```

也可使用图像增强模块来实现上面的功能,例如:

```
>>> from PIL import ImageEnhance
>>> enh = ImageEnhance.Brightness(im)
>>> enh.enhance(1.3).show()
```

(17) 图像冷暖色调调整。

```
>>> r, g, b = im.split()
>>> r = r.point(lambda i: i*1.3)
>>> g = g.point(lambda i: i*0.9)
>>> b = b.point(lambda i: 0)
>>> im10 = Image.merge(im.mode, (r,g,b))
>>> im10.show()
```

(18) 图像对比度增强。

```
>>> im = Image.open('sample.jpg')
>>> im.show()
>>> from PIL import ImageEnhance
>>> enh = ImageEnhance.Contrast(im)
>>> enh.enhance(1.3).show()
```

15.2.2 使用 Pillow 计算椭圆中心

本节案例用于计算和确定任意形状椭圆的中心，使用 Pillow 扩展库实现。

例 15-1 计算椭圆中心。

例 15-1

```
from PIL import Image
import os

def searchLeft(width, height, im):
    for w in range(width):                      #从左向右扫描
        for h in range(height):                 #从下向上扫描
            color = im.getpixel((w, h))         #获取图像指定位置的像素颜色
            if color != (255, 255, 255):
                return w                        #遇到并返回椭圆边界最左端的 x 坐标

def searchRight(width, height, im):
    for w in range(width-1, -1, -1):            #从右向左扫描
        for h in range(height):
            color = im.getpixel((w, h))
            if color != (255, 255, 255):
                return w                        #遇到并返回椭圆边界最右端的 x 坐标

def searchTop(width, height, im):
    for h in range(height-1, -1, -1):
        for w in range(width):
            color = im.getpixel((w,h))
            if color != (255, 255, 255):
                return h                        #遇到并返回椭圆边界最上端的 y 坐标

def searchBottom(width, height, im):
    for h in range(height):
        for w in range(width):
            color = im.getpixel((w, h))
            if color != (255, 255, 255):
                return h                        #遇到并返回椭圆边界最下端的 y 坐标
```

```python
#遍历指定文件夹中所有bmp图像文件,假设图像为白色背景,椭圆为其他任意颜色
images = [f for f in os.listdir('testimages') if f.endswith('.bmp')]
for f in images:
    f = 'testimages\\' + f
    im = Image.open(f)
    width, height = im.size                          #获取图像大小
    x0, x1 = searchLeft(width, height, im), searchRight(width, height, im)
    y0, y1 = searchBottom(width, height, im), searchTop(width, height, im)
    center = ((x0+x1)//2, (y0+y1)//2)
    im.putpixel(center, (255, 0, 0))                 #把椭圆中心像素画成红色
    im.save(f[0:-4]+'_center.bmp')                   #保存为新图像文件
    im.close()
```

15.2.3 使用 Pillow 动态生成比例分配图

本节使用 Pillow 实现另一个案例。功能:使用 3 种颜色填充横条矩形区域,并在每段中分别居中输出字母 A、B、C,要求 A、B、C 各自所占比例可动态调整。

例 15-2 动态生成比例分配图。

```python
from PIL import Image, ImageDraw, ImageFont

def redraw(f, v1, v2):
    start, end = int(600*v1), int(600*v2)            #600为图像宽度
    im = Image.open(f)
    for w in range(start):                            #绘制红色区域
        for h in range(36, 61):                       #36和60为下面一条的起止纵坐标
            im.putpixel((w,h), (255,0,0))
    for w in range(start, end):                       #绘制绿色区域
        for h in range(36, 61):
            im.putpixel((w,h), (0,255,0))
    for w in range(end, 600):                         #绘制品红色区域
        for h in range(36, 61):
            im.putpixel((w,h), (255,0,255))
    draw = ImageDraw.Draw(im)
    font = ImageFont.truetype('simsun.ttc', 18)
    draw.text((start//2,38), 'A', (0, 0, 0), font=font)       #在各自区域内居中显示字母
    draw.text(((end-start)//2+start,38), 'B', (0,0,0), font=font)
    draw.text(((600-end)//2+end, 38), 'C', (0,0,0), font=font)
    im.save(f)                                        #保存图片

redraw(r'd:\biaotou1.png', 0.1, 0.9)
```

程序运行结果如图 15-4 所示。图中上面浅蓝色部分的百分比是提前做好的,下面 3 种颜色的矩形区域和 A、B、C 是由程序动态生成,可以根据图片大小修改代码中的数值。

图 15-4 比例分配图

15.2.4 使用 Pillow 生成验证码图片

验证码在网络应用开发中占有重要地位,广泛应用于用户注册、登录、留言、购物、网络

支付等场合,可以有效阻止恶意用户频繁地提交非法数据。图片验证码是比较传统的验证码形式,图片中除了经过平移、旋转、错切、缩放等基本变换的字母和数字之外,还有一些线条或其他干扰因素。另外,还有问答型验证码,验证码是一个简单的问题,用户需要输入正确的答案才能进行后续的操作。某些系统的验证码系统更加复杂,实现了基于内容的图像识别功能或者拼图功能,题目难度较大,在一定程度上也阻碍了用户的正常使用。

例 15-3 生成验证码图片。

例 15-3

```
import random
import string
from PIL import Image, ImageDraw, ImageFont

#所有可能的字符,主要是英文字母和数字
characters = string.ascii_letters + string.digits
#获取指定长度的字符串,也可以使用字符串方法 join()和 random.choices()函数实现
def selectedCharacters(length):
    return random.choices(characters, k=length)

def getColor():
    '''get a random color'''
    r = random.randint(0, 255)
    g = random.randint(0, 255)
    b = random.randint(0, 255)
    return(r, g, b)

def main(size=(200,100), characterNumber=6, bgcolor=(255,255,255)):
    imageTemp = Image.new('RGB', size, bgcolor)
    #设置字体和字号
    font = ImageFont.truetype('c:\windows\fonts\TIMESBD.TTF', 48)
    draw = ImageDraw.Draw(imageTemp)
    text = selectedCharacters(characterNumber)
    width, height = draw.textsize(text, font)
    #绘制验证码字符串
    offset = 2
    for i in range(characterNumber):
        offset += width//characterNumber
        position = (offset, (size[1]-height)//2+random.randint(-10,10))
        draw.text(xy=position, text=text[i], font=font, fill=getColor())
    #对验证码图片进行简单变换,这里采用简单的点运算
    imageFinal = Image.new('RGB', size, bgcolor)
    pixelsFinal = imageFinal.load()
    pixelsTemp = imageTemp.load()
    for y in range(0, size[1]):
        offset = random.randint(-1, 1)
        for x in range(0, size[0]):
            newx = x +offset
            if newx >= size[0]:
                newx = size[0]-1
            elif newx < 0:
                newx = 0
            pixelsFinal[newx,y] = pixelsTemp[x,y]
```

```python
        draw = ImageDraw.Draw(imageFinal)
        for i in range(int(size[0]*size[1]*0.07)):        #绘制干扰噪点像素
            draw.point((random.randint(0,size[0]), random.randint(0, size[1])),
                       fill=getColor())
        for i in range(8):                                #绘制干扰线条
            start = (0, random.randint(0, size[1]-1))
            end = (size[0], random.randint(0, size[1]-1))
            draw.line([start, end], fill=getColor(), width=1)
        for i in range(8):                                #绘制干扰弧线
            start = (-50, -50)
            end = (size[0]+10, random.randint(0, size[1]+10))
            draw.arc(start+end, 0, 360, fill=getColor())
        imageFinal.save('result.jpg')                     #保存验证码图片
        imageFinal.show()

if __name__ == '__main__':
    main((200, 100), 8, (255, 255, 255))
```

将上面的程序保存并运行，即可生成验证码图片，如图 15-5 所示。

 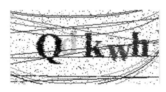 

(a) 验证码图片（一）　　　　(b) 验证码图片（二）　　　　(c) 验证码图片（三）

图 15-5　验证码图片

15.3　音乐编程

15.3.1　音乐播放

（1）扩展库 pygame 除了提供 mixer 模块支持音乐播放之外，还包含了大量其他支持游戏编程的模块，如表 15-2 所示。

表 15-2　pygame 主要模块

模　　块	说　　明	模　　块	说　　明
display	屏幕显示	time	时间控制
event	事件处理	cursors	控制鼠标指针
image	图像处理	transform	修改和移动图像
mouse	鼠标消息处理	key	读取键盘按键
movie	视频文件播放，需要安装 PyMedia	font	使用字体
surface	绘制屏幕		

扩展库 pygame 中用于音乐播放有关的函数主要在 mixer 模块中，如表 15-3 所示。

表 15-3 pygame.mixer 的主要函数

函数	说明
init()	初始化,必须最先调用
music.load(filename)	打开音乐文件
music.play(count,start)	播放音乐文件
music.stop()	停止播放
music.pause()	暂停播放
music.unpause()	继续播放
music.get_busy()	检测声卡是否正被占用

下面的代码使用 pygame.mixer 模块编写了一个简单的音乐播放器。程序运行后,将会自动随机播放指定文件夹中所有 MP3 音乐文件,并自动打印显示当前正在播放的音乐文件名。当然,可以修改这段代码以支持其他类型音乐文件的播放,或者还可以结合第 9 章 GUI 编程的知识编写一个漂亮美观的音乐播放器程序。

```
import os
import time
import random
import pygame

folder = r'h:\music'
musics = [folder+'\\'+music for music in os.listdir(folder) if music.endswith('.mp3')]
total = len(musics)
pygame.mixer.init()
while True:
    if not pygame.mixer.music.get_busy():
        nextMusic = random.choice(musics)              #随机选择一个音乐文件
        pygame.mixer.music.load(nextMusic.encode())    #打开音乐文件
        pygame.mixer.music.play(1)                     #播放音乐文件
        print('playing…', nextMusic)
    else:
        time.sleep(1)
```

(2)除了可以使用 pygame 播放 MP3 音乐文件之外,也可以使用标准库 wave 和扩展库 pyaudio 播放 WAV 波形音乐文件,下面的代码演示了这两个库的用法。

```
import wave
import pyaudio

buffer_size = 10240
wf = wave.open('北国之春.wav', 'rb')
audio = pyaudio.PyAudio()
stream = audio.open(format=audio.get_format_from_width(wf.getsampwidth()),
                    channels=wf.getnchannels(),
                    rate=wf.getframerate(),
                    output=True)
while True:
```

```
        data = wf.readframes(buffer_size)
        if not data:
            break
        stream.write(data)
stream.stop_stream()
stream.close()
audio.terminate()
```

15.3.2　WAV 波形音乐文件处理

在扩展库 SciPy 的 io 包中提供了 wavfile 模块支持对未压缩的 WAV 波形音乐文件进行操作，本节通过几个案例演示 wavfile 模块的用法。在运行这些代码之前，应首先安装扩展库 NumPy 和 SciPy，关于这两个扩展库的更多用法见第 17 章。

1. 让音乐内容重复两次

```
import numpy as np
from scipy.io import wavfile

def doubleMusic(srcMusicFile, dstMusicFile):
    #data[0]为采样频率,data[1]为声音数据
    data = wavfile.read(srcMusicFile)
    data12 = np.array(list(data[1])*2)      #可改为 data12 = np.append(data[1],data[1])
    wavfile.write(dstMusicFile, data[0], data12)

doubleMusic('北国之春.wav', 'result.wav')
```

2. 让音乐的音量减小一半

```
from scipy.io import wavfile

def halfMusic(srcMusicFile, dstMusicFile):
    #读取 WAV 声音文件
    #其中,data[0]是采样频率,data[1]是 numpy.array 格式的声音数据
    data = wavfile.read(srcMusicFile)
    #声音数据变为原来的 1/2,写入新文件
    wavfile.write(dstMusicFile, data[0], data[1]//2)
halfMusic('北国之春.wav', 'result.wav')
```

3. 修改音乐文件的音量实现淡入淡出效果

```
import numpy as np
from scipy.io import wavfile

def fadeInOutMusic(srcMusicFile, dstMusicFile):
    sampleRate, musicData = wavfile.read(srcMusicFile)
    #截取中间 1/3
    start = len(musicData) // 3
    musicData = musicData[start:-start]
    #前 1/10 淡入,后 1/10 淡出,中间不变
    length = len(musicData)
    n = 10
```

```
    start = length//n
    #通过调整 round()函数的第二个参数,可以控制淡入淡出的速度,也可使用 NumPy 生成系数
    factors = tuple(map(lambda num: round(num/start, 1), range(start)))
    factors = factors +(1,)*(length-start*2) +factors[::-1]
    musicData = np.array(tuple(map(lambda data, factor: [np.int16(data[0]*factor),
                                                         np.int16(data[1]*factor)],
                                    musicData, factors)))    #可使用 NumPy 简化,见 17.1 节
    #写入结果文件
    wavfile.write(dstMusicFile, sampleRate, musicData)
fadeInOutMusic('北国之春.wav', 'result.wav')
```

4. 分离音乐文件的左右声道

```
import numpy as np
from scipy.io import wavfile

def splitChannel(srcMusicFile):
    sampleRate, musicData = wavfile.read(srcMusicFile)
    #提取左右声道数据,写入结果文件
    wavfile.write('left.wav', sampleRate, musicData[:, 0])
    wavfile.write('right.wav', sampleRate, musicData[:, 1])
splitChannel('北国之春.wav')
```

15.4 语 音 识 别

（1）使用 Python 编写语音识别程序需要用到 speech 模块,并且需要安装扩展库 Pywin32 和 Microsoft Speech SDK。

speech 模块支持的主要功能：文本合成语音,将键盘输入的文本信息以语音信号方式输出；语音识别,将输入的语音信号识别为文本；特定词的识别,对输入的语音信号进行特定词的捕捉；特定用户、特定词的识别,能够对不同人、不同特定词进行识别。

speech 模块的主要成员如表 15-4 所示。

表 15-4 speech 模块的主要成员

成 员	说 明
say(phrase)	读出给定的文本
input(prompt=None, 　　 phraselist=None)	打印信息 prompt 提示用户使用语音录入在 phraselist 中列出的文本,并返回用户录入的内容。该函数会阻塞当前线程直至得到用户录入或者按 Ctrl+C 组合键结束
listenfor(phraselist, callback)	如果用户语音录入 phraselist 中的任何文本,则自动调用回调函数 callback,并返回 Listener 对象
listenforanything(callback)	得到用户语音录入的内容后自动执行回调函数 callback(spoken_text, listener),并返回 Listener 对象
Listener.islistening(self)	当 Listener 对象处于监听状态时返回 True
Listener.stoplistening(self)	停止监听,当 Listener 对象处于监听状态时返回 True

成　　员	说　　明
islistening()	只要有 Listener 对象正在监听就返回 True
stoplistening()	停止所有 Listener 对象的监听状态,如果有 Listener 对象处于监听状态则返回 True

例如,下面的代码让计算机读出用户输入的内容,当用户输入 stop 时结束。

```
>>> while True:
        words = input("Please input some words:")
        if words.lower() == 'stop':
            break
        speech.say(words)
```

下面的代码让计算机接收用户语音输入,并重复一遍用户语音输入的内容,以文字形式显示用户语音输入的内容。在微信公众号"Python 小屋"发送消息"语音提问"可学习更多。

```
>>> contents = speech.input()
>>> speech.say(contents)
>>> print(contents)
```

(2) 扩展库 pyaudio 提供了录音功能,支持从麦克风拾取声音并写入声音文件,下面的代码使用这个扩展库实现了一个简单的录音软件,界面使用 tkinter 开发,可以参考第 9 章的介绍。在微信公众号"Python 小屋"发送消息"翻录音乐"可学习更多案例。

```
import wave
import threading
import tkinter
import tkinter.filedialog
import tkinter.messagebox
import pyaudio

CHUNK_SIZE = 1024
CHANNELS = 2
FORMAT = pyaudio.paInt16
RATE = 44100
RECORD_SECONDS = 5
fileName = None
allowRecording = False

def record():
    global fileName
    p = pyaudio.PyAudio()
    stream = p.open(format=FORMAT, channels=CHANNELS, rate=RATE,
                    input=True, frames_per_buffer=CHUNK_SIZE)
    wf = wave.open(fileName, 'wb')
    wf.setnchannels(CHANNELS)
    wf.setsampwidth(p.get_sample_size(FORMAT))
    wf.setframerate(RATE)
```

```python
    while allowRecording:
        #从录音设备读取数据,直接写入 WAV 文件
        data = stream.read(CHUNK_SIZE)
        wf.writeframes(data)
    wf.close()
    stream.stop_stream()
    stream.close()
    p.terminate()
    fileName = None

#创建 tkinter 应用程序
root = tkinter.Tk()
root.title('录音机——董付国')
root.geometry('280x80+400+300')
root.resizable(False, False)

#开始按钮
def start():
    global allowRecording, fileName
    fileName = tkinter.filedialog.asksaveasfilename(filetypes=[('未压缩波形文件',
                                                                '*.wav')])
    if not fileName:
        return
    if not fileName.endswith('.wav'):
        fileName = fileName + '.wav'
    allowRecording = True
    lbStatus['text'] = '正在录音……'
    threading.Thread(target=record).start()
btnStart = tkinter.Button(root, text='开始录音', command=start)
btnStart.place(x=30, y=20, width=100, height=20)

#结束按钮
def stop():
    global allowRecording
    allowRecording = False
    lbStatus['text'] = '准备就绪'
btnStop = tkinter.Button(root, text='停止录音', command=stop)
btnStop.place(x=140, y=20, width=100, height=20)

lbStatus = tkinter.Label(root, text='准备就绪', anchor='w', fg='green')
lbStatus.place(x=30, y=50, width=200, height=20)

#关闭程序时检查是否正在录制
def closeWindow():
    if allowRecording:
        tkinter.messagebox.showerror('正在录制', '请先停止录制')
        return
    root.destroy()
root.protocol('WM_DELETE_WINDOW', closeWindow)

root.mainloop()
```

15.5 视频处理和摄像头接口调用

15.5.1 OpenCV 应用

扩展库 opencv-python 提供了大量用于视频处理和摄像头接口调用的功能,本节通过几个案例演示相关的用法。

1. 把 AVI 文件分离成静态图像

```python
import cv2
from PIL import Image

def splitFrames(videoFileName):
    cap = cv2.VideoCapture(videoFileName)
    num = 1
    while True:
        success, data = cap.read()
        if not success:
            break
        cv2.imwrite(str(num)+'.png', data)
        num = num + 1
    cap.release()
splitFrames('__name__变量的用法.avi')
```

2. 调用摄像头接口进行拍照并保存为图像文件

```python
from os import mkdir
from os.path import isdir
import datetime
from time import sleep
import cv2

while True:
    #参数 0 表示笔记本计算机自带摄像头
    cap = cv2.VideoCapture(0)
    #获取当前日期时间,例如 2023-11-24 23:11:00
    now = str(datetime.datetime.now())[:19].replace(':', '_')
    if not isdir(now[:10]):
        mkdir(now[:10])
    #捕捉当前图像,ret = True 表示成功,ret= False 表示失败
    ret, frame = cap.read()
    if ret:
        #保存图像,以当前日期时间为文件名
        fn = now[:10] + '\\' + now + '.jpg'
        cv2.imwrite(fn, frame)
    cap.release()
    #每 5 秒捕捉一次图像
    sleep(5)
```

3. 调用摄像头接口进行录像并保存为视频文件

```python
from os import mkdir
from os.path import isdir
```

```python
import datetime
from time import sleep
from threading import Thread
import cv2

cap = cv2.VideoCapture(0)                                  #参数 0 表示笔记本计算机自带摄像头
#获取当前日期时间,例如 2023-11-24 23:11:00
now = str(datetime.datetime.now())[:19].replace(':', '_')
dirName = now[:10]
tempAviFile = dirName + '\\' + now + '.avi'
if not isdir(dirName):
    mkdir(dirName)

#创建视频文件
aviFile = cv2.VideoWriter(tempAviFile, cv2.VideoWriter_fourcc(*'MJPG'),
                          25, (640,480))                   #帧速和视频宽度、高度

def write():
    while cap.isOpened():
        #捕捉当前图像,ret = True 表示成功,ret = False 表示失败
        ret, frame = cap.read()
        if ret:
            aviFile.write(frame)                           #写入视频文件
    aviFile.release()
Thread(target=write).start()

input('按任意键结束.')
cap.release()
```

15.5.2 moviepy 应用

扩展库 moviepy 提供了更加强大的视频处理功能,可以使用 pip 命令安装该扩展库及其依赖库。

1. 提取视频中的音频

```
from moviepy.editor import *

aviFileName = r'G:\录屏测试文件\LP_20190809110515.avi'
mp3FileName = r'G:\录屏测试文件\提取出的音频.mp3'
video = VideoFileClip(aviFileName)
video.audio.write_audiofile(mp3FileName)
```

2. 删除视频中的音频

```
from moviepy.editor import *

aviFileName = r'G:\录屏测试文件\LP_20190809110515.avi'
silenceFileName = r'G:\录屏测试文件\删除声音后的视频.mp4'
video = VideoFileClip(aviFileName)
#删除声音
video = video.without_audio()
video.write_videofile(silenceFileName)
```

3. 视频剪辑、合成、添加字幕

```python
from moviepy.editor import *

aviFileName1 = r'G:\录屏测试文件\LP_20190809105619.avi'
aviFileName2 = r'G:\录屏测试文件\LP_20190809110515.avi'
aviFileNameResult = r'G:\录屏测试文件\合成并添加字幕.mp4'
#从第 8 秒开始,剪到 6 分 51 秒
video1 = VideoFileClip(aviFileName1).subclip(t_start=8, t_end=(6,51))
video2 = VideoFileClip(aviFileName2).cutout(0, 5)         #剪掉 0~5s
video3 = concatenate_videoclips([video1, video2])         #拼接两段视频
#创建并添加、合成字幕
text_clip = TextClip('董付国老师系列课程', fontsize=50,
                    font=r'C:\Windows\fonts\STXINGKA.TTF',
                    color='black', bg_color='transparent', transparent=True
                    ).set_position(('right', 'top')).set_duration(1200)
                    .set_start(0)
video = CompositeVideoClip([video3, text_clip])
video.write_videofile(aviFileNameResult)
```

4. 旋转视频

```python
from moviepy.editor import *

aviFileName = r'G:\录屏测试文件\LP_20190809110515.avi'
resultFileName = r'G:\录屏测试文件\旋转 90 度.mp4'
video = VideoFileClip(aviFileName).rotate(90)
video.write_videofile(resultFileName)
```

5. 为视频设置异形遮罩窗口

```python
from copy import deepcopy
import numpy as np
from PIL import Image
from moviepy.editor import *

#图片尺寸应与视频尺寸相同
video = VideoFileClip('测试视频.mp4')
mask = np.array(Image.open('视频 mask.png'))
#防止图片有 alpha 通道,只保留 rgb 分量
mask = mask[:,:,:3]
invisible = mask!=[255,255,255]

def add_mask(frame):
    image_new = deepcopy(frame)
    #把不可见区域设置为该区域内遮罩图片的像素颜色
    image_new[invisible] = mask[invisible]
    return image_new
video.fl_image(add_mask).write_videofile('测试视频_图片 mask.mp4')
```

6. 批量视频添加滚动字幕

```python
from itertools import count
from os import listdir
from random import randint
```

```python
from os.path import splitext
from moviepy.editor import VideoFileClip, TextClip, CompositeVideoClip

#要添加的飘动文字,以及创建的视频剪辑大小
txt_clip = TextClip('董付国老师免费微课\n配套教材:《Python 程序设计》(第 4 版)',
                   fontsize=30, align='center',
                   font='SimHei', color='#aaaaaa')
t_width, t_height = txt_clip.w, txt_clip.h
#遍历当前目录下所有 MP4 文件,逐个添加滚动字幕
fns = [fn for fn in listdir() if fn.endswith('.mp4')]
for fn in fns:
    video = VideoFileClip(fn)
    #原始视频大小和时长
    v_width, v_height, v_duration = video.w, video.h, video.duration
    #把所有要合成的视频剪辑放在这个列表中,原始视频放在第一个
    #然后把滚动字幕叠加在原始视频之上
    clips = [video]
    #文字的位置,从右向左移动
    x, y, span = v_width, 0, 0.2
    #创建多个不同位置的文字剪辑,放到列表中原始视频文件的后面
    for start in count(0, span):
        if start >= v_duration:
            break
        clips.append(txt_clip.set_start(start).set_position((x,y))
                     .set_duration(span))
        #移出视频左边界之后,再重新从右边界进来,垂直位置随机生成
        if x <= 0- t_width:
            x = v_width
            y = randint(0, v_height- t_height)
        #向左移动的步幅,和变量 span 一起可以控制字幕滚动速度
        x = x - 10
    #合成所有视频剪辑,输出为最终文件
    CompositeVideoClip(clips).write_videofile('新'.join(splitext(fn)),
                                              codec='libx264')
```

本 章 小 结

(1) Python 的跨平台扩展模块 PyOpenGL 封装了 OpenGL API,支持图形编程所需要的所有功能。

(2) OpenGL 采用的是"状态机"工作方式,一旦设置了某种状态之后,除非显式修改,否则该状态将一直保持,例如图形顶点的颜色、法向量和纹理坐标等。

(3) Pillow 是支持 Python 的图像处理模块,提供了强大的图像处理功能。

(4) 可以使用 pygame.mixer、Phonon、DirectSound 或 WMPlayer.ocx 等多种方式进行音乐文件播放。

(5) 使用 Python 编写语音识别程序需要用到 speech 模块,并且需要安装扩展库 Pywin32 和 Microsoft Speech SDK。

(6) pyaudio 提供了录音和播放的功能。

(7) 扩展库 opencv-python 支持视频文件处理以及摄像头接口调用等功能。

（8）扩展库 moviepy 提供了强大的视频处理功能。

（9）在微信公众号"Python 小屋"发送消息"pyopengl 配置"了解计算机图形学开发环境的配置方法。

习　题

1. 编写程序，使用扩展库 PyOpenGL 在窗口上绘制一个三角形，设置 3 个顶点为不同的颜色，并对内部进行光滑着色。

2. 编写程序，使用扩展库 Pillow 读取两幅大小一样的图片，然后将两幅图像的内容叠加到一幅图像，结果图像中每个像素值为原来两幅图像对应位置像素值的平均值。

3. 编写程序，使用扩展库 Pillow 读取一幅图像的内容，将其按象限分为 4 等份，然后 1、3 象限内容交换，2、4 象限内容交换，生成一幅新图像。

4. 结合 GUI 编程知识，编写一个程序，创建一个窗口并在上面放置两个按钮，分别为"开始播放"和"暂停播放"，将 15.3 节中的音乐播放程序进行封装。

5. 运行 15.4 节和 15.5 节中的代码并查看运行结果。

6. 编程程序，实现录屏软件的基本功能，包括录制屏幕画面和麦克风声音。

第 16 章　逆向工程与软件分析

对于大多数程序员而言，或许并不关心关于硬件与操作系统底层或者软件运行机制的细节，只需要也只希望把更多的精力放在高层的业务逻辑实现上面。但是毫无疑问，如果对底层细节了解或熟悉，就能够对自己开发的软件进行更好的把握和控制。对硬件和系统底层的深刻理解有利于写出更好的应用程序，对于程序员的职业发展也是非常有帮助的。在某些领域，逆向工程是解决问题非常重要的方式，甚至可能是唯一的方式，例如软件安全测试、加密和解密、软件汉化、漏洞挖掘、计算机取证、恶意软件分析和版权保护等。在这些领域中，一般很难获得软件源代码，只能对二进制可执行文件进行分析，而可执行文件由于编译器的优化一般变得非常难以理解，甚至很多恶意软件根本没有可独立运行的文件，而是将代码注入其他正常进程中。另外，最近几年提出的 ROP、JOP 攻击甚至没有注入任何代码，仅仅通过精心选择和重新组合进程中已有的指令序列就可以实现恶意功能。所有这些都给安全分析人员造成很大困难和挑战，这要求分析人员对逆向工程有更全面而准确的理解和把握，并且能够熟练运用各种成熟的工具，必要的时候甚至需要自己编写程序来完成分析任务。从另一个角度来讲，从源代码级别对软件进行分析，无法获知编译器对最终可执行文件造成的影响，或者说，很难保证编译器能够忠实地、毫无错误地工作。不幸的是，编译器本身也是软件的一种，同样也有可能存在漏洞。从底层对最终可执行文件进行分析，可以综合考虑各方面的因素（包括加载过程、进程管理和内存管理等），虽然难度相对较大，但是可以得到更加全面和准确的信息，甚至可以控制和修改软件的运行过程。

在本章中，重点介绍 Windows 平台上 PE 文件的分析。PE 的全称是 Portable Executable，指可移植的可执行文件，包括 EXE 文件、COM 文件、DLL 文件、OCX 文件、SYS 文件、SCR 文件等 Windows 平台上所有可执行文件类型，可以说 PE 文件是 Windows 操作系统和 Windows 平台上所有软件和程序能够正常运行的重要基础。

需要说明的是，应尽量避免直接在本地物理主机上分析恶意软件，以免被恶意软件感染而造成不必要的损失。为了保证物理主机安全，同时也为了能够在分析环境被恶意软件感染之后快速恢复系统，建议使用 VirtualBox、VMware、QEMU 等虚拟机系统或沙箱系统进行保护。如果没有条件使用虚拟机或沙箱系统，也可使用 Deep Freeze、Truman、FPG 或其他类似软件来保护物理主机以防止系统被感染。

16.1　主流项目与插件简介

在软件安全和逆向工程领域，有大量的成熟工具以及针对不同工具和目的开发的各种插件，例如 IDA Pro、OllyDbg、WinDbg、W32DASM、PEid、ssdeep、DiStorm、DisView、LordPE、PIN、Universal PE Unpacker 和 Sample Chart Builder 等，可以说是数不胜数。本书中主要介绍使用 Python 开发或可以使用 Python 进行二次开发的工具和插件，以及如何使用 Python 开发 PE 文件逆向分析工具。

16.1.1 主流项目

目前已有大量使用 Python 作为主要语言开发的软件逆向分析工具，下面列出了知名度较高的几款。

（1）PyEmu：可编写脚本的模拟器，对恶意软件分析非常有用。

（2）Immunity Debugger：著名的调试器，是在 OllyDbg 的源代码基础上建立起来的，外观非常相似，并且两者共享很多底层功能和控制。Immunity Debugger 带有内置的 Python 接口和专门用于研究漏洞和执行恶意软件分析的强大 API，是可编写脚本的 GUI 和命令行软件调试器，支持 exploit 编写、二进制可执行文件逆向工程等各种应用。

（3）Paimei：完全使用 Python 编写，是非常成熟的逆向工程框架，包括 PyDBG、PIDA 和 pGRAPH 等多个可扩展模块，可以执行大量静态分析和动态分析，如模糊测试、代码覆盖率跟踪和数据流跟踪等。

（4）ropper：比较成熟的 ROP Gadgets 查找与可执行文件分析工具，其反汇编部分使用了成熟的 Capstone 框架。

（5）WinAppDbg：纯 Python 调试器，没有本机代码，使用 ctypes 封装了许多与调试器有关的 Win32 API 调用，并且为操作线程、库和进程提供了强有力的抽象。利用该工具可以将自己编写的脚本附加为调试器、跟踪执行、拦截 API 调用，以及在待调试进程中处理事件，并且可以设置各种断点。

（6）YARA：恶意软件识别和分类引擎，也可以利用 YARA 创建规则以检测字符串、入侵序列、正则表达式和字节模式等。既可以使用命令行模式下的 YARA 工具扫描文件，也可以利用 YARA 提供的 API 函数将 YARA 扫描引擎集成到 C 或 Python 语言程序中。

16.1.2 常用插件

经过多年努力，不同的研究人员和公司分别推出了用于不同软件分析需要的 IDA、OllyDbg 以及 Immunity Debugger 插件，大大简化了分析人员的工作。除了以下几种常用插件，还可以通过 SDK 编写自己的插件，或者通过一定技术将 OllyDbg 插件转换为 Immunity Debugger 插件，大大提高了插件的应用范围和生命力。

（1）IDAPython：IDAPython 是运行于交互式反汇编器 IDA 的插件，用于实现 IDA 的 Python 编程接口。IDA 在逆向工程领域具有广泛的应用，尤其是二进制文件静态分析，其强大的反汇编功能一直在业内处于领先水平。IDAPython 插件使得 Python 脚本程序能够在 IDA 中运行并实现自定义的软件分析功能，通过该插件运行的 Python 脚本程序可以访问整个 IDA 数据库，并且可以方便地调用所有 IDC 函数和使用所有已安装的 Python 模块中的功能。较高版本的 IDA 中集成了 IDAPython 插件，如果需要安装或升级，需要登录其官方网站下载安装适合当前已安装 Python 和 IDA 版本的 IDAPython 插件。

（2）Hex-Rays Decompiler：IDA 插件，非常成熟的反编译插件。

（3）PatchDiff2：IDA 插件，主要用于补丁对比。

（4）BinDiff：IDA 插件，主要用于二进制文件差异比较。

（5）hidedebug：Immunity Debugger 插件，可以隐藏调试器的存在，用来对抗某些通用的反调试技术。

（6）IDAStealth：IDA 插件，可隐藏 IDA Debugger 的存在，用来对抗某些通用的反调试技术。

16.2　IDAPython 与 Immunity Debugger 编程

16.2.1　IDAPython 编程

安装 IDAPython 插件时，一定要正确选择适合已安装 Python 和 IDA 的版本，否则可能无法在 IDA 中加载 IDAPython 和运行 Python 程序。安装成功以后，启动 IDA 会看到软件界面最下端有个 Python 标志，在后面的文本框中可以直接输入并运行 Python 代码，如图 16-1 所示。另外，编写 Python 程序后，在 IDA 主界面中单击菜单 File→Script file 命令，在弹出的"脚本文件选择"对话框中，可以看到文件类型为.idc 和.py 两种，如图 16-2 所示。也就是说，现在可以在 IDA 中运行 Python 程序了，最后选择并运行自己编写的 Python 程序来实现自定义的二进制文件分析任务。

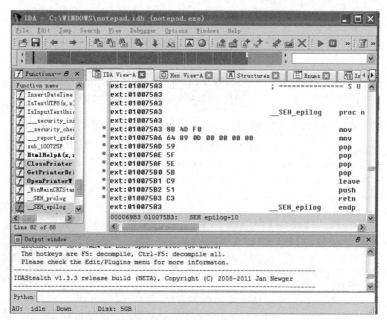

图 16-1　IDAPython 插件安装成功的 IDA 软件界面

接下来，主要通过几个示例来演示如何使用 Python 编程，并在 IDA 中运行来实现 PE 文件分析。详细的 IDC 库函数可以查阅 IDA 官方网址：

（1）查看 PE 文件中所有段的名字、起始地址以及结束地址。

```
for seg in Segments():
    print(SegName(seg), '(', hex(SegStart(seg)), ',', hex(SegEnd(seg)),')')
```

（2）查看 PE 文件中所有段的名字与长度。

```
segments = dict()
for seg_ea in Segments():
    segments[SegName(seg_ea)] = SegEnd(seg_ea)-seg_ea
```

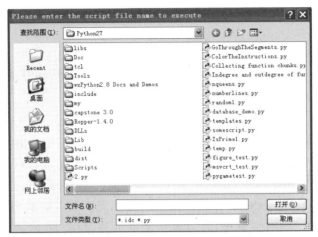

图 16-2 "脚本文件选择"对话框

```
for seg_name, seg_data in segments.items():
    print(seg_name, seg_data)
```

(3) 查看 PE 文件中所有函数信息。

```
for segment in Segments():
    for function_ea in Functions(SegStart(segment), SegEnd(segment)):
        print(hex(function_ea), GetFunctionName(function_ea))
```

(4) 查找 PE 文件中指定函数调用,并将该行设置为红色进行高亮显示。

```
from idaapi import *
danger_functions = ['strcpy','sprintf','strncpy','memcpy']
for func in danger_functions:
    addr = LocByName(func)
    if addr != BADADDR:
        cross_refs = CodeRefsTo(addr, 0)
        print('Cross References to %s'%func)
        print('----------------------------')
        for ref in cross_refs:
            print('%08x'%ref)
            SetColor(ref, CIC_ITEM, 0x0000ff)
        print('----------------------------')
```

(5) 遍历函数 chunk。

```
function_chunks = []
for ea in Functions():
    func_iter = idaapi.func_tail_iterator_t(idaapi.get_func(ea))
    status = func_iter.main()
    while status:
        chunk = func_iter.chunk()
        function_chunks.append((chunk.startEA, chunk.endEA))
        status = func_iter.next()
for chunk in function_chunks:
    print(hex(chunk[0]), hex(chunk[1])), 'belongs to function:',
        GetFunctionName(chunk[0])
```

（6）统计函数入度与出度。

```
from sets import Set
ea = ScreenEA()
callers = dict()
callees = dict()
for function_ea in Functions(SegStart(ea), SegEnd(ea)):  #遍历当前段中的函数
    f_name = GetFunctionName(function_ea)               #获取函数名字
    callers[f_name] = Set(map(GetFunctionName, CodeRefsTo(function_ea, 0)))
                                                        #调用该函数的所有函数
    for ref_ea in CodeRefsTo(function_ea, 0):           #遍历调用该函数的所有函数
        caller_name = GetFunctionName(ref_ea);
        callees[caller_name] = callees.get(caller_name, Set())
        callees[caller_name].add(f_name)
functions = Set(callees.keys() + callers.keys())
for f in functions:
    print('%-4d::%s::%4d'%(len(callers.get(f, [])), f,
        len(callees.get(f, []))))
```

（7）统计 PE 文件中的指令频度。

```
mnemonics = dict()
ea = ScreenEA()
for head in Heads(SegStart(ea), SegEnd(ea)):
    if isCode(GetFlags(head)):
        mnem = GetMnem(head)
        mnemonics[mnem] = mnemonics.get(mnem, 0) +1
mnem_list = map(lambda x:(x[1], x[0]),
                mnemonics.items())
mnem_list.sort()
for cnt, mnem in mnem_list:
    print(mnem, cnt)
```

针对某动态链接库文件，上面的代码运行结果如图 16-3 所示。

（8）查找潜在的 ROP Gadgets。

ROP（Return-Oriented Programming）是近几年来流行的一种攻击方式。在早些年，黑客通过各种溢出漏洞和保护机制的缺陷来实现任意代码注入和执行，后来由于数据执行保护（Data Execution Prevention，DEP）和地址空间布局随机化（Address Space Layout Randomization，ASLR）等保护技术的部署，实现代码注入攻击的难度越来越大，于是聪明的黑客又发明了 ROP 攻击及其各种变种，其主要思想是通过精确控制进程的执行流程，重新组合和复用可执行文件中已经存在的代码，实现恶意目的并绕过特定的防护技术和系统。目前针对 ROP 攻击较为有效的防护技术有不定期 ASLR 与控制流完整性（Control Flow Integrity，CFI）约束技术。粗粒度 ASLR 已经被确认不安全，而细粒度 ASLR 会导致代码膨胀从而增加了潜在的 Gadgets，并且很可能会使得库函数无法共享，严重影响了细粒度 ASLR 的实用性。

图 16-3　指令频度统计结果

ROP 攻击首先要利用特定漏洞来实现栈上内容的覆盖,通过覆盖栈上的函数返回地址来实现控制流的修改,重新组合已有的代码并构造 Gadgets 链来实现恶意目的。ROP Gadgets 是指较短的汇编指令序列,一般以 ret 指令或其他间接跳转指令(如 jmp eax 或 call eax 等)结束,每个 Gadget 仅仅实现功能非常有限的运算,但大量的 Gadgets 链接起来却可以实现任意功能。ROP 已被证明是图灵完备的。

下面的代码演示了在 PE 可执行文件中搜索 ROP Gadgets 的基本原理。这只是个基本的演示,并没有考虑更加复杂的情况。例如,如果控制流能够跳转到指令中间开始执行,就会打乱原有的指令序列而产生新的指令,从而产生原本不存在的 Gadget,这种情况这里没有考虑。可以根据需要对下面的代码进行修改,或者阅读 ropper 工具的源代码以了解更多知识。

```python
import time
import re
instructions = []
controlInstructions = ('call','ret','retn','jmp','jz','je','jnz','jne',
                       'js','jns','jo','jno','jp','jpe','jnp','jpo','jc',
                       'jb','jnae','jbe','jna','jnc','jnb','jae','jnbe',
                       'ja','jl','jnge','jnl','jge','jle','jng','jnle',
                       'jg','jcxz','jecxz')
def ReadInstructions():
    for seg_ea in Segments():
        for head in Heads(seg_ea,SegEnd(seg_ea)):
            if isCode(GetFlags(head)):
                #here using GetMnem(head) can get only mnemonic
                instruction = GetDisasm(head)
                instructions.append((hex(head), instruction))

    #print all the direct or indirect control instructions
    print('The number of all instructions found is:', len(instructions))
    print('And the direct or indirect control instructions are:')
    allControlInstructionsCount = 0 #the number of control instructions

    #get all the mnemonics from instructions
    mnemonics = [t[1].split()[0] for t in instructions]

    for ins in controlInstructions:
        if ins in mnemonics:
            print(ins, mnemonics.count(ins))
            allControlInstructionsCount =+ mnemonics.count(ins)
    print('The number of all control instructions is:', allControlInstructionsCount)

#check if given instruction is a indirect control diversion instruction
def Check(instruction):
    if instruction.startswith('ret') or instruction.startswith('retn'):
        return True
    else:
        for instr in controlInstructions:
            if instr in ('ret', 'retn'):
                continue
            if instruction.startswith(instr+'e'):#like call edi
```

```python
            return True
    return False

#output the potential gadgets
def Output(start, end):
    print('='*30)
    for i in range(start, end+1):
        print(instructions[i])

#find potential gadgets
def FindGadgets():
    total = len(instructions)
    gadgetNumber = 0
    index = total - 1
    while index >= 0:
        instruction = instructions[index]
        if Check(instruction[1]):
            gadgetNumber += 1
            for i in range(1, 20):
                if Check(instructions[index-i][1]):
                    Output(index-i+1, index)
                    index = index - i
                    break
            else:
                Output(index-19, index)
                index -= 19
        else:
            index -= 1
    print('='*30)
    print('Total number of gadgets:', gadgetNumber)

start = time.time()
ReadInstructions()
FindGadgets()
print(time.time() - start)
```

16.2.2 Immunity Debugger 编程

Immunity Debugger 是一款使用 Python 开发的非常成熟的调试器软件，支持软件调试几乎所有的功能，可以用于实现漏洞利用编写、模糊测试、恶意软件分析以及可执行文件的逆向工程分析，并且支持 PyCommand 接口以支持 Python 编程进行二次开发，其启动界面如图 16-4 所示。

在主界面中的工具栏上单击左边第二个带有符号>>>的工具按钮，打开 Immunity Debugger Python Shell 窗口，在该窗口中可以直接执行 Python 语句，并通过对象 imm 来访问 Immlib 库的所有成员，如图 16-5 所示。也可以通过在 Immunity Debugger 主界面下面的命令框中输入"!"符号来执行编写好的 Python 程序完成分析任务，如图 16-6 所示。

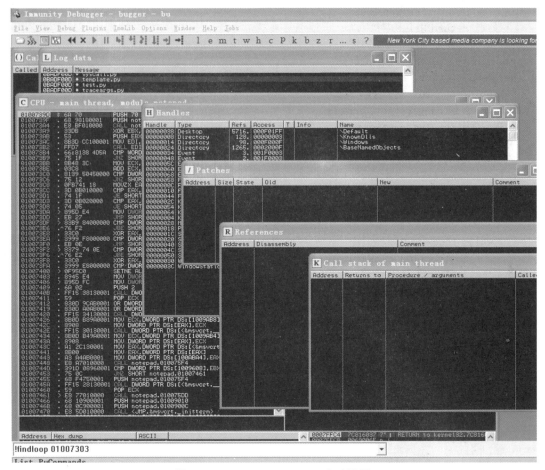

图 16-4 Immunity Debugger 启动界面

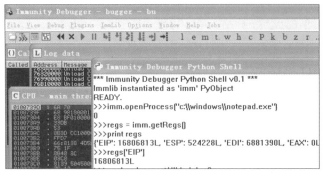

图 16-5 在 Immunity Debugger 中执行 Python 语句

编写自己的插件时，可以使用 IDLE 或记事本等任意文本编辑器编写 Python 源程序，然后将程序文件存放至 Immunity Debugger 安装目录下的 PyCommands 目录中，最后通过在 Immunity Debugger 主界面下方命令框中输入"!"后加上程序文件名称即可执行，如图 16-6 所示。下面再通过几个示例来演示如何利用 Python 编程实现 PE 文件分析，也可以参考 Immunity Debugger 安装目录下的 PyCommands 目录中的文件来了解更多分析技

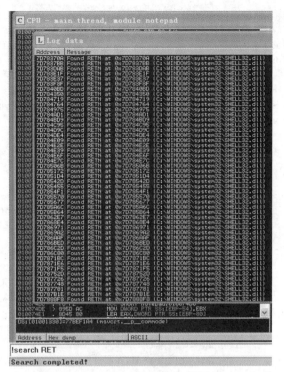

图 16-6　在 Immunity Debugger 中执行 Python 程序

巧，或者根据自己的软件分析需求来编写相应的插件。

1. 寻找可执行文件中的循环

```
import getopt
from immlib import *
from immutils import *

DESC=""" Find natural loops given a function start address """

def usage(imm):
    imm.log("!findloop -a <address>")
    imm.log("-a (function start address)")
    imm.log("-h This help")

def main(args):
    imm = Debugger()
    try:
        opts, argo = getopt.getopt(args, "a:")
    except:
        return usage(imm)
    for o,a in opts:
        if o == "-a":
            loops = imm.findLoops(int(a, 16))
            for loop in loops:
                imm.log("LOOP! from:0x%08x, to:0x%08x"%(loop[0], loop[1]), loop[0])
                func = imm.getFunction(int(a, 16))
                bbs = func.getBasicBlocks()
```

```python
                #寻找第一个和最后一个节点
                first = 0xffffffff
                last = 0
                for node in loop[2]:
                    if node < first: first = node
                    if node > last: last = node
                #标记循环节点,但如果存在任何形式的注释就不做任何改变
                for node in loop[2]:
                    imm.log("Loop node:0x%08x"%node,node)
                    for bb in bbs:
                        if bb.getStart() == node:
                            instrs = bb.getInstructions(imm)
                            for op in instrs:
                                if not imm.getComment(op.getAddress()) and\
                                    op.getAddress() != node:
                                    if node == last and\
                                        op.getAddress() == instrs[-1].getAddress():
                                        #最后一个节点的最后一个指令
                                        imm.setComment(op.getAddress(), "/")
                                    else:
                                        imm.setComment(op.getAddress(), "|")
                        if not imm.getComment(node):
                            if node == first:
                                imm.setComment(node, "\ Loop 0x%08X Node"%(loop[0]))
                            else:
                                imm.setComment(node, "| Loop 0x%08X Node"%(loop[0]))
        return "Done!"
    if o == "-h":
        return usage(imm)
```

2. 寻找可执行文件中的打包器

```python
import immlib
import getopt
import struct

DESC = """Find a Packer/Cryptor on a Module (Note: It might take some times due to the
amount of signature on our db)"""

def usage(imm):
    imm.log("!findpacker [-f] -m filename/module Get the RPC information of a loaded
    dll or for all loaded DLL's",focus=1)
    imm.log("-m filename/module File or Module to search for")
    imm.log("-f When set, it look in the file instead of the loaded module")
    imm.log("ex: !findpacker -m notepad")
    imm.log("NOTE: It might take some times due to the amount of signature on our db")

def main(args):
    imm = immlib.Debugger()
    if not args:
        usage(imm)
```

```
        return "No args"
try:
    opts, argo = getopt.getopt(args, "m:f")
except getopt.GetoptError:
    usage(imm)
    return "Bad heap argument %s" %args[0]
module = None
OnMemory = 1
for o,a in opts:
    if o == "-m":
        module = a
    elif o == '-f':
        OnMemory = 0
if not module:
    usage(imm)
    return "No module provided, see the Log Window for details of usage"
try:
    ret = imm.findPacker(module, OnMemory=OnMemory)
except Exception, msg:
    return "Error: %s" %msg
if not ret:
    return "No Packer found"
for (addr, name) in ret:
    imm.log("Packer found!: %s at 0x%08x" %(name, addr), address=addr)
return "Packers found on %s: %d" %(module, len(ret))
```

3. 寻找可执行文件中的指令

```
import immlib

DESC = "Search code in memory"

def usage(imm):
    imm.log("!searchcode Search code in memory")
    imm.log("!searchcode <asm code>")

def main(args):
    imm = immlib.Debugger()
    look = " ".join(args)
    ret = imm.search(imm.assemble(look))
    for a in ret:
        module = imm.findModule(a)
        if not module:
            module = "none"
        else:
            module = module[0]
        #Grab the memory access type for this address
        page = imm.getMemoryPageByAddress(a)
        access = page.getAccess(human=True)
        imm.log("Found %s at 0x%08x [%s] Access: (%s)" %(look, a, module, access),
                address=a)
```

```
    if ret:
        return "Found %d address (Check the Log Window for details)" %len(ret)
    else:
        return "Sorry, no code found"
```

16.3 Windows 平台软件调试原理

不论使用什么语言开发软件调试器,其基本原理都是一致的,最终都是调用操作系统自身提供的调试接口,设置断点,并对被调试软件的执行过程以及有关事件进行跟踪和处理。在本节中,主要介绍 Windows 平台上的调试接口、调试事件和断点等基本概念与调试原理。

16.3.1 Windows 调试接口

在 Win32 子系统中自带了大量支持不同类型应用开发的 API 函数,其中一部分被称为 Win32 调试 API(Win32 Debug API),提供了编写软件调试器所需要的大部分功能。利用这些 API 可以加载一个程序或将调试器捆绑到一个正在运行的进程上以供调试;可以获得被调试进程的深层信息,如进程 ID、入口地址和映像基址等;甚至可以对被调试的程序进行任意修改,包括进程的内存和线程的运行环境等。表 16-1 列出了常用的 Windows 调试 API。

表 16-1 常用的 Windows 调试 API

API 函数	功 能 说 明
ContinueDebugEvent()	恢复先前由于调试事件而挂起的线程
DebugActiveProcess()	将调试器捆绑到一个正在运行的进程上
DebugActiveProcessStop()	将调试器从一个正在运行的进程上卸载
DebugBreak()	在当前进程中产生一个断点异常,如果当前进程不处在被调试状态,那么这个异常将被系统例程接管,多数情况下会导致当前进程被终止。与在程序中直接插入 INT 3 的效果一样
DebugBreakProcess()	在指定进程中产生一个断点异常
FatalExit()	使调用进程强制退出,将控制权转移至调试器,在退出前会先调用一个 INT 3 断点
FlushInstructionCache()	刷新指令高速缓存
GetThreadContext()	获取指定线程的执行环境
GetThreadSelectorEntry()	返回指定选择器和线程的描述符表的入口地址
IsDebuggerPresent()	判断调用进程是否处于被调试环境中
OutputDebugString()	将一个字符串传递给调试器显示
ReadProcessMemory()	读取指定进程的某区域内的数据
SetThreadContext()	设置指定线程的执行环境
WaitForDebugEvent()	等待被调试进程发生调试事件
WriteProcessMemory()	在指定进程的某区域内写入数据

16.3.2 调试事件

调试器的主要工作是监视目标进程的执行并对目标进程执行过程中发生的每一个调试事件进行相应的响应和处理。当目标进程发生一个调试事件后，系统将会通知调试器来处理这个事件，调试器利用 WaitForDebugEvent() 函数来获取目标进程中发生的调试事件信息。常用的调试事件如表 16-2 所示。

表 16-2 调试事件

调 试 事 件	含　　义
CREATE_PROCESS_DEBUG_EVENT	进程被创建。当调试的进程刚被创建（还未运行）或调试器开始调试已经激活的进程时，就会生成这个事件
CREATE_THEAD_DEBUG_EVENT	在调试进程中创建一个新的进程或调试器开始调试已经激活的进程时，就会生成这个调试事件。要注意的是，当调试的主线程被创建时不会收到该通知
EXCEPTION_DEBUG_EVENT	在调试的进程中出现了异常，就会生成该调试事件
EXIT_PROCESS_DEBUG_EVENT	每当退出调试进程中的最后一个线程时，产生这个事件
EXIT_THREAD_DEBUG_EVENT	调试中的线程退出时事件发生，调试的主线程退出时不会收到该通知
LOAD_DLL_DEBUG_EVENT	每当被调试的进程装载 DLL 文件时，就生成这个事件。当 PE 装载器第一次解析出与 DLL 文件有关的链接时，将收到这一事件。调试进程使用了 LoadLibrary 时也会发生。每当 DLL 文件装载到地址空间中去时，都要调用该调试事件
OUTPUT_DEBUG_STRING_EVENT	当调试进程调用 DebugOutputString 函数向程序发送消息字符串时该事件发生
UNLOAD_DLL_DEBUG_EVENT	每当调试进程使用 FreeLibrary 函数卸载 DLL 文件时，就会生成调试事件。仅当最后一次从过程的地址空间卸载 DLL 文件时，才出现该调试事件（也就是说 DLL 文件的使用次数为 0 时）
RIP_EVENT	只有 Windows 98 检查过的构件才会生成该调试事件。该调试事件用于报告错误信息

当 WaitForDebugEvent() 接收到一个调试事件时，会把调试事件的信息填写入 DEBUG_EVENT 结构中返回，然后检查 dwDebugEventCode 字段中的值，根据它来判断被调试的进程中发生了哪种类型的调试事件。在调试事件结构体中，dwProcessId 的值是调试事件所发生的进程的标识符，dwThreadId 的值是调试事件所发生的线程的标识符，最后一个是与调试事件类型 dwDebugEventCode 对应的共用体成员。

```
typedef struct _DEBUG_EVENT {
    DWORD dwDebugEventCode;
    DWORD dwProcessId;
    DWORD dwThreadId;
    union {
        EXCEPTION_DEBUG_INFO Exception;
        CREATE_THREAD_DEBUG_INFO CreateThread;
        CREATE_PROCESS_DEBUG_INFO CreateProcessInfo;
```

```
            EXIT_THREAD_DEBUG_INFO ExitThread;
            EXIT_PROCESS_DEBUG_INFO ExitProcess;
            LOAD_DLL_DEBUG_INFO LoadDll;
            UNLOAD_DLL_DEBUG_INFO UnloadDll;
            OUTPUT_DEBUG_STRING_INFO DebugString;
            RIP_INFO RipInfo;
        } u;
    } DEBUG_EVENT;
```

16.3.3 进程调试

实际应用时,根据不同的需要,可以创建一个新的进程进行调试,也可以调试一个正在运行的进程,不管哪种方式,都需要建立循环不断地接收和处理调试事件,并进行相应的处理,然后等待下一个调试事件触发。

1. 创建一个新进程以供调试

通过 `CreateProcess()` 创建新进程时,如果在 `dwCreationFlags` 标志字段中设置了 `DEBUG_PROCESS` 或 `DEBUG_ONLY_THIS_PROCESS` 标志,将创建一个用于调试的新进程。

2. 调试一个已有进程

利用 `DebugActiveProcess()` 函数可以将调试器捆绑到一个正在运行的进程上,如果执行成功,则效果类似于利用 `DEBUG_ONLY_THIS_PROCESS` 标志创建的新进程。需要注意的是,在 NT 内核下当试图通过 `DebugActiveProcess()` 函数将调试器捆绑到一个创建时带有安全描述符的进程上时,将被拒绝。

3. 建立事件监视循环

使用 `WaitForDebugEvent()` 和 `ContinueDebugEvent()` 函数建立循环来不断地监视调试事件。`WaitForDebugEvent()` 在一段时间内等待目标进程中调试事件的发生,如果在这段时间没有调试事件发生,那么函数将返回 False;如果在指定时间内调试事件发生了,那么函数将返回 True,并且把所发生的调试事件及其相关信息填写入一个 DEBUG_EVENT 结构中。然后调试器会检查这些信息,并据此进行相应处理。在对这些事件进行相应的操作后,就可以使用 `ContinueDebugEvent()` 函数来恢复线程的执行,并等待下一个调试事件的发生。需要注意的是,`WaitForDebugEvent()` 只能使用在创建以供调试的或是已被捆绑调试器的进程中的某个线程上。下面的代码以 C 语言形式演示了该循环的构建方式。

```
PROCESS_INFORMATION pi;
STARTUP_INFO si;
DEBUG_EVENT devent;
if(CreateProcess(0, "target.exe", 0, 0, FALSE, DEBUG_ONLY_THIS_PROCESS, 0, 0, &si, pi))
{
    while(TRUE)
    {
        if(WaitForDebugEvent(&devent, 100))        //在100ms内等待调试事件
        {
            switch (devent.dwDebugEventCode)
            {
                case CREATE_PROCESS_DEBUG_EVENT:
```

```
                        //在此处编写自己的处理代码
                        break;
                case EXIT_PROCESS_DEBUG_EVENT:
                        //在此处编写自己的处理代码
                        break;
                case EXCEPTION_DEBUG_EVENT:
                        //在此处编写自己的处理代码
                        break;
                //其他类型调试事件处理代码
            }
            ContinueDebugEvent(devent.dwProcessId, devent.dwThreadId,
                        DBG_CONTINUE);
        }
        else
        {
            //其他一些操作
        }
    }
}
else
{
    MessageBox(0, "Unexpected load error", "Fatal Error", MB_OK);
}
```

16.3.4 线程环境

每个进程都有一个最初的主线程,通过主线程可以创建在同一地址空间中运行的其他线程。进程并不执行代码,真正执行代码的是线程。同一个进程中的所有线程共享相同的地址空间和相同的系统资源,但是每个线程又有不同的执行环境。

Windows 分配给每个线程一个很短的时间片,时间片用完之后,系统将暂停当前线程并切换到下一个具有最高优先级的待调度线程。在切换之前,系统会把当前线程执行状态保存到一个名为 CONTEXT 的结构体中,包括线程执行所用寄存器、系统堆栈和用户堆栈、线程所用的描述符表等其他状态信息。当该线程再次被调度进入 CPU 运行时,系统将恢复上次保存的上下文,以便线程可以继续上一次未完成的工作。

在调试时,为了满足某些特定调试目的,也可以根据需要来读取和修改线程环境,具体步骤有 4 个。

(1) 调用 `SuspendThread()` 函数暂停线程。
(2) 调用 `GetThreadContext()` 函数读取线程环境。
(3) 修改读取到的数据,再调用 `SetThreadContext()` 函数设置线程新的执行环境。
(4) 调用 `ResumeThread()` 函数恢复线程执行。

16.3.5 断点

断点是最常用的软件调试技术之一,其基本思想是在某一个位置设置一个"陷阱",当 CPU 执行到这个位置时停止被调试的程序并中断到调试器中,让调试者进行分析和调试,调试者分析结束后,可以让被调试程序恢复执行。通过设置断点可以暂停程序执行,并可以观察和记录指令信息、变量值、堆栈参数和内存数据,还可以深入了解和把握程序执行的内部原理和详细过程,断点对于软件调试具有重要的意义和作用。

断点可以分为软件断点、硬件断点和内存断点三大类,也有的分为代码断点、数据断点和 I/O 断点三类,这里只介绍前一种分类标准。

1. 软件断点

软件断点是一个单字节指令(INT 3,字节码为 0xCC),可以在程序中设置多个软件断点,使得程序执行到该处时能够暂停执行,并将控制权转移给调试器的断点处理函数。

当调试器被告知在目标地址设置一个断点,它首先读取目标地址的第一字节的操作码,然后保存起来,同时把地址存储在内部的中断列表中。接着,调试器把一字节操作码 0xCC 写入刚才的地址。当 CPU 执行到 0xCC 操作码时就会触发一个 INT 3 中断事件,此时调试器就能捕捉到这个事件。调试器继续判断这个发生中断事件的地址(通过指令指针寄存器 EIP)是不是自己先前设置断点的地址。如果在调试器内部的断点列表中找到了这个地址,就将设置断点前存储起来的操作码写回到目标地址,这样进程被调试器恢复后就能正常执行。

2. 硬件断点

硬件断点通过调试寄存器实现,设置在 CPU 级别上,当需要调试某个指定区域而又无法修改该区域时,硬件断点非常有用。

一个 CPU 一般会有 8 个调试寄存器(DR0～DR7),用于管理硬件断点。其中,调试寄存器 DR0 到调试寄存器 DR3 存储硬件断点地址,同一时间内最多只能设置 4 个硬件断点;DR4 和 DR5 保留,DR6 是状态寄存器,说明被断点触发的调试事件的类型;DR7 本质上是一个硬件断点的开关寄存器,同时也存储了断点的不同类型。通过在 DR7 寄存器里设置不同标志,能够创建以下几种断点:当特定的地址上有指令执行时中断、当特定的地址上有数据写入时中断、当特定的地址上有数据读或者写但不执行时中断。

硬件断点使用 INT 1 实现,该中断负责硬件中断和步进事件。步进是指根据预定的流程一条一条地执行指令,每执行完一条指令后暂停下来,从而可以精确地观察关键代码并监视寄存器和内存数据的变化。在 CPU 每次执行代码之前,都会先确认当前将要执行代码的地址是不是硬件断点的地址,同时也要确认是否有代码要访问被设置了硬件断点的内存区域。如果任何储存在 DR0～DR3 中的地址所指向的区域被访问了,就会触发 INT 1 中断,同时暂停 CPU;如果不是中断地址则 CPU 执行该行代码,到下一行代码时,CPU 继续重复上面的过程。

3. 内存断点

内存断点是通过修改内存中指定块或页的访问权限来实现的。通过将指定内存块或页的访问权限属性设置为受保护的,则任何不符合访问权限约束的操作都将失败,并抛出异常,导致 CPU 暂停执行,使得调试器可以查看当前执行状态。

一般来说,每个内存块或页的访问权限都由 3 种不同的访问权限组成:是否可执行、是否可读、是否可写。每个操作系统都提供了用于查询和修改内存页访问权限的函数,在 Windows 操作系统中可以使用 `VirtualProtect()` 函数来修改主调进程虚拟地址空间中已提交页面的保护属性,使用 `VirtualProtectEx()` 函数可以修改其他进程虚拟地址空间页面的保护属性。

16.4 案例精选

本节通过一些示例来演示如何使用 Python 编写程序分析 PE 文件，其中用到了不同的扩展库，可根据需要下载安装。

1. 利用 pefile 模块查看 PE 文件详细信息

```
>>> import pefile
>>> f = pefile.PE(r'C:\windows\notepad.exe')
>>> print(f)                          #略去输出结果
>>> print(f.FILE_HEADER)
>>> print(f.OPTIONAL_HEADER)
>>> for k in f.sections:
    print(k)
>>> f.is_dll()
False
>>> f.is_exe()
True
```

2. 利用 pefile 模块枚举 DLL 的导出项

```
>>> import pefile
>>> pe = pefile.PE(r'C:\windows\glut32.dll')
>>> if hasattr(pe, 'DIRECTORY_ENTRY_EXPORT'):
    for exp in pe.DIRECTORY_ENTRY_EXPORT.symbols:
        print(hex(pe.OPTIONAL_HEADER.ImageBase+exp.address),exp.name, exp.ordinal)
```

3. 利用 pefile 和 pydasm 模块从 PE 文件入口点开始反汇编

```
import sys
import pefile
import pydasm

pe = pefile.PE(r"C:\windows\notepad.exe")
console = sys.stdout
f = open('pe_dasm.txt', 'w')
sys.stdout = f

ep = pe.OPTIONAL_HEADER.AddressOfEntryPoint
ep_ava = ep+pe.OPTIONAL_HEADER.ImageBase
data = pe.get_memory_mapped_image()[ep:]
offset = 0
while offset < len(data):
    i = pydasm.get_instruction(data[offset:], pydasm.MODE_32)
    instruction = pydasm.get_instruction_string(i, pydasm.FORMAT_INTEL,
                                                ep_ava+offset)
    if instruction != None:
        print(hex(ep+offset), '\t', instruction)
    else:
        break
    try:
```

```
            offset += i.length
    except BaseExceptionase:
        break

f.close()
sys.stdout = console
```

4. 利用 WinAppDbg 监视 Windows API 调用

```
import os
import sys
from winappdbg import Debug, EventHandler        #首先需要使用 pip 安装 winappdbg

class MyEventHandler(EventHandler):
    #Add the APIs you want to hook
    apiHooks = {'kernel32.dll': [('CreateFileW', 7)]}

    #The pre_ functions are called upon entering the API
    def pre_CreateFileW(self, event, ra, lpFileName, dwDesiredAccess,
                        dwShareMode, lpSecurityAttributes, dwCreationDisposition,
                        dwFlagsAndAttributes, hTemplateFile):
        fname = event.get_process().peek_string(lpFileName, fUnicode=True)
        print("CreateFileW: %s" %(fname))

    #The post_ functions are called upon exiting the API
    def post_CreateFileW(self, event, retval):
        if retval:
            print('Suceeded (handle value: %x)' %(retval))
        else:
            print('Failed!')

if __name__ == "__main__":
    if len(sys.argv) < 2 or not os.path.isfile(sys.argv[1]):
        print("\nUsage: %s <File to monitor>[arg1, arg2, ...]\n" %sys.argv[0])
        sys.exit()

    #Instance a Debug object, passing it the MyEventHandler instance
    debug = Debug(MyEventHandler())
    try:
        #Start a new process for debugging
        p = debug.execv(sys.argv[1:], bFollow=True)
        #Wait for the debugged process to finish
        debug.loop()
    #Stop the debugger
    finally:
        debug.stop()
```

将上面的代码保存为 simpleapi.py，使用方法与运行结果如图 16-7 所示。

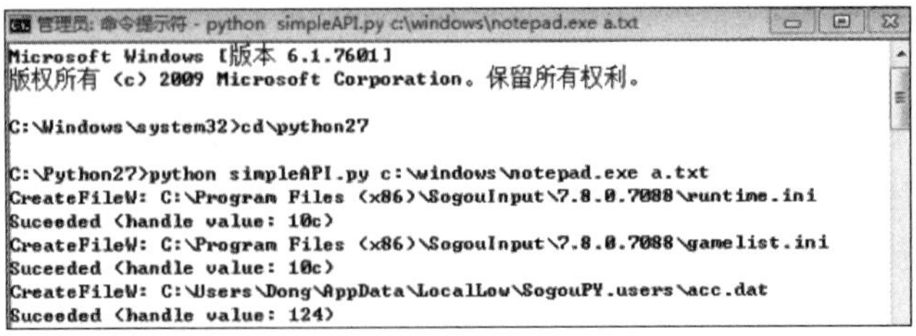

图 16-7　Windows API 监视器

本 章 小 结

（1）在 Windows 平台上，EXE 文件、COM 文件、DLL 文件、OCX 文件、SYS 文件和 SCR 文件等都属于 PE 文件。

（2）PE 文件规范最新版本是 2013 年 2 月 6 日发布的 8.3 版。

（3）在分析软件尤其是恶意软件时，应尽量使用虚拟机或沙箱系统，避免本地物理主机系统被感染而造成不必要的损失。

（4）IDA、W32DASM 是成熟的可执行文件反汇编工具，OllyDbg、WinDbg 和 Immunity Debugger 是成熟的软件调试工具。

（5）通过 IDAPython 插件可以在 IDA 中运行 Python 程序实现自定义的软件测试与分析功能。

（6）ROP、JOP 是近几年流行的攻击方式，目前比较有效的防范技术是 CFI。

（7）软件调试时经常需要设置断点，常见的断点类型有软件断点、硬件断点和内存断点。

（8）调试器的主要工作就是监视目标进程的运行并对目标执行过程中发生的每一个调试事件都进行相应的反应和处理。

习 题

1. 下载 PE 文件规范 8.3 版本，并尝试了解 PE 文件的基本结构。
2. 下载并安装 IDA Pro 与 Immunity Debugger，并简单了解 PE 文件反汇编和调试步骤。
3. 安装并配置 IDAPython 插件，然后运行 16.2.1 节的 Python 代码。
4. 在 Immunity Debugger 调试器中运行 16.2.2 节中的代码。
5. 叙述软件调试断点的概念、作用及其分类。
6. 运行 16.4 节中的代码并查看运行结果。

第 17 章 数据分析、科学计算与可视化

用于数据分析、科学计算与可视化的扩展库非常多,例如 NumPy、SciPy、SymPy、Pandas、Matplotlib、Traits、TraitsUI、Chaco、TVTK、Mayavi、VPython 和 OpenCV。其中,NumPy 是科学计算、数据分析、可视化以及机器学习库依赖的扩展库,提供了 Python 中没有的数组对象,支持 N 维数组运算、大型矩阵、成熟的广播函数库、矢量运算、线性代数、傅里叶变换以及随机数生成等功能,并可与 C++、FORTRAN 等语言无缝结合。SciPy 模块依赖于 NumPy,提供了更多的数学工具,包括矩阵运算、线性方程组求解、积分和优化等。Pandas 是非常成熟的数据分析库。Matplotlib 是比较常用的绘图模块,可以快速地将计算结果以不同类型的图形展示出来。由于篇幅限制,本章略去了大部分运行结果,并且在配套课件中补充了更多内容,在作者另一本教材《Python 数据分析与数据可视化》中对扩展库 NumPy、Pandas、Matplotib 进行了全面的讲解。

17.1 NumPy 数组运算与矩阵运算

根据 Python 社区的习惯,一般使用下面的方式来导入 NumPy 模块:

```
>>> import numpy as np
```

1. 创建数组

```
>>> a = np.array((1, 2, 3, 4, 5))                      #一维数组
>>> b = np.array(([1, 2, 3], [4, 5, 6], [7, 8, 9]))    #二维数组
>>> x = np.linspace(0, 5, 10)                          #一维数组
>>> y = np.logspace(0, 100, 10)                        #一维数组
```

2. 数组与数值的算术运算

```
>>> a = np.array((1, 2, 3, 4, 5))
>>> a * 2                                              #每个数字乘以 2
array([2, 4, 6, 8, 10])
>>> a // 2
array([0, 1, 1, 2, 2])
>>> a / 2.0
array([0.5, 1. , 1.5, 2. , 2.5])
>>> a ** 2                                             #每个数字的平方
array([1, 4, 9, 16, 25])
>>> 2 ** a                                             #2 的每个数字次方
array([2, 4, 8, 16, 32], dtype=int32)
```

3. 数组与数组的算术运算

```
>>> a = np.array((1, 2, 3))
>>> b = np.array(([1, 2, 3], [4, 5, 6], [7, 8, 9]))
```

```
>>> c = a * b                           #广播,a 中数字乘以 b 中对应列
>>> print(c)                            #使用 print()输出与直接查看的形式不同
[[ 1  4  9]
 [ 4 10 18]
 [ 7 16 27]]
>>> print(c/b)
[[1 2 3]
 [1 2 3]
 [1 2 3]]
>>> a = np.array((1, 2, 3))
>>> b = np.array((1, 2, 3))
>>> a * b                               #数组形状相同,对应位置的元素进行运算
array([1, 4, 9])
>>> a + b
array([2, 4, 6])
```

4. 二维数组转置

```
>>> b = np.array(([1, 2, 3], [4, 5, 6], [7, 8, 9]))
>>> print(b.T)
[[1 4 7]
 [2 5 8]
 [3 6 9]]
```

5. 向量点积

```
>>> import numpy as np
>>> a = np.array((5, 6, 7))
>>> b = np.array((6, 6, 6))
>>> print(np.dot(a, b))                 #对应位置分量乘积的和
108
```

向量点积

6. 数组元素访问

```
>>> import numpy as np
>>> b = np.array(([1, 2, 3], [4, 5, 6], [7, 8, 9]))
>>> b[0, 0]
1
>>> b[0][0]                             #两种形式的功能一样
1
```

数组还支持多元素同时访问,例如:

```
>>> x = np.arange(0, 100, 10, dtype=np.floating)
>>> x
array([ 0., 10., 20., 30., 40., 50., 60., 70., 80., 90.])
>>> index = np.random.randint(0, len(x), 5)    #5 个随机数
>>> index
array([9, 6, 3, 9, 7])
>>> noise = np.random.standard_normal(5) * 0.3
>>> noise                               #可以使用 np.set_printoptions(precision=16)设置显示精度
array([ 0.43460475, 0.57262955, -0.15114837, 0.02738525, -0.01063617])
>>> x[index]                            #使用数组作索引,访问多个元素
array([ 90., 60., 30., 90., 70.])
```

```
>>> x[index] += noise                                    #同时修改多个元素的值
>>> x[index]
array([90.02738525, 60.57262955, 29.84885163, 90.02738525, 69.98936383])
>>> x[[1,3,5]]                                           #使用列表作下标访问多个元素
array([10.    , 29.84885163, 50.    ])
```

7. 三角函数运算

```
>>> b = np.array(([1, 2, 3], [4, 5, 6], [7, 8, 9]))
>>> print(np.sin(b))                                     #输出结果(略),更多函数见 PPT
```

8. 四舍五入

```
>>> print(np.round(np.sin(b)))                           #NumPy 数组支持大量的类似函数
[[ 1.  1.  0.]
 [-1. -1.  0.]
 [ 1.  1.  0.]]
```

9. 对二维数组不同维度上的元素进行求和

```
>>> x = np.arange(0, 10).reshape(2, 5)
>>> x
array([[0, 1, 2, 3, 4],
       [5, 6, 7, 8, 9]])
>>> np.sum(x)                                            #所有元素之和
45
>>> np.sum(x, axis=0)                                    #纵向求和,二维数组的第一个维度
array([5, 7, 9, 11, 13])
>>> np.sum(x, axis=1)                                    #横向求和,二维数组的第二个维度
array([10, 35])
```

10. 计算二维数组不同维度上元素的均值

```
>>> x = np.arange(0, 10).reshape(2,5)                    #二维数组,2 行 5 列
>>> np.average(x, axis=0)
array([2.5, 3.5, 4.5, 5.5, 6.5])
>>> np.average(x, axis=1)
array([2., 7.])
```

11. 计算数据的标准差与方差

```
>>> x = np.random.randint(0, 10, size=(3, 3))
>>> x
array([[4, 2, 8],
       [0, 8, 9],
       [0, 2, 7]])
>>> np.std(x)                                            #默认计算所有数值的标准差
3.4029761846919007
>>> np.std(x, axis=1)                                    #每行数值分别计算标准差
array([2.49443826, 4.02768199, 2.94392029])
>>> np.var(x)
11.580246913580245
```

12. 对二维数组不同维度上的元素求最大值

```
>>> np.max(x)                                            #默认返回所有元素的最大值
```

9
```
>>>np.max(x, axis=1)                           #每行的最大值
array([8, 9, 7])
```

13. 对二维数组不同维度上的元素进行排序

```
>>>np.sort(x)                                  #默认沿数组的 shape 属性中最后一个维度排序
array([[2, 4, 8],
       [0, 8, 9],
       [0, 2, 7]])
>>>np.sort(x, axis=0)
array([[0, 2, 7],
       [0, 2, 8],
       [4, 8, 9]])
```

14. 生成特殊数组

```
>>>print(np.zeros((3, 3)))                     #全 0 数组,3 行 3 列
[[0. 0. 0.]
 [0. 0. 0.]
 [0. 0. 0.]]
>>>print(np.ones((3, 3)))                      #全 1 数组
[[1. 1. 1.]
 [1. 1. 1.]
 [1. 1. 1.]]
>>>print(np.identity(3))                       #单位数组
[[1. 0. 0.]
 [0. 1. 0.]
 [0. 0. 1.]]
>>>np.empty((3, 3))            #只申请空间,不初始化,速度很快,每次运行结果可能不同
array([[ 4.24510694e+175,  5.03061214e+223,  4.72100120e+164],
       [ 2.63551414e-144, -1.00000000e+000,  0.00000000e+000],
       [ 0.00000000e+000,  0.00000000e+000,  1.00000000e+000]])
```

15. 改变数组形状

```
>>>a = np.arange(1, 11, 1)                     #一维数组
>>>a
array([1, 2, 3, 4, 5, 6, 7, 8, 9, 10])
>>>a.shape = 2, 5                              #改为二维数组,2 行 5 列
>>>a
array([[1, 2, 3, 4, 5],
       [6, 7, 8, 9, 10]])
>>>a.shape = 5, -1                             #-1 表示自动计算,这里相当于 2
>>>a
array([[1, 2],
       [3, 4],
       [5, 6],
       [7, 8],
       [9, 10]])
>>>b = a.reshape(2, 5)                         #不能修改数组中元素的数量
>>>b
```

改变数组形状

```
array([[1, 2, 3, 4, 5],
       [6, 7, 8, 9, 10]])
```

16. 切片操作

切片操作

```
>>> a = np.arange(10)
>>> a
array([0, 1, 2, 3, 4, 5, 6, 7, 8, 9])
>>> a[::-1]
array([9, 8, 7, 6, 5, 4, 3, 2, 1, 0])
>>> a[::2]
array([0, 2, 4, 6, 8])
>>> a[:5]
array([0, 1, 2, 3, 4])
>>> c = np.array([[j*10 +i for i in range(6)] for j in range(6)])
>>> c[0, 3:5]                                  #行下标为 0,列下标为[3,5)区间的元素
array([3, 4])
>>> c[0]
array([0, 1, 2, 3, 4, 5])
>>> c[2:5, 2:5]                                #逗号前是行下标,逗号后是列下标
array([[22, 23, 24],
       [32, 33, 34],
       [42, 43, 44]])
```

17. 布尔运算

布尔运算

```
>>> x = np.random.rand(10)
>>> x
array([0.93874098, 0.97312716, 0.45264749, 0.74117525, 0.89758246,
       0.29755703, 0.2182093, 0.5673035, 0.90745768, 0.71920431])
>>> x > 0.5                                    #测试每个数字是否大于 0.5
array([True, True, False, True, True, False, False, True, True, True], dtype=bool)
>>> x[x>0.5]                                   #返回与 True 对应位置上的数字组成的数组
array([0.93874098, 0.97312716, 0.74117525, 0.89758246, 0.5673035,
       0.90745768, 0.71920431])
>>> np.array([1, 2, 3]) < np.array([3, 2, 1])  #对应位置上的数字进行比较
array([True, False, False], dtype=bool)
>>> np.array([1, 2, 3]) == np.array([3, 2, 1])
array([False, True, False], dtype=bool)
```

18. 取整运算

```
>>> x = np.random.rand(10) * 50
>>> x
array([ 0.69708323, 14.99931488, 15.04431214, 24.60547929,
       12.12020273, 42.72638176, 16.01128916, 38.91558471,
       39.6877989, 21.98678429])
>>> np.array([t-int(t) for t in x])            #也可以直接使用 x- np.int32(x)
array([0.69708323, 0.99931488, 0.04431214, 0.60547929, 0.12020273,
       0.72638176, 0.01128916, 0.91558471, 0.6877989, 0.98678429])
```

19. 广播

```
>>> a = np.arange(0, 60, 10).reshape(-1, 1)
>>> b = np.arange(0, 6)
```

```
>>> a + b                                          #更详细的用法用PPT或作者公众号
array([[ 0,  1,  2,  3,  4,  5],
       [10, 11, 12, 13, 14, 15],
       [20, 21, 22, 23, 24, 25],
       [30, 31, 32, 33, 34, 35],
       [40, 41, 42, 43, 44, 45],
       [50, 51, 52, 53, 54, 55]])
```

20. 分段函数

分段函数

```
>>> x = np.random.randint(0, 10, size=(1, 10))
>>> x
array([[0, 4, 3, 3, 8, 4, 7, 3, 1, 7]])
>>> np.where(x<5, 0, 1)                            #小于5的变为0,其他变为1
array([[0, 0, 0, 0, 1, 0, 1, 0, 0, 1]])
>>> x = np.random.randint(0, 10, size=(1, 10))
>>> x
array([[3, 6, 5, 1, 0, 7, 3, 9, 6, 0]])
>>> np.piecewise(x, [x>7,x<4], [lambda x:x*2, lambda x:x*3, 0])
array([[9, 0, 0, 3, 0, 0, 9, 18, 0, 0]])           #大于7的乘以2,小于4的乘以3,其他变为0
```

21. 计算唯一值以及出现次数

```
>>> x = np.random.randint(0, 10, 10)
>>> x
array([4, 7, 3, 6, 7, 4, 1, 9, 4, 8])
>>> np.bincount(x)                                 #结果数字分别表示 0, 1, 2, …, max(x)的出现次数
array([0, 1, 0, 1, 3, 0, 1, 2, 1, 1])
>>> np.unique(x)                                   #唯一元素,升序排序
array([1, 3, 4, 6, 7, 8, 9])
```

22. 计算加权平均值

```
>>> x = np.random.randint(0, 10, 10)
>>> x
array([7, 8, 5, 8, 0, 7, 9, 9, 9, 7])
>>> y = np.round_(np.random.random(10), 1)
>>> y
array([0.6, 0.8, 0.8, 0. , 0.6, 0.1, 0. , 0.2, 0.8, 0.7])
>>> np.sum(x*y) / np.sum(np.bincount(x))
2.9199999999999999
>>> x = np.array((5, 10))
>>> np.average(x, weights=(0.3, 0.7))
8.5
```

23. 矩阵运算

```
>>> import numpy as np
>>> a_list = [3, 5, 7]
>>> a_mat = np.matrix(a_list)                      #创建矩阵
>>> a_mat
matrix([[3, 5, 7]])
>>> np.shape(a_mat)                                #矩阵一定是二维的
(1, 3)
>>> b_mat = np.matrix((1, 2, 3))
```

矩阵运算

```
>>> b_mat
matrix([[1, 2, 3]])
>>> a_mat * b_mat.T                                #T 表示转置
matrix([[34]])
>>> a_mat.argsort()                                #返回每个元素的排序序号
matrix([[0, 1, 2]])
>>> a_mat.mean()                                   #所有元素的平均值
5.0
>>> a_mat.sum()                                    #所有元素的和
15
>>> a_mat.max()                                    #所有元素的最大值
7
>>> d_mat = np.matrix([[1, 2, 3], [4, 5, 6], [7, 8, 9]])
>>> d_mat.diagonal()                               #矩阵对角线元素
matrix([[1, 5, 9]])
>>> d_mat.flatten()                                #矩阵平铺,行优先
matrix([[1, 2, 3, 4, 5, 6, 7, 8, 9]])
>>> d_mat.flatten('F')                             #矩阵平铺,列优先
matrix([[1, 4, 7, 2, 5, 8, 3, 6, 9]])
>>> e, v = np.linalg.eig([[1, 1],[2, 2]])          #特征值与特征向量
>>> x = np.matrix([[1, 2], [3, 4]])
>>> y = np.linalg.inv(x)                           #计算逆矩阵
>>> a = np.matrix([[1, 2, 3], [4, 5, 6]])
>>> q, r = np.linalg.qr(a)                         #矩阵 QR 分解
>>> a = [[1, 2], [3, 4]]
>>> np.linalg.det(a)                               #计算行列式
>>> U, s, V = np.linalg.svd(a, full_matrices=False) #奇异值分解
>>> a = np.array([[3, 1], [1, 2]])
>>> b = np.array([9, 8])
>>> x = np.linalg.solve(a, b)                      #线性方程组求解
>>> x = np.matrix([[1, 2], [3, -4]])
>>> np.linalg.norm(x)                              #norm()函数用于计算范数
5.4772255750516612                                 #(1**2+2**2+3**2+(-4)**2)**0.5
>>> np.linalg.norm(x, -2)                          #最小奇异值
1.9543950758485487
>>> np.linalg.norm(x, -1)                          #min(sum(abs(x), axis=0))
4.0
>>> np.linalg.norm(x, 1)                           #max(sum(abs(x), axis=0))
6.0
>>> np.linalg.norm(np.array([1, 2, 3, 4]), 3)
4.6415888336127784
>>> x.std(axis=1)                                  #横向标准差
>>> x.std(axis=0)                                  #纵向标准差
>>> x.var(axis=0)                                  #纵向方差
>>> x.var(axis=1)                                  #横向方差
```

17.2 SciPy 简单应用

扩展库 SciPy 在 NumPy 的基础上增加了大量用于数学计算、科学计算以及工程计算的模块,包括线性代数、常微分方程数值求解、信号处理、图像处理和稀疏矩阵等。SciPy 主要模块如表 17-1 所示。

表 17-1　SciPy 主要模块

模　　块	说　　明
constants	常数
special	特殊函数
linalg	增强的线性代数模块
optimize	数值优化算法，如最小二乘拟合（leastsq）、函数最小值（fmin 系列）、非线性方程组求解（fsolve）等
interpolate	插值（interp1d、interp2d 等）
integrate	数值积分
signal	信号处理
ndimage	图像处理，包括 filters 滤波器模块、fourier 傅里叶变换模块、interpolation 图像插值模块、measurements 图像测量模块、morphology 形态学图像处理模块等
stats	统计
sparse	稀疏矩阵
datasets	数据集，包含了几个图像的数据

17.2.1　常数与特殊函数

SciPy 的 constants 模块包含了大量用于科学计算的常数，详情可以查看 http://docs.scipy.org/doc/scipy/reference/constants.html。

可以使用下面的方法来访问该模块中预定义的常数：

```
>>> from scipy import constants as C
>>> C.c                     #真空中的光速
299792458.0
>>> C.h                     #普朗克常数
6.62606896e-34
>>> C.mile                  #一英里等于多少米
1609.3439999999998
>>> C.inch                  #一英寸等于多少米
0.0254
>>> C.degree                #一度等于多少弧度
0.017453292519943295
>>> C.minute                #一分钟等于多少秒
60.0
```

此外，SciPy 的 special 模块包含大量函数库，包括基本数学函数、特殊函数及 NumPy 中的所有函数。

```
>>> from scipy import special as S
>>> x = [0, np.pi/2, np.pi, np.pi*1.5, np.pi*2]
>>> S.sin(x)
array([  0.00000000e+00,   1.00000000e+00,   1.22464680e-16,
        -1.00000000e+00,  -2.44929360e-16])
>>> x = [1, 2+3j, 4-5j]
```

```
>>> S.conjugate(x)                  #共轭复数
array([1.-0.j, 2.-3.j, 4.+5.j])
>>> S.gamma(4)                      #gamma 函数
6.0
>>> from scipy import special as S
>>> S.cbrt(8)                       #立方根
2.0
>>> S.exp10(3)                      #10**3
1000.0
>>> S.sindg(90)                     #正弦函数,参数为角度
1.0
>>> S.comb(5, 3)                    #从 5 个中任选 3 个的组合数,Python 3.8 的 math 模块也提供了
10.0
>>> S.perm(5, 3)                    #排列数,Python 3.8 的 math 模块也提供了
60.0
>>> S.beta(10, 200)                 #beta 函数
2.839607777781333e-18
>>> S.sinc(0)                       #sinc 函数
1.0
>>> S.sinc(1)
3.8981718325193755e-17
```

17.2.2 SciPy 中值滤波

中值滤波是数字信号处理、数字图像处理中常用的预处理技术之一。该技术的特点是将信号中每个值都替换为其邻域内的中值,即邻域内所有值排序后中间位置上的值。下面通过两个示例来演示 SciPy 模块中中值滤波的实现和应用。

```
>>> import random
>>> import numpy as np
>>> import scipy.signal as signal
>>> x = np.arange(0, 100, 10)
>>> random.shuffle(x)
>>> x
array([40, 0, 60, 20, 50, 70, 80, 90, 30, 10])
>>> signal.medfilt(x, 3)            #中值滤波,邻域大小为 3
array([0., 40., 20., 50., 50., 70., 80., 80., 30., 10.])
```

下面的代码使用中值滤波实现了信号去噪,并将处理前后的信号值进行了对比:

```
import numpy as np
import scipy.signal as signal

x = np.arange(0, 6, 0.1)
y = np.sin(x)
z = y.copy()                                        #浅复制
print('='*20, 'y:', y, sep='\n')
print('before adding noise. z-y:', z-y, sep='\n')
index = np.random.randint(0, len(x), 20)
noise = np.random.standard_normal(20) * 0.8
z[index] += noise                                   #添加噪声
print('='*20, 'after adding noise. z-y:', z-y, sep='\n')
```

```
result = signal.medfilt(z, 3)                    #中值滤波,邻域大小为 3
print('='*20, 'after median filtering. z-y:', result-y, sep='\n')
```

运行程序可以看到,经过中值滤波处理之后,信号整体更加接近原始值,但同时也导致了一些微小的失真。

17.2.3 使用 SciPy 进行多项式计算

```
>>> from scipy import poly1d                     #也可以使用 NumPy 的 poly1d
>>> p1 = poly1d([1, 2, 3, 4])
>>> print(p1)                                    #输出结果中,第一行的数字为第二行对应位置项中 x 的指数
   3     2
1x + 2x + 3x + 4
>>> p2 = poly1d([1, 2, 3, 4], True)              #等价于 p2 = (x-1)(x-2)(x-3)(x-4)
>>> print(p2)
   4      3      2
1x - 10x + 35x - 50x + 24
>>> p3 = poly1d([1, 2, 3, 4], variable='z')      #使用 z 作为变量
>>> print(p3)
   3     2
1z + 2z + 3z + 4
>>> p1(0)                                        #把多项式中的变量替换为指定的值
4
>>> p1(1)
10
>>> p1.r                                         #计算多项式对应方程的根
array([-1.65062919+0.j, -0.17468540+1.54686889j, -0.17468540-1.54686889j])
>>> p1(p1.r[0])
(-8.8817841970012523e-16+0j)
>>> p1.c                                         #查看和修改多项式的系数
array([1, 2, 3, 4])
>>> print(p3)
   3     2
1z + 2z + 3z + 4
>>> p3.c[0] = 5
>>> print(p3)
   3     2
5z + 2z + 3z + 4
>>> p1.order                                     #查看多项式最高阶
3
>>> print(p1.deriv())                            #一阶导数
   2
3x + 4x + 3
>>> print(p1.deriv(2))                           #二阶导数
6x + 4
>>> print(p1.integ(m=1, k=0))                    #一重不定积分,设常数项为 0
     4        3       2
0.25x + 0.6667x + 1.5x + 4x
>>> print(p1.integ(m=2, k=3))                    #二重不定积分,设常数项为 3
     5        4       3     2
0.05x + 0.1667x + 0.5x + 2x + 3x + 3
```

17.2.4 数理统计与随机变量

扩展库 SciPy 的模块 stats 提供了 80 多种连续随机变量和 10 多种离散随机变量，可以导入模块后使用内置函数 dir()查看成员并使用内置函数 help()查看使用帮助。下面的代码演示了正态分布的用法。

```
#设置正态分布参数,其中 loc 是期望值参数,scale 是标准差参数
X = stats.norm(loc=1.0, scale=2.0)
#计算随机变量的期望值和方差
print(X.stats())
#计算累积分布函数值
print(X.cdf(3.14))
```

下面的代码演示了随机变量的用法。

```
import numpy as np
from scipy import stats

#骰子 6 个面的值
x = range(1, 7)
#每个面出现的概率
p = (0.2, 0.2, 0.2, 0.2, 0.1, 0.1)

#创建表示这个骰子的随机变量 dice
dice = stats.rv_discrete(values=(x, p))
#获取随机变量的 20 个随机值,模拟投掷骰子 20 次
values = dice.rvs(size=20)
print(values)
#最大值,最小值,中值
print(values.max(), values.min(), np.median(values))
#方差,标准差
print(values.var(), values.std())
#几何平均值(n 个值连乘结果的 n 次方根),算术平均值
print(stats.gmean(values), values.mean())
```

17.3 Matplotlib 可视化案例精选

Matplotlib 扩展库依赖于 NumPy 扩展库和 tkinter 标准库,可以绘制多种形式的图形,包括折线图、柱状图、饼状图、散点图和雷达图等,是数据可视化与科学计算可视化的重要工具。由于篇幅限制,本节只介绍了很小一部分应用,更多可视化案例和应用可关注微信公众号"Python 小屋"进行学习,或参考作者另一本教材《Python 数据分析与数据可视化》（微课版）。

17.3.1

17.3.1 绘制折线图

```
import numpy as np
import pylab as pl                                      #pylab 是扩展库 Matplotlib 的一个模块
import matplotlib.font_manager as fm
```

```
myfont = fm.FontProperties(fname=r'C:\Windows\Fonts\STKAITI.ttf')   #设置字体
t = np.arange(0.0, 2.0*np.pi, 0.01)                                  #自变量取值范围
s, z = np.sin(t), np.cos(t)                                          #计算正弦、余弦函数值
pl.plot(t, s, label='正弦')
pl.plot(t, z, label='余弦')
pl.xlabel('x-变量', fontproperties='STKAITI', fontsize=24)           #设置 x 轴标签
pl.ylabel('y-正弦余弦函数值', fontproperties='STKAITI', fontsize=24)
pl.title('sin-cos 函数图像', fontproperties='STKAITI', fontsize=32)  #图形标题
pl.legend(prop=myfont)                                               #设置图例
pl.show()                                                            #显示图形
```

运行结果如图 17-1 所示。

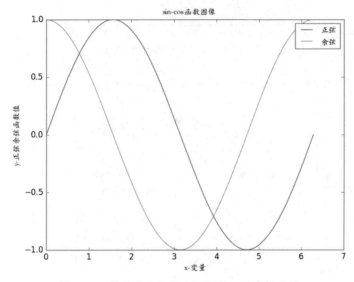

图 17-1　带有中文标签和图例的正弦余弦曲线

17.3.2　绘制散点图

```
import pylab as pl

x = pl.arange(0, 2.0*pl.pi, 0.1)        #在 pylab 模块中可使用 NumPy 的函数
y = pl.cos(x)
pl.scatter(x, y)
pl.show()
```

运行结果如图 17-2 所示。

下面的代码使用随机数生成数值，并根据数值大小来计算散点的大小。

```
import numpy as np
import matplotlib.pylab as pl

x, y = np.random.random(100), np.random.random(100)
pl.scatter(x, y, s=x*500, c='r', marker='*')    #s 指大小,c 指颜色,marker 指符号形状
pl.show()
```

运行结果如图 17-3 所示。

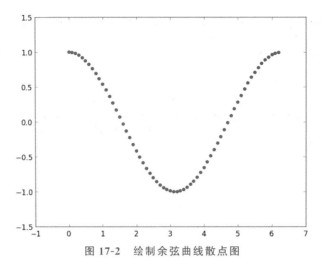

图 17-2　绘制余弦曲线散点图

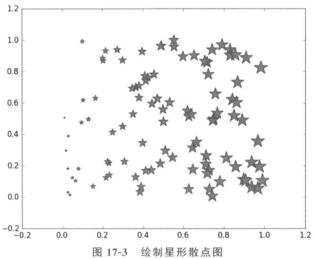

图 17-3　绘制星形散点图

17.3.3　绘制饼状图

17.3.3

```
import numpy as np
import matplotlib.pyplot as plt               #pyplot 是绘图的核心模块

labels = ('Frogs', 'Hogs', 'Dogs', 'Logs')    #标签,逆时针绘制扇形
sizes = (15, 30, 45, 10)
colors = ('yellowgreen', 'gold', '#FF0000', 'lightcoral')
explode = (0, 0.1, 0, 0.1)                    #第二、四个扇形向外裂出

fig = plt.figure()                            #创建图形
ax = fig.gca()                                #获取轴域
ax.pie(np.random.random(4), explode=explode, labels=labels, colors=colors,
       autopct='%1.1f%%', shadow=True, startangle=90,
       radius=0.25, center=(0, 0), frame=True)
```

```
ax.pie(np.random.random(4), explode=explode, labels=labels, colors=colors,
       autopct='%1.1f%%', shadow=True, startangle=90,
       radius=0.25, center=(1, 1), frame=True)
ax.pie(np.random.random(4), explode=explode, labels=labels, colors=colors,
       autopct='%1.1f%%', shadow=True, startangle=90,
       radius=0.25, center=(0, 1), frame=True)
ax.pie(np.random.random(4), explode=explode, labels=labels, colors=colors,
       autopct='%1.1f%%', shadow=True, startangle=90,
       radius=0.25, center=(1, 0), frame=True)
ax.set_xticks([0, 1],['Sunny', 'Cloudy'])        #设置显示刻度的位置和文本
ax.set_yticks([0, 1],['Dry', 'Rainy'])
ax.set_xlim((-0.5, 1.5))                         #设置坐标轴范围
ax.set_ylim((-0.5, 1.5))
ax.set_aspect('equal')                           #设置纵横比相等
plt.show()
```

程序运行结果如图 17-4 所示。

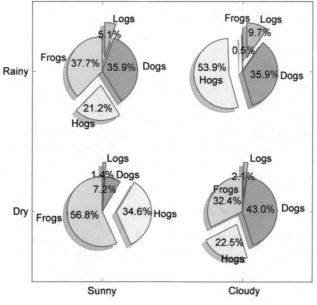

图 17-4　绘制饼状图

17.3.4　在图例中显示公式

```
import numpy as np
import matplotlib.pyplot as plt

x = np.linspace(0, 2*np.pi, 500)
y, z = np.sin(x), np.cos(x*x)
plt.figure(figsize=(8, 4))
#标签前后加$将使用内嵌的 LaTex 引擎将其显示为公式
plt.plot(x, y, label='$sin(x)$', color='red', linewidth= 2)    #红色,2 像素宽
plt.plot(x, z, 'b--', label='$cos(x^2)$')                      #蓝色,虚线
plt.xlabel('Time(s)')
```

```
plt.ylabel('Volt')
plt.title('sin and cos figure using pyplot')
plt.ylim(-1.2, 1.2)
plt.legend()                                    #显示图示
plt.show()                                      #显示绘图窗口
```

上面的代码运行结果如图 17-5 所示。

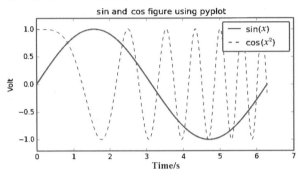

图 17-5 同时绘制多个图形并在图例中显示公式

17.3.5 创建和使用子图

```
import numpy as np
import matplotlib.pyplot as plt

x = np.linspace(0, 2*np.pi, 500)                #创建自变量数组
y1, y2, y3 = np.sin(x), np.cos(x), np.sin(x*x)  #创建函数值数组
plt.figure(1)                                   #创建图形
ax1 = plt.subplot(2, 2, 1)                      #第一行第一列子图
ax2 = plt.subplot(2, 2, 2)                      #第一行第二列子图
ax3 = plt.subplot(212, facecolor='y')           #第二行
plt.sca(ax1)                                    #选择 ax1
plt.plot(x, y1, color='red')                    #绘制红色曲线
plt.ylim(-1.2, 1.2)                             #限制 y 坐标轴范围
plt.sca(ax2)                                    #选择 ax2
plt.plot(x, y2, 'b--')                          #绘制蓝色曲线
plt.ylim(-1.2, 1.2)
plt.sca(ax3)                                    #选择 ax3
plt.plot(x, y3, 'g--')
plt.ylim(-1.2, 1.2)
plt.show()
```

运行结果如图 17-6 所示。

17.3.6 绘制有描边和填充效果的柱状图

```
import numpy as np
import matplotlib.pyplot as plt

x = np.linspace(0, 10, 11)
y = 11 - x
```

```
plt.bar(x, y,
        color='#772277',                    #柱的颜色
        alpha=0.8,                          #透明度
        edgecolor='blue',                   #边框颜色
        linestyle='--',                     #边框样式为虚线
        linewidth=1,                        #边框线宽
        hatch='*')                          #内部使用五角星填充
for xx, yy in zip(x, y):                    #为每个柱形添加文本标注
    plt.text(xx-0.2, yy+0.1, '%2d'%yy)
plt.show()
```

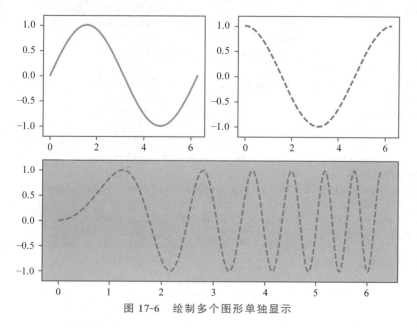

图 17-6　绘制多个图形单独显示

运行结果如图 17-7 所示。

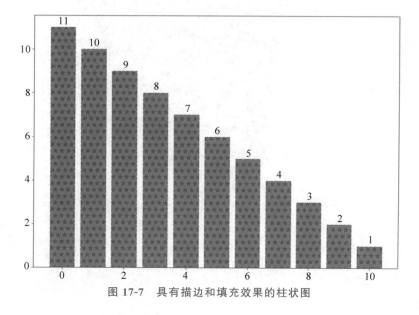

图 17-7　具有描边和填充效果的柱状图

17.3.7 使用雷达图展示学生成绩

```python
import numpy as np
import matplotlib.pyplot as plt

courses = ['C++', 'Python', '高数', '大学英语', '软件工程',
           '组成原理', '数字图像处理', '计算机图形学']
scores = [80, 95, 78, 85, 45, 65, 80, 60]
dataLength = len(scores)                              #数据长度
#angles 数组把圆周等分为 dataLength 份
angles = np.linspace(0,                               #数组第一个数据
                     2*np.pi,                         #数组最后一个数据
                     dataLength,                      #数组中数据数量
                     endpoint=False)                  #不包含终点
scores.append(scores[0])
angles = np.append(angles, angles[0])                 #闭合
#绘制雷达图
plt.polar(angles,                                     #设置角度
          scores,                                     #设置各角度上的数据
          'rv--',                                     #设置颜色、线型和端点符号
          linewidth=2)                                #设置线宽
#设置角度网格标签
plt.thetagrids(angles[:8]*180/np.pi, courses, fontproperties='simhei')
#填充雷达图内部
plt.fill(angles, scores, facecolor='r', alpha=0.6)
plt.show()
```

运行结果如图 17-8 所示。

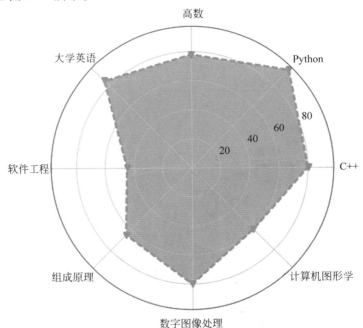

图 17-8 学生成绩雷达图

17.3.8 绘制三维曲面

```
import numpy as np
import mpl_toolkits.mplot3d
import matplotlib.pyplot as plt

x, y = np.mgrid[-2:2:20j, -2:2:20j]          #生成二维坐标网络
z = 50 * np.sin(x+y)
ax = plt.subplot(111, projection='3d')        #创建三维子图
ax.plot_surface(x, y, z, rstride=2, cstride=1, cmap=plt.cm.Blues_r)
ax.set_xlabel('X')
ax.set_ylabel('Y')
ax.set_zlabel('Z')
plt.show()
```

运行结果如图 17-9 所示,在绘图窗口中可用鼠标来旋转所绘制图形。

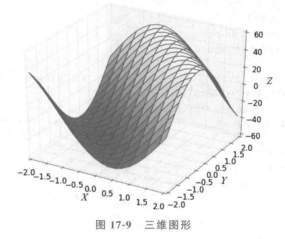

图 17-9 三维图形

下面的代码绘制了另一个略加复杂的三维图形:

```
import pylab as pl
import numpy as np
import mpl_toolkits.mplot3d

rho, theta = np.mgrid[0:1:40j, 0:2*np.pi:40j]
z = rho ** 2
x = rho * np.cos(theta)
y = rho * np.sin(theta)
ax = pl.subplot(111, projection='3d')
ax.plot_surface(x, y, z)
pl.show()
```

运行结果如图 17-10 所示。

17.3.9 绘制三维曲线

下面的代码演示了如何绘制三维曲线。

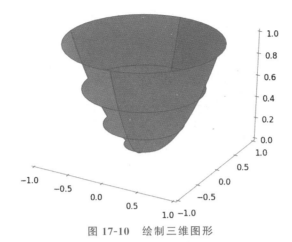

图 17-10　绘制三维图形

```
import numpy as np
import matplotlib as mpl
import matplotlib.pyplot as plt
from mpl_toolkits.mplot3d import Axes3D

mpl.rcParams['legend.fontsize'] = 10           #设置图例中的字号
ax = plt.subplot(111, projection='3d')
theta = np.linspace(-4*np.pi, 4*np.pi, 100)
z = np.linspace(-4, 4, 100)*0.3
r = z**3 +1
x = r * np.sin(theta)
y = r * np.cos(theta)
ax.plot(x, y, z, label='parametric curve')
ax.legend()
plt.show()
```

程序运行结果如图 17-11 所示。

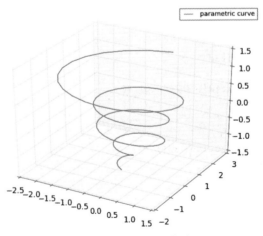

图 17-11　绘制三维曲线

17.3.10 设置图例样式

```python
import numpy as np
import matplotlib.pyplot as plt
import matplotlib.font_manager as fm

t = np.arange(0.0, 2*np.pi, 0.01)
s, z = np.sin(t), np.cos(t)
plt.plot(t, s, label='正弦')
plt.plot(t, z, label='余弦')
plt.title('sin-cos 函数图像',              #标题文本
          fontproperties='STLITI',         #标题字体
          fontsize=24)                     #标题字号
myfont = fm.FontProperties(fname=r'C:\Windows\Fonts\STKAITI.ttf')
plt.legend(prop=myfont,                    #图例字体
           title='Legend',                 #图例标题
           loc='lower left',               #图例左下角位于图形(0.43,0.75)的位置
           bbox_to_anchor=(0.43,0.75),
           shadow=True,                    #显示阴影
           facecolor='yellowgreen',        #图例背景色
           edgecolor='red',                #图例边框颜色
           ncol=2,                         #显示为两列
           markerfirst=False)              #图例文字在前,符号在后

plt.show()
```

17.4 数据分析扩展库 Pandas 用法精要

Pandas(Python Data Analysis Library)是基于 NumPy 的数据分析扩展库,提供了大量数据模型和高效操作大型数据集所需要的功能,可以说 Pandas 是使 Python 能够成为高效且强大的数据分析语言的重要因素之一。由于篇幅限制,本节只介绍很小一部分应用,更多数据分析方面的应用可关注微信公众号"Python 小屋"进行学习,或参考作者另一本教材《Python 数据分析与数据可视化》(微课版)。

Pandas 主要提供了 3 种数据结构:①Series,带标签的一维数组;②DataFrame,带标签的二维表格结构;③DatetimeIndex,日期时间索引数组。

可以使用 conda 或者 pip 工具下载和安装 Pandas,然后按照 Python 社区的习惯,使用下面的语句导入:

```
>>> import pandas as pd
```

1. 生成一维数组

```
>>> import numpy as np
>>> x = pd.Series([1, 3, 5, np.nan])       #np.nan 表示缺失值
```

2. 生成二维数组

```
>>> dates = pd.date_range(start='20240101', end='20241231', freq='D')   #间隔为天
>>> dates = pd.date_range(start='20240101', end='20241231', freq='M')   #间隔为月
```

```
>>> df = pd.DataFrame(np.random.randn(12,4), index=dates, columns=list('ABCD'))
>>> df = pd.DataFrame([[np.random.randint(1,100) for j in range(4)] for i in range(12)],
                     index=dates, columns=list('ABCD'))        #4列随机数
>>> df = pd.DataFrame({'A':[np.random.randint(1,100) for i in range(4)],
                     'B':pd.date_range(start='20200101', periods=4, freq='D'),
                     'C':pd.Series([1, 2, 3, 4], index=list(range(4)),dtype='float32'),
                     'D':np.array([3]*4, dtype='int32'),
                     'E':pd.Categorical(['test','train','test','train']),
                     'F':'foo'})
>>> df = pd.DataFrame({'A':[np.random.randint(1, 100) for i in range(4)],
                     'B':pd.date_range(start='20200101', periods=4, freq='D'),
                     'C':pd.Series([1, 2, 3, 4], index=['zhang', 'li', 'zhou', 'wang'],
                                  dtype='float32'),
                     'D':np.array([3]*4, dtype='int32'),
                     'E':pd.Categorical(['test', 'train', 'test', 'train']),
                     'F':'foo'})
```

3. 二维数据查看

```
>>> df.head()            #默认显示前5行,为节约篇幅,略去输出结果
>>> df.head(3)           #查看前3行
>>> df.tail(2)           #查看最后2行
```

4. 查看二维数据的索引、列名和数据

```
>>> df.index
>>> df.columns
>>> df.values
```

5. 查看数据的统计信息

```
>>> df.describe()        #平均值、标准差、最小值、最大值等信息
```

6. 二维数据转置

```
>>> df.T
```

7. 排序

```
>>> df.sort_index(axis=0, ascending=False)       #按行标签进行降序排序
>>> df.sort_index(axis=1, ascending=False)       #按列标签进行降序排序
>>> df.sort_values(by='A')                       #按数据进行排序,by参数可以为列表
>>> df.sort_values(by=['A','B'], ascending=[False,True])    #按A列的值降序排列
```

8. 数据选择

数据选择

```
>>> df['A']                                      #选择列
>>> df[0:2]                                      #使用切片选择多行,左闭右开区间
>>> df['zhang':'zhou']                           #选择多行,闭区间
>>> df.loc[:, ['A', 'C']]                        #选择多列
>>> df.loc[['zhang', 'zhou'], ['A', 'D', 'E']]   #同时指定多行与多列进行选择
>>> df.loc['zhang', ['A', 'D', 'E']]
>>> df.at['zhang', 'A']                          #查询指定行、列位置的数据值
>>> df.at['zhang', 'D']
>>> df.iloc[3]                                   #查询第3行数据,下标从0开始
```

第 17 章 数据分析、科学计算与可视化

```
>>> df.iloc[0:3, 0:4]                           #查询前 3 行、前 4 列数据
>>> df.iloc[[0, 2, 3], [0, 4]]                  #查询指定的多行、多列数据
>>> df.iloc[0, 1]                               #查询指定行、列位置的数据值
>>> df.iloc[2, 2]
>>> df[df.A>50]                                 #按给定条件查询
>>> df[df['E']=='test']                         #E 列为 'test' 的数据
>>> df[df['A'].isin([20,69])]                   #A 列为 20 或 69 的数据
>>> df.nlargest(3, ['C'])                       #返回 C 列最大的前 3 行
>>> df[df.sum(axis=1)==76]                      #每行数值相加等于 76 的数据
```

9. 数据修改

```
>>> df.iat[0, 2] = 3                            #修改指定行、列位置的数据值
>>> df.loc[:, 'D'] = np.random.randint(50, 60, 4)  #修改某列的值
>>> df['C'] = -df['C']                          #对 C 列数据取反
>>> df.replace(23, 5)                           #把所有 23 替换为 5
>>> df.replace({23:5, 'test':'TEST'})           #使用字典指定替换关系
>>> df.drop('E', axis=1)        #删除指定列,可以使用参数 inplace=True 设置原地删除
>>> df['E'] = df['E'].map(str.upper)            #把 E 列变为大写
>>> df['C'] = df['C'].map(lambda x:x+5)         #C 列数值加 5
```

10. 缺失值、重复值、异常值处理

```
>>> df1 = df.reindex(index=['zhang','li','zhou','wang'],columns=list(df.columns)+['G'])
>>> df1.iat[0, 6] = 3                       #修改指定位置元素值,该列其他元素为缺失值 NaN
>>> pd.isnull(df1)                          #测试缺失值,返回值为 True/False 阵列
>>> df1.dropna()                            #返回不包含缺失值的行
>>> df1['G'].fillna(5, inplace=True)        #使用指定值填充缺失值
>>> df.duplicated()                         #检查重复行
>>> df.drop_duplicates()                    #返回新数组,删除重复行
>>> df.drop_duplicates(['E'])               #删除 E 列重复的数据
>>> data = pd.DataFrame(np.random.randn(500, 4))
>>> data.describe()                         #查看数据的统计信息
>>> data[np.abs(data)>2.5] = np.sign(data) * 2.5   #异常值处理
                                            #把所有数据都限定在[-2.5, 2.5]
```

缺失值、重复值、异常值处理

11. 其他操作

```
>>> df1.mean()                              #平均值,自动忽略缺失值
>>> df.mean(1)                              #横向计算平均值
>>> df1.shift(1)                            #数据移位
>>> df1['D'].value_counts()                 #直方图统计
>>> df2 = pd.DataFrame(np.random.randn(10, 4))
>>> p1 = df2[:3]                            #数据行拆分
>>> p2 = df2[3:7]
>>> p3 = df2[7:]
>>> df3 = pd.concat([p1, p2, p3])           #数据行合并
>>> df2 == df3                              #测试两个二维数据是否相等,返回 True/False 阵列
>>> df4 = pd.DataFrame({'A':[np.random.randint(1, 5) for i in range(8)],
                        'B':[np.random.randint(10, 15) for i in range(8)],
                        'C':[np.random.randint(20, 30) for i in range(8)],
                        'D':[np.random.randint(80, 100) for i in range(8)]})
```

```python
>>> df4.groupby('A').sum()                                    #数据分组计算
>>> df4.groupby(['A','B']).mean()
>>> df4.groupby(by=['A', 'B'], as_index=False).mean()
>>> df4.groupby(by=['A', 'B']).aggregate({'C':np.mean, 'D':np.min})
                                                              #分组后,C列使用平均值,D列使用最小值
>>> data = np.random.randint(0, 100, 10)
>>> category = [0, 25, 50, 100]
>>> pd.cut(data, category)                                    #离散化
>>> pd.cut(data, category, right=False)                       #左闭右开区间
>>> labels = ['low', 'middle', 'high']
>>> pd.cut(data, category, right=False, labels=labels)        #指定标签
>>> pd.cut(data, 4)                                           #四分位数区间
>>> pd.qcut(data, 4)                                          #四分位数区间
>>> df = pd.DataFrame({'a':[1, 2, 3, 4], 'b':[2, 3, 4, 5], 'c':[3, 4, 5, 6], 'd':[3, 3, 3, 3]})
>>> df.pivot(index='a', columns='b', values='c')              #透视表
>>> pd.crosstab(index=df.a, columns=df.b)                     #交叉表
>>> pd.crosstab(index=df.a, columns=df.b, margins=True)
>>> pd.crosstab(index=df.a, columns=df.b, values=df.c, aggfunc='sum', margins=True)
>>> pd.crosstab(index=df.a, columns=df.b, values=df.c, aggfunc='mean', margins=True)
>>> df.diff()                                                 #纵向一阶差分
>>> df.diff(axis=1)                                           #横向一阶差分
>>> df.diff(periods=2)                                        #纵向二阶差分
>>> df.corr()                                                 #pearson 相关系数
>>> df.corr('kendall')                                        #kendall Tau 相关系数
>>> df.corr('spearman')                                       #spearman 秩相关
```

12. 结合 Matplotlib 绘图

```python
>>> import pandas as pd
>>> import numpy as np
>>> import matplotlib.pyplot as plt
>>> df = pd.DataFrame(np.random.randn(1000, 2), columns=['B','C']).cumsum()
>>> df['A'] = pd.Series(list(range(len(df))))
>>> plt.figure()
>>> df.plot(x='A')                                            #折线图,以 A 列的值为 x 坐标
>>> plt.show()
>>> df = pd.DataFrame(np.random.rand(10, 4), columns=['a', 'b', 'c', 'd'])
>>> df.plot(kind='bar')                                       #柱状图,默认以 index 为 x 坐标
>>> plt.show()
>>> df.plot(kind='barh', stacked=True)                        #堆叠的水平柱状图
>>> plt.show()
>>> df = pd.DataFrame({'height':[180, 170, 172, 183, 179, 178, 160],
                       'weight':[85, 80, 85, 75, 78, 78, 70]})
>>> df.plot(x='height', y='weight', kind='scatter',
            marker='*', s=60, label='height-weight')          #绘制散点图
>>> plt.show()
>>> df['weight'].plot(kind='pie', autopct='%.2f%%', labels=df['weight'].values,
            shadow=True)                                      #饼状图
>>> plt.show()
>>> df.plot(kind='box')                                       #箱图
```

```
>>> plt.show()
>>> df['weight'].plot(kind='kde', style='r-.')        #密度图
>>> plt.show()
```

13. 文件读写

文件读写

```
>>> df.to_excel('d:\test.xlsx', sheet_name='dfg')     #将数据保存为 Excel 文件
>>> df = pd.read_excel('d:\test.xlsx', 'dfg', index_col=None, na_values=['NA'])
>>> df.to_csv('d:\test.csv')                          #将数据保存为 CSV 文件
>>> df = pd.read_csv('d:\test.csv')                   #读取 CSV 文件中的数据
```

17.5　统计分析模块 statistics 常用函数

（1）计算平均数函数 mean()。

```
>>> import statistics
>>> statistics.mean([1, 2, 3, 4, 5, 6, 7, 8, 9])      #使用包含整数的列表作为参数
5.0
>>> statistics.mean(range(1, 10))                     #使用 range 对象作为参数
5.0
>>> import fractions
>>> x = [(3, 7), (1, 21), (5, 3), (1, 3)]
>>> y = [fractions.Fraction(*item) for item in x]
>>> y
[Fraction(3, 7), Fraction(1, 21), Fraction(5, 3), Fraction(1, 3)]
>>> statistics.mean(y)                                #使用包含分数的列表作为参数
Fraction(13, 21)
>>> import decimal
>>> x = ('0.5', '0.75', '0.625', '0.375')
>>> y = map(decimal.Decimal, x)
>>> statistics.mean(y)
Decimal('0.5625')
```

（2）中位数函数 median()、median_low()、median_high()、median_grouped()。

```
>>> statistics.median([1, 3, 5, 7])                   #偶数个样本时取中间两个数的平均数
4.0
>>> statistics.median_low([1, 3, 5, 7])               #偶数个样本时取中间两个数的较小者
3
>>> statistics.median_high([1, 3, 5, 7])              #偶数个样本时取中间两个数的较大者
5
>>> statistics.median(range(1, 10))                   #奇数个样本时取排序后中间位置的数
5
>>> statistics.median_low([5, 3, 7]), statistics.median_high([5, 3, 7])
(5, 5)
>>> statistics.median_grouped([5, 3, 7])
5.0
>>> statistics.median_grouped([52, 52, 53, 54])
52.5
>>> statistics.median_grouped([1, 3, 3, 5, 7])
3.25
>>> statistics.median_grouped([1, 2, 2, 3, 4, 4, 4, 4, 5])
3.7
```

```
>>> statistics.median_grouped([1, 2, 2, 3, 4, 4, 4, 4, 4, 5], interval=2)
3.4
```

(3) 返回最常见数据或出现次数最多的数据(most common data)的函数 mode()。

```
>>> statistics.mode([1, 3, 5, 7])              #无法确定出现次数最多的唯一元素,失败
statistics.StatisticsError: no unique mode; found 4 equally common values
>>> statistics.mode([1, 3, 5, 7, 3])           #返回出现次数最多的元素
3
>>> statistics.mode(['red', 'blue', 'blue', 'red', 'green', 'red', 'red'])
'red'
```

(4) pstdev()：返回总体标准差(population standard deviation，the square root of the population variance)。

```
>>> statistics.pstdev([1.5, 2.5, 2.5, 2.75, 3.25, 4.75])
0.986893273527251
>>> statistics.pstdev(range(20))
5.766281297335398
```

(5) pvariance()：返回总体方差(population variance)或二次矩(second moment)。

```
>>> statistics.pvariance([1.5, 2.5, 2.5, 2.75, 3.25, 4.75])
0.9739583333333334
>>> x = [1, 2, 3, 4, 5, 10, 9, 8, 7, 6]
>>> mu = statistics.mean(x)
>>> mu
5.5
>>> statistics.pvariance([1, 2, 3, 4, 5, 10, 9, 8, 7, 6], mu)
8.25
>>> statistics.pvariance(range(20))
33.25
>>> statistics.pvariance((random.randint(1,10000) for i in range(30)))
10903549.933333334
```

(6) variance()、stdev()，计算样本方差(sample variance)和样本标准差(sample standard deviation，the square root of the sample variance，也称均方差)。

```
>>> statistics.variance(range(20))
35.0
>>> statistics.stdev(range(20))
5.916079783099616
>>> _ * _
35.0
>>> statistics.variance([3, 3, 3, 3, 3, 3]), statistics.stdev([3, 3, 3, 3, 3, 3])
(0.0, 0.0)
```

<h1 style="text-align:center">本 章 小 结</h1>

(1) 比较常用的科学计算与可视化扩展库有 NumPy、SciPy 和 Matplotlib。
(2) NumPy 支持数组与标量的运算、数组与数组的运算、向量内积、数组的函数运算、数组多元素操作、不同维度的计算、布尔运算、切片操作、计算标准差与方差以及特殊数组生

成和矩阵运算等功能。

（3）SciPy 依赖于 NumPy，在其基础上增加了大量用于数学计算、科学计算以及工程计算的模块，包括线性代数、常微分方程数值求解、信号处理、图像处理和稀疏矩阵等。

（4）Matplotlib 依赖于 NumPy 和 tkinter，可以绘制多种形式的图形，包括折线图、直方图、饼状图和散点图等，是科学计算可视化的重要工具。

（5）Pandas 基于 NumPy，提供了大量数据模型和高效操作大型数据集所需要的功能，可以说 Pandas 是使得 Python 能够成为高效且强大的数据分析语言的重要因素之一。

（6）statistics 提供了平均数、中位数、总体标准差、总体方差等与统计相关的功能函数。

习　题

1. 运行本章所有代码并查看运行结果。

2. 使用 Python 内置函数 `dir()` 查看 SciPy 模块中的对象与方法，并使用 Python 内置函数 `help()` 查看其使用说明。

3. 编写程序，使用 Pandas 把多个结构相同的 Excel 文件合并为一个 Excel 文件。

4. 假设有个 Excel 文件"电影导演演员.xlsx"，有三列数据分别为电影名称、导演和演员列表（同一个电影可能会有多个演员，每个演员姓名之间使用逗号分隔），要求统计每个演员的参演电影数量，并统计最受欢迎的前三个演员。

5. 编写程序，使用 Pandas 模拟转盘抽奖游戏，统计 10000 次游戏中各奖项的中奖次数。假设转盘角度归一化之后 `[0,0.08)` 区间对应一等奖，`[0.08,0.3)` 区间对应二等奖，`[0.3,1)` 区间对应三等奖。

6. 查阅资料，使用 Matplotlib 绘制动态折线图、柱状图、散点图。

7. 查阅资料，使用 Matplotlib 显示图像文件，响应并处理鼠标的单击、移动、释放等事件，能够使用鼠标在图像中绘制直线，并在鼠标按下移动时实时显示鼠标当前位置与初始单击位置的直线距离。

8. 查阅资料，使用 Matplotlib 绘制正弦曲线和余弦曲线，设置图例的字体、背景色、分栏、位置等属性。

9. 查阅资料，使用 Matplotlib 绘制尼哥米德蚌线。

10. 假设当前文件夹中 data.csv 文件中存放了 2023 年某饭店营业额，第一列为日期（如 2023-09-10），第二列为每天交易额（如 3510），文件中第一行为表头，其余行为实际数据。查阅资料，编写程序，完成下面的任务，要求对结果图形进行适当的美化。

（1）使用 Pandas 读取文件 data.csv 中的数据，创建 DataFrame 对象，并删除其中所有缺失值。

（2）绘制折线图，反应该饭店每天的营业额情况，并把图形保存为本地文件 first.jpg。

（3）按月份进行统计，绘制柱状图显示每月的营业额，并把图形保存为本地文件 second.jpg。

（4）按月份进行统计，找出相邻两个月最大涨幅，并把涨幅最大的月份写入文件 maxMonth.txt。

（5）按季度统计该饭店 2023 年的营业额数据，绘制饼状图显示 4 个季度的营业额分布情况，并把图形保存为本地文件 third.jpg。

第 18 章 密码学编程

信息加密和信息隐藏是实现信息安全与保密的主要技术。其中,信息隐藏或隐写术具有悠久的历史,常用于版权保护和信息保密等相关领域,近几年来与之有关的研究呈上升趋势。作为传统的信息安全技术,加密和解密算法一直都是业内研究的重点。

Python 标准库 hashlib 实现了 SHA1、SHA224、SHA256、SHA384、SHA512 以及 MD5 等多个安全哈希算法,标准库 zlib 提供了 adler32 和 crc32 算法的实现,标准库 hmac 实现了 HMAC 算法。在众多扩展库中,pycryptodome 可以说是密码学编程模块中最成功也是最成熟的,cryptography 也有一定数量的用户在使用。扩展库 pycryptodome 和 cryptography 提供了 SHA 系列算法和 RIPEMD160 等多个安全哈希算法,以及 DES、AES、RSA、DSA、ElGamal 等多个加密算法和数字签名算法的实现。

18.1 安全哈希算法

安全哈希算法也称报文摘要算法,对任意长度的消息可以计算得到固定长度的唯一指纹。理论上,即使内容非常相似的消息也不会得到相同的指纹。安全哈希算法是不可逆的,无法从指纹还原得到原始消息,属于单向变换算法。安全哈希算法常用于数字签名领域,很多管理信息系统把用户密码的哈希值存储到数据库中而不直接存储密码。另外,文件完整性检查也经常用到 MD5 或其他安全哈希算法,见例 7-14。

下面的代码使用 Python 标准库 hashlib 计算字节串的安全哈希值。

```
>>> import hashlib
>>> hashlib.md5('abcdefg'.encode()).hexdigest()          #使用 MD5 算法,略去输出结果
>>> hashlib.sha512('abcdefg'.encode()).hexdigest()       #使用 SHA512 算法
>>> hashlib.sha256('Python 小屋'.encode()).hexdigest()   #使用 SHA256 算法
```

Python 扩展库 pycryptodome 也提供了 MD2、MD4、MD5、HMAC、RIPEMD、SHA、SHA224、SHA256、SHA384、SHA512 等多个安全哈希算法的实现。

```
>>> from Crypto.Hash import SHA256
>>> h = SHA256.SHA256Hash('Python 程序设计(第 4 版)'.encode())
>>> h.hexdigest()
```

18.2 对称密钥密码算法 DES 和 AES

作为经典的对称密钥密码算法,DES 早在 1976 年就被美国政府采用,随后得到美国国家标准局和美国国家标准协会的认可,并成为全球范围内事实上的工业标准。DES 算法使用 56 位密钥对 64 位的数据块加密,并对 64 位的数据块进行 16 轮编码,最终完成变换。下面的代码演示了 Python 扩展库 pycryptodome 中 DES 算法的用法。

```
>>> from Crypto.Cipher import DES
>>> des_encrypt_decrypt = DES.new(b'ShanDong', DES.MODE_ECB)
>>> p = 'Beautiful is better than ugly.'
>>> pp = p.encode()
#按8字节对齐
>>> c = des_encrypt_decrypt.encrypt(pp.ljust((len(pp)//8+1)*8, b'0'))    #加密
>>> cp = des_encrypt_decrypt.decrypt(c)                                   #解密
>>> cp
b'Beautiful is better than ugly.00'
>>> cp[0:len(pp)].decode()
'Beautiful is better than ugly.'
```

高级加密标准(Advanced Encryption Standard，AES)又称 Rijndael 算法，是美国联邦政府采用的一种区块加密标准，用于替代 DES 算法，AES 算法使用代换-置换网络，在软件和硬件上都能快速地加解密。AES 加密数据块分组长度必须为 128b，密钥长度可以是 128/192/256b 中的任意一个(数据块及密钥长度不足时需要补齐)。

例 18-1 使用 Python 扩展库 pycryptodome 提供的 AES 算法实现消息加密和解密。

```python
import string
import random
from Crypto.Cipher import AES

#生成指定长度的密钥,单位为字节
def keyGenerater(length):
    if length not in (16, 24, 32):
        return None
    x = string.ascii_letters + string.digits
    return ''.join(random.choices(x, k=length))

def encryptor_decryptor(key, mode):
    return AES.new(key, mode, b'0000000000000000')

#使用指定密钥和模式对给定信息加密
def AESencrypt(key, mode, text):
    encryptor = encryptor_decryptor(key, mode)
    return encryptor.encrypt(text)

#使用指定密钥和模式对给定信息解密
def AESdecrypt(key, mode, text):
    decryptor = encryptor_decryptor(key, mode)
    return decryptor.decrypt(text)

if __name__ == '__main__':
    text = 'Python 小屋是一个非常棒的微信公众号,由董付国老师维护.'
    key = keyGenerater(16).encode()
    #随机选择 AES 的模式
    mode = random.choice((AES.MODE_CBC, AES.MODE_CFB, AES.MODE_ECB, AES.MODE_OFB))
    if not key:
        print('Something is wrong.')
    else:
        print('key:', key)
```

```
print('mode:', mode)
print('Before encryption:', text)
#明文必须是字节串形式,且长度为 16 的倍数
text_encoded = text.encode()
text_length = len(text_encoded)
padding_length = 16 - text_length%16
text_encoded = text_encoded + b'0'*padding_length
text_encrypted = AESencrypt(key, mode, text_encoded)
print('After encryption:', text_encrypted)
text_decrypted = AESdecrypt(key, mode, text_encrypted)
print('After decryption:', text_decrypted.decode()[:-padding_length])
```

18.3 非对称密钥密码算法 RSA 与数字签名算法 DSA

18.3.1 RSA

RSA 是一种经典的非对称密钥密码体制,试图从加密密钥和解密密钥中的任何一个推导出另一个在计算上是不可行的。RSA 的安全性建立在"大数分解和素性检测"这一著名数论难题的基础上。公钥可以完全公开,不需要保密,但必须提供完整性检测机制以保证不受篡改;私钥由用户自己保存。通信双方无须事先交换密钥就可以进行保密通信。

RSA 密码体制算法如下。

(1) 由用户选择两个互异并且距离较远的大素数 p 和 q。

(2) 计算 $n=p\times q$ 和 $f(n)=(p-1)\times(q-1)$。

(3) 选择正整数 e,使其与 $f(n)$ 的最大公约数为 1;然后计算正整数 d,使得 $e\times d$ 对 $f(n)$ 的余数为 1,即 $e\times d \equiv 1 \bmod f(n)$,最后销毁 p 和 q。

经过以上步骤,得出公钥 (n,e) 和私钥 (n,d)。设 M 为明文,C 为对应的密文,则加密变换为 $C=M^e \bmod n$;解密变换为 $M=C^d \bmod n$。

Python 扩展库 rsa 封装了 RSA,可以方便地使用该算法生成密钥以及加解密。

```
>>> import rsa
>>> key = rsa.newkeys(3000)              #随机生成密钥,3000 为 n 的二进制位数
>>> private = key[1]                     #查看私钥分量,输出结果略
>>> print(private.d, private.e, private.n, private.p, private.q)
```

例 18-2　使用 rsa 模块来实现消息加密和解密。

```
import rsa

key = rsa.newkeys(3000)                              #生成随机密钥
privateKey = key[1]                                  #私钥
publicKey = key[0]                                   #公钥

message = '中国山东烟台.Now is better than never.'
print('Before encrypted:', message)
message = message.encode()
cryptedMessage = rsa.encrypt(message, publicKey)     #加密
```

```
print('After encrypted:\n', cryptedMessage)

message = rsa.decrypt(cryptedMessage, privateKey)  #解密
message = message.decode()
print('After decrypted:', message)
```

Python 扩展库 pycryptodome（以 3.18.0 为例）也封装了 RSA、DSA 和 ElGamal 算法，可用于数字签名和其他相关领域。

下面的代码演示了如何使用 pycryptodome 提供的 RSA 模块进行加密和解密。

```
>>> from Crypto.PublicKey import RSA
>>> from Crypto.Cipher import PKCS1_OAEP
>>> key = RSA.generate(2048)                          #生成密钥,长度越大越安全,攻击难度越大
>>> print(key.n, key.p, key.q, key.e, key.d, sep='\n') #查看密钥各分量,结果略
>>> p = 'Python 小屋,董付国,开通于 2016 年 6 月 29 日'
>>> encryptor = PKCS1_OAEP.new(key)
>>> c = encryptor.encrypt(p.encode('utf8'))           #编码,加密
>>> encryptor.decrypt(c).decode('utf8')               #解密,解码,得到原始信息
'Python 小屋,董付国,开通于 2016 年 6 月 29 日'
>>> with open('key.pem', 'wb') as fp:                 #导出密钥,保存
        fp.write(key.exportKey('PEM'))                #显示写入文件的字节串长度

1678
>>> with open('key.pem', 'rb') as fp:                 #从文件中导入密钥
        key1 = RSA.importKey(fp.read())

>>> encryptor1 = PKCS1_OAEP.new(key1)                 #使用导入的密钥进行解密
>>> encryptor1.decrypt(c).decode('utf8')
'Python 小屋,董付国,开通于 2016 年 6 月 29 日'
```

18.3.2 DSA

DSA 是基于公钥机制的数字签名算法，其安全性基于离散对数问题（DLP），即给定一个循环群中的元素 g 和 h，很难找到一个整数 x 使得 $g^x = h$。下面的代码简单演示了 pycryptodome 扩展库中 DSA 算法的用法。

```
from Crypto.Random import random
from Crypto.PublicKey import DSA
from Crypto.Signature import DSS
from Crypto.Hash import SHA256

message = 'Simple is better than complex.'
key = DSA.generate(1024)                              #生成密钥
h = SHA256.new(message.encode())                      #计算消息的哈希值
signer = DSS.new(key, 'fips-186-3')
sig = signer.sign(h)                                  #签名
try:
    signer.verify(h, sig)                             #验证
    print('通过验证')
except:
```

```
        print('验证失败')

h1 = SHA256.new(message.encode()+b'3')           #伪造另一个消息的哈希值
try:
    signer.verify(h1, sig)                       #验证失败,引发异常
    print('通过验证')
except:
    print('验证失败')
```

本 章 小 结

（1）信息加密和信息隐藏是实现信息安全和保密的主要技术。

（2）Python 标准库 hashlib 实现了 SHA1、SHA224、SHA256、SHA384、SHA512 以及 MD5 等多个安全哈希算法,标准库 zlib 提供了 adler32 和 crc32 的实现,标准库 hmac 实现了 HMAC 算法。

（3）安全哈希算法也称报文摘要算法,属于单向变换算法,对任意长度的消息可以计算得到固定长度的唯一指纹,即使是非常相似的消息也不会得到相同的指纹。

（4）扩展库 pycryptodome 不仅提供了 SHA256 和 RIPEMD160 等多个安全哈希算法,还提供了 DES、AES、RSA、ElGamal 等多个密码学算法的实现。

（5）安全哈希算法常用于数字签名、用户密码管理、文件完整性检查等领域。

（6）Python 扩展库 Rsa 和 pycryptodome 都提供了 RSA 算法的实现。

习　　题

1. 根据安全哈希值可以还原原始的消息内容(判断题,对或错)。
2. RSA 的安全性主要取决于密钥的长度(判断题,对或错)。
3. 查看 Python 扩展库 pycryptodome 文件夹结构,了解其提供的各种模块和库。
4. 编写程序,实现凯撒算法加密与解密。

参考文献

[1] 董付国. Python程序设计基础(微课版)[M]. 3版. 北京:清华大学出版社,2023.
[2] 董付国. Python可以这样学[M]. 北京:清华大学出版社,2017.
[3] 董付国. Python程序设计开发宝典[M]. 北京:清华大学出版社,2017.
[4] 董付国,应根球. 中学生可以这样学Python(微课版)[M]. 北京:清华大学出版社,2020.
[5] 董付国. 玩转Python轻松过二级[M]. 北京:清华大学出版社,2018.
[6] 董付国. Python程序设计基础与应用[M]. 2版. 北京:机械工业出版社,2022.
[7] HORSTMANN C,NECAISE R. Python程序设计[M]. 董付国,译. 北京:机械工业出版社,2018.
[8] 董付国. Python程序设计实验指导书[M]. 2版. 北京:清华大学出版社,2024.
[9] 董付国,应根球. Python编程基础与案例集锦(中学版)[M]. 北京:电子工业出版社,2019.
[10] 董付国. 大数据的Python基础(微课版)[M]. 2版. 北京:机械工业出版社,2023.
[11] 董付国. Python数据分析、挖掘与可视化(慕课版)[M]. 2版. 北京:人民邮电出版社,2023.
[12] 董付国. Python网络程序设计(微课版)[M]. 北京:清华大学出版社,2021.
[13] 董付国. Python数据分析与数据可视化(微课版)[M]. 北京:清华大学出版社,2023.